# Taxonomy of Mycotoxigenic Fungi

**S. Girisham**
**V. Koteswara Rao**
**S.M. Reddy**
Department of Microbiology
Kakatiya University
Warangal-506009, Telangana

SCIENTIFIC
PUBLISHERS

*Published by:*

**SCIENTIFIC PUBLISHERS (INDIA)**

Jodhpur
–
*5 A, New Pali Road*
*P.O. Box 91*
*Jodhpur - 342 001*
*INDIA*

Delhi
–
*4806/24, Ansari Road*
*Daryaganj*
*New Delhi - 110 002*
*INDIA*

E-mail: info@scientificpub.com
Website: www.scientificpub.com



Printed in India

# Preface

Mycotoxins are extremely toxic chemical substances produced by certain moulds growing naturally in many agricultural commodities especially cereals and oil seeds both in the field and as well as in storage and later when processed into food and feed concentrations. These toxins are toxic to warm blooded animals including man. Infact term mycotoxin is a misnomer as the etymological meaning is toxic to fungi by analogy to zootoxins and phytotoxins which are toxic to animals and plants respectively. Though human mycotoxicoses are less well understood, they are increasingly diagnosed and studied. Human intake of these toxins mainly occurs from plant based food and animal-derived foods such as milk, cheese and certain meat products. Mycotoxins are in general low molecular weight, non-antigenic fungal secondary metabolites formed by way of several metabolic pathways and still some mycotoxins are derived from a combination of two or more of the principle pathways.

Though wide varieties of mycotoxins are elaborated their number may range more than 300 which are elaborated by not more than 100-150 fungal species. The potential of mycotoxin production is complex and needs to be studied in depth in the literature as there are multimode number of fungi. This is mainly due to misidentification. The gap between mycologists and toxicologists is increasing.

In view of their pleomorphic nature or large production of spores, it is becoming difficult to make proper identification of mycotoxin producing fungi. Most of the times the toxigenic moulds are misidentified. The concept of mycological identification is fast changing as a result only trained mycologist are in a position to make correct identification. Most of the times, identification of mycotoxin producing fungi is critical in the diagnosis of mycotoxicosis. Global concern is given to the mycotoxin contamination of food products. Numerous activities at National, Regional and International levels have been carried out in order to prevent mycotoxin monitor and control mycotoxin contamination, and to train various staff involved in research prevention, *surveillance*, and control activities in both public and private sectors. The present authors have studied these fungi for more than 20 years with in depth knowledge on different features of these fungi. Such manuals for identification of these fungi are also not available.

Detailed aspects of identification have been discussed. It is profusely illustrated which would facilitate easy identification. A key is also provided to make it easy for correct identification. An attempt has also been made to refer the recent nomenclature.

The authors would like to thank to all those who helped directly or indirectly to successfully complete the herculean task. We thank Head Department of Microbiology and authors for providing necessary encouragement. We are also thankful to UGC for financial assistance.

**S. Girisham**
**V. Koteswara Rao**
**S.M. Reddy**

April, 2016

# Contents

*Chapter 1*

# Introduction

## General Introduction

As a sequel to the report of aflatoxin production by *Aspergillus flavus* in 1960 (Sargeant *et al.*, 1961), a great interest has been developed to know the harmful effects of fungi associated with stored foods and feeds. Large assembly of fungi are now known to elaborate wide variety of metabolites which exhibit varying toxicity, ranging from chronic to acute toxicity which are called as mycotoxins. These fungi mainly exists in anamorphic state and have ascomycete's connection. These fungi which are comparatively xerophilic and have the potential to produce conidial mass in abundance. They need warmer conditions to elaborate these toxic metabolites on nutritionally rich substrates, making them unsuitable and hazardous to health of a man and animal. These are popularly called moulds. More than fifty genera of fungi are known to contain mycotoxigenic species. Most of these moulds produce their sexual states from groups of ascomycetous orders Eurotiales and Hypocreales, with a smaller number distributed among other orders of Ascomycetes. The genera which are reported to contain mycotoxigenic species are precised in Table 1.1.

Perusal of Table 1.1 leads to the conclusion that toxigenic ability is sporadically distributed and that any fungus is as likely as any other to be mycotoxigenic. However, it is logical to suggest that these fungi are phylogenically related as these mycotoxins are synthesized at the end of long enzyme mediated reaction sequences. Since most of these fungi do not reproduce sexually or produce rarely, their taxonomic identification is based on anamorphic state which makes determination of their phylogeny difficult. Consequently there is little correlation between the pattern of toxin production by a particular fungus and its classificatory position. However, connections between teleomorphs and their anamorphs are now known in some (Tubaki, 1981) which enables us to predict the likely teleomorphic genus (or groups or teleomorph genera), for most anamorphic genera and only a few genera with toxigenic species cannot be linked to teleomorphs.

Majority and most significant mycotoxigenic species can be assigned to two orders, Eurotiales and Hypocreales. Probably, the toxins produced by these fungi share the same basic biosynthetic pathways but specific toxins are not shared between the orders. Species of other orders of Ascomycetes produce various toxins but these toxins are also produced by species of either Eurotiales or Hypocreales. However, Claviceptales produce only ergot alkaloids.

**Table 1.1.** Genera of fungi containing Mycotoxigenic species

| | | | |
|---|---|---|---|
| *Acremonium* | *Dendrodochium* | *Magnaparthe* | *Rosellinia** |
| *Alterneria* | *Dichotomomyces** | *Microdochium* | *Sclerotinia** |
| *Arthrinium* | *Diaparthe* | *Monographella ** | *Sclerotium* |
| *Aschochyta* | *Diplodia* | *Myrothecium* | *Sepadonium* |
| *Aspergillus* | *Drechslera* | *Nigrosabulum** | *Sphacelia* |
| *Beauveria* | *Epicoccum* | *Nigrospora* | *Stachbotrys* |
| *Bipolaris* | *Epichloe** | *Paecilomyces* | *Talaromyces** |
| *Byssochlamys** | *Fusarium* | *Papulaspora* | *Thanatephorous* |
| *Ceratocystis** | *Fusicoccum* | *Penicillium* | *Thielavia** |
| *Chaetomium** | *Gerlachia* | *Periconia* | *Torulopsis* |
| *Chalara* | *Gibberella** | *Pestalotiopsis* | *Trichoderma* |
| *Cladosporium* | *Gliocladium* | *Phoma* | *Trichothecium* |
| *Clavariopsis* | *Gloeotinia** | *Phomopsis* | *Verticillium* |
| *Claviceps** | *Hormicium* | *Piricauda* | *Verticimonosporium* |
| *Colletotrichum* | *Hyalodendron* | *Pithomyces* | *Zygosporium* |
| *Curvularia* | *Khuskia** | *Pteroconium* | |
| *Cylindrocarpon* | *Lasidiplodia* | *Pyricularia* | |
| *Dematophora* | *Metarrhizium* | *Rhizoctonia* | |

* - Teleomorphic stage

Members of Eurotiales dominate toxigenic species represented by *Aspergillus* and *Penicillium*. Other toxigenic members of this group are species of *Byssochlamys* and *Paecilomyces*. Toxins produced by members of this order are diverse representing at least six of the main biosynthetic categories (Table 1.2).

**Table 1.2** Mycotoxins and their Biosynthetic Categories of Secondary Metabolites

| Biosynthetic Pathway | Representative Mycotoxins |
|---|---|
| **Poly ketides** | |
| **Tetra-** | Patulin, Penicillic Acid, Chlorflavonin |
| **Penta-** | Citrinin, Ochratoxins |
| **Hexa-** | Maltoryzine |
| **Hepta-** | Viriditoxin, Cytochalasins, Rugulosin |
| **Octa-** | Ergochromes |
| **Nona-** | Zearalenone, Viridicatum toxin |
| **Deca-** | Aflatoxins, Austocystins, Erythroskyrine |
| **Tetronic/ Tetramic acid** | Tenuazonic Acid, Cyclopiazonic Acid, Cytochalasins |
| **Diketopiperazines Simple** | Aspergillic Acid, Echinulins |
| **Modified** | Brevianamides, Sporidesmins, Fumitremorgins, Oxaline |

| Biosynthetic Pathway | Representative Mycotoxins |
|---|---|
| **Peptides** | Tentoxin, Ergotamine, Tryptoquivaline |
| **$C_6C_3$ Products** | Chlorflavonin, Xanthocillin, Terphenyllin |
| **Terpenes Mono-** | Viridicatum Toxin |
| **Sesqui-** | Trichothecenes |
| **Di-** | Paspaline |

These include among others:

(1) Polyketides- Patulin, Citrinin, Ochratoxins, Maltoryzine, Viriditoxin, Cytochalasins, Rugulosin, Ergochromes and Aflatoxins

(2) Tetramic Acid-Cyclopiazonic Acid;

(3) Diketopiperzines-Aspergillic Acid, Fumitrogens and Oxaline

(4) Neurotrophic Peptide-Tryptoquivaline.

(5) $C_6$ $C_3$ product-Xanthocillin and

(6) Nonadride - Rubratoxin a compound of mixed fatty acid–TCA cycle origin.

However, a few of them are produced by Non-Eurotialean genera such as a ergochromes are also formed by *Claviceps, Phoma* and Lichens. Cyclopiazonic acid is produced by species of *Alternaria, Pyricularia* and *Phoma*. Similarly, sterigmatocystins are produced by species of *Bipolaris* and *Chaetomium*. Thus, some mycotoxins are being elaborated by several fungi (Table 1.3). On the other hand, single fungus produce more than one mycotoxins which are quite unrelated (Table 1.4).

**Table 1.3.** Mycotoxins produced by more than one fungus

| Mycotoxin | Fungal species |
|---|---|
| Aflatoxin B1 | *Aspergillus* (10), *Emericella* (3) |
| Sterigmatocystin | *Aspergillus* (5), *Chaetomium* (5), *Bipolaris* (1), *Emericella* (20), *Drechslera* sp. |
| Patulin | *Penicillium* (17) |
| Penitrem A | *Penicillium* (9) |
| Fumitrogen B | *Aspergillus* (2), *Penicillium lanosum* |
| Ochratoxin A | *Aspergillus* (17), *Penicillium* (2) |
| Citrinin | *Penicillium* (10), *Aspergillus* (4) |
| Cycloplazonic acid | *Penicillium* (5), *Aspergillus* (6) |
| Penicillic acid | *Penicillium* (15), *Aspergillus* (4) |
| Zearalenone | *Fusarium* (7) |
| T2 toxin | *Fusarium* (5), *Trichoderma viride* |
| Roridin A | *Myrothecium roridum, Stachybotrys atra* |
| Cytochalasins (14 compounds) | *Metarrhizlum anisopollae*, *Zygosporium* (2), *Phomopsis C. paspali, Rosellinia necatrix, Aspergillus clavatus* |

| Mycotoxin | Fungal species |
|---|---|
| Rubratoxin | *Penicillium rubrum* |
| Skyrins (Hydroxy anthroquinones) | *Penicillium* (7), *Pachbastum bamatum, Sepedonium ampullosporum, Chaetomium elatum, Cladosporium fulvum, Talaromyces avellaneus, Endothia parasitica* and *Preussia multispora* |
| Gliotoxin | *Trichoderma* (2), *Gliocladium (*2), *Aspergillus* (2) and *Penicillium* (3) |
| Secalonic acid | *Aspergillus* (3), *Penicillium* (6), *Eupenicillium egyptiacum, Parmelia entotheichroa, Claviceps purpurea, Phoma terrestris* |
| Cytochalsisns | *Drechslera dematioidea, Phoma* sp, *Metarhizium anisopliae, Zygosporium masonii, Aspergillus clavatus, Rosellinia necatrix, Phomopsis paspali, Nigrosabulum* sp. |

**Table 1.4.** Fungi producing more than one mycotoxin

| Organism | Mycotoxin |
|---|---|
| *Aspergillus terreus* | Patulin, Citrinin, Territrems, Aranotins, Citroviridin |
| *A. ochraceus* | Viomellin, Penicillic acid, Ochratoxins, Vioxanthin, Islandiotoxin, Emodin, Rugulosin |
| *A. flavus* | Aflatoxin B1, B2, Aflatrems, CPA, 3-nitropionic acid, Paspaline, Paspalicine, Paspalinine |
| *A. parasiticus* | Aflatoxin B1, B2, Aflatoxin G1,G2 |
| *A. fumigtus* | Fumigaclavines A, B, C, Gliotoxin, Tryptoquivalins, Tryptoquivalons |
| *Penicillium islandicum* | Cyclochlorotine, Erythoskyrin, Luicoskyrin |
| *P. viridicatum* | Viridiotoxin, Penicillic acid, Verruculotoxin, Viomellenin, Vioxanthin |
| *P. expansum* | Chaetoglobosins A-J, Citrinin, Communesins, Patulin, Griseofulvin |
| *Fusarium sporotrichoides* | Zearalenone, $T_2$ toxin, Beauvricin, Trichothecenes (TCTC) A type |

Toxigenic species are found in all of the six subdivisions of *Penicillium* (Pitt, 1979) and out of 18 groups of *Penicillia,* 10 contain toxigenic species. *Penicillia* are found in four subgenera but the production of specific toxin is not localized to any subgenus. For example, patulin is produced by teleomorphic genera *Byssochlamys* and *Eupenicillium* and by two sections each of *Penicillium* and *Aspergillus*. On the other hand, aflatoxins are produced by only two species of *Aspergillus* (*A. flavus* and *A. parasiticus*), but structurally similar Sterigmatocystins are produced by species belonging to four groups including the *A. flavus* group, one species of *Penicillium* and other Non-Eurotialean fungi. The ability to produce particular toxin by a species is not necessarily to produce toxins by other closely related species. Variation in the ability to produce a particular toxin also rests among different strains of the same species. For instance, *Aspergillus flavus, A. parasticus, A. sojae* and *A. oryzae* are difficult to distinguish morphologically but *A. flavus* and *A. parasticus* produce aflatoxins, while no strain of *A. oryzae* and *A. sojae* which are used in food fermentation are not known to produce toxins. Wicklow and Cole (1984) feel that these species evolved from *A. flavus* and *A. parasiticus* and lost toxigenicity under domestication.

Over 40 species of anamorphs of Hypocrealean fungi are toxigenic and belong to *Acremonium, Fusarium, Gliocladium, Myrothecium, Trichoderma* and *Verticillium*. Teleomorphs are not known for *Metarrhizium* or *Verticimonosporium* but their morphology and biology suggests their hypocrealean nature. These fungi are known to produce comparatively less number of toxins which include zearalenone (polyketide), trichothecenes (sesquiterpenes) and butenolides. Their biosynthetic pathway is as yet unknown when Butenolides are produced by only these fungi, Zearalenone is also produced by species of *Curvularia*, a genus of Pleosporales. On the other hand, trichothecenes are produced exclusively by species of *Fusarium, Myrothecium, Acremonium*, *Trichothecium* and *Trichoderma*. However, *Stachybotrys* which has affiliation to Sordariales also produce macrocyclic trichothecenes. Similarly *Microdochium nivale,* which was formerly known as *Fusarium nivale* now included in Hyphonectriaceae of polystigmales and the flowering plant genus *Baccharis* are reported to elaboraty these foscius.

*Claviceps* with anamorphic state *Sphacelia,* is the only genus of *Claviceptales* positively identified as toxigenic. The pharmacologically most active components of ergots alkaloids which are amides of D-lysergic acid. No lysergic acid is found outside the *Claviceps*. The other two alkaloids produced by *Claviceps* are Clavine alkaloids and Ergolines. Clavine alkaloids are also formed by species of *Aspergillus* and *Penicillium*, while ergolines are found in *Aspergillus, Penicillium* and *Phoma* which is a genus of Pleosporiales, and also Lichens.

*Acremonium typhium,* the anamorph of *Epichloe typhina,* a pasture grass endophyte, has been linked to the Fescue-Foot Toxic syndrome. Butanolides produced by species of *Fusarium* have also been implicated in this condition.

Some of the anamorphic genera like *Phoma, Curvularia, Bipolaris, Drechslera, Alternaria, Diplodia* and *Periconia* are reported to have their teleomorphs in bitunicate ascomycetes belonging to Pleosporales. Of these fungi, *Alternaria* is pharmacologically important and produces polyketides, alternarioles and related altertoxins by its one of the species, *A. alternata*. This species also produces tetramic acid, tenuazonic acid, in common with *Pyricularia* and *Phoma*. The other tetramic acid, the cyclopiazonic acid is elaborated by species of *Aspergillus* and *Penicillium*.

The other toxigenic fungi belong to four unrelated orders of Ascomycetes and represented by one or two species. *Chaetomium* belonging to order Sordariales is reported to produce Cytochalasins and Sterigmatocystins with several other unrelated fungi. Several species of *Cladosporium,* whose teleomorph stage is in the dothideales produces epi- and fagicladosporic acids, fatty acid derivatives which can induce Alimentary Toxic Aleukia (ATA) in common with *Fusarium. Stachybotrys* an anamorph of Sordariales which produces trichothecenes in common with *Monographella* and flowering plant, *Baccharis*.

Some anamorphic fungi such as *Pithomyces, Zygosporium* and *Epicoccum* are reported to be toxigenic. Teleomorphs are not known and it is not possible to speculate their affinities. Though conidial ontogeny in *Pithomyces* and *Epicoccum* are similar, they are not related. *P. chartarum* produces Sporidesmins, the cause of photosensitivity, liver damage and death in sheep, cattle and deer. These toxins include gliotoxins produced by hypocrealean genus *Gliocladium* and flavipine produced by *Epicoccum purpurascens.* Species of *Aspergillus* and *Zygosporium* in common with many unrelated fungi produce cytochalasins.

Most of the known mycotoxins are derived either from polyketides or from sesquiterpenes, two basically different categories of secondary metabolites (Bulouc, 1980). Both of these are common in Ascomycetes and in fungi imperfecti. Thus, the basic pathways to most mycotoxins are generally resident within the Ascomycetes and their anamorphs.

There are many techniques which have been used in the classification of bacteria and yeasts for many years, which include DNA hybridization electrophoresis, GC content data, serological techniques, cytology and chromosome mapping. There is no doubt that much more can be achieved in fungal classification using traditional techniques simply because morphological taxonomy has suffered from a shortage of labour for years. But the wider use of novel methods will undoubtedly accelerate the production of really satisfactory classification.

There has been a great deal of interest during last 100 years in the classification of Deuteromycetes. Several mycologists considered the classification of deuteromycetes on anamorphic structures in isolation with the obvious benefit of including those anamorphs without known teleomorphs. Until the work of Hughes (1953), most classifications of deuteromycetes were based largely on the artificial system of conidium colour, shape and septation as suggested by Saccardo (1886). The work of Hughes (1953) on conidiophore and conidium development provided inputs to others. Kendrick (1971), Ellis (1976) and Cole and Samson (1979) have tried to developed a system where conidia produced by fragmentation of hyphae (Thallic) or swelling of the hypal tip (blastic) development. The blastic division was further divided into enteroblastic where conidia are produced singly from different loci on a single conidiogenous cell. However, this system of classification has been challenged by Minter *et al.* (1982) and Minter *et al.* (1983) by saying that the division of Deuteromycetes established by previous workers are not fundamentally important. They suggest that careful description rather than attempting to shoehorn extremely plastic organisms into a simple series of classification boxes. The work on these lines will certainly help to develop a more stable and phylogenetic classification. Key for the identification of these mycotoxigenic fungi is proposed as given below.

**Key to the identification of mycotoxin producing fungi**

1. Mycelium present in life cycle ........................................ 4

 1a. Mycelium unknown, asexual reproduction by budding ........................ 2

2. Asexual reproduction by budding (Fig. 1) ........................ *Torulopsis* (1)

 2a. Reproduces by basidia and basidiospores ........................ 3

3. Basidia arise from sclerotia formed from agglomeration on moniliform hyphae, basidia barrel shaped, stigmata straight longer than metabasidium (Fig. 57) .......... *Thanatophorus* (2)

3a. Basidia are formed from elongated yeast like cells (Fig. 65) .......... *Filobasidium* (3)

4. Reproduction by ascus and ascospores production, anamorph may have separate name .......... 47

4a. Ascus and ascospores are not produced and may reproduce asexually or form sclerotia .......... 5

5. Conidia formed on simple or branched conidiophores or arising directly from vegetative mycelium .......... 9

5a. Conidia are not produced and survive by sclerotium formation .......... 6

6. Sclerotia slender-fusiform, cylindrical with tapering ends, brown, black or purple in the mature inflorescence of grasses (Fig. 30) .......... *Claviceps* (4)

6a. Sclerotia not in the inflorescence of grasses .......... 7

7. Sclerotia homogeneous structures i.e. no specialized tissues .......... 8

7a. Sclerotia not homogeneous organized into tissues, rind, cortex medulla, globose to concave, pinkish to dark- purple –brown, clamp connections present (Fig. 3) .......... *Sclerotium* (5)

8. Sclerotia black, 1.0 to 2.5 mm diameter, mycelium dark-green, becoming black, constricted at the septate Teleomorph *Thanatephorus* (Fig. 6) .......... *Rhizoctonia* (6)

8a. Sclerotia not black, very small, consisting of regular to irregular groups of cells, sclerotia referred to a bulbils (Fig. 5) .......... *Papulaspora* (7)

9. Conidia formed in pycnidia .......... 10

9a. Conidia not formed in pycnidia .......... 15

10. Pycnidia separate, wall distinct, yellowish-brown to dark-brown with a circular short protruding opening, globose to sub-globose .......... 11

10a. Pycnidia without a distinct wall, but reduced to cavities in a stroma, conidia, hyaline, fusiform, large in bark (Fig. 4) .......... *Fusicoccum* (8)

11. Conidia one-celled .......... 12

11a. Conidia two-celled .......... 13

12. Pycnidia separate and wall glaberous, conidia elliptical, oval, cylindrical, hyaline (Fig. 7) .......... *Phoma* (9)

12a. Pycnidia immersed in tissues with a short protruding ostiole, conidia oval to lanceolate, hyaline, β-conidia (curved, needle like) present (Fig. 8) .......... *Phomopsis* (10)

13. Conidia hyaline and 2-celled .......... 14

13a. Conidia dark-brown, cylindrical or slightly curved with rounded apex, 2-celled (Fig. 9) .......... *Diplodia* (11)

14. Conidia 2-celled and hyaline (Fig. 45) .......... *Ascochyta* (12)

14a. Conidia hyaline, when young becoming dark-brown, 2-celled with longitudinal striations with age (Fig. 10)............................ *Lasiodiplodia* (13)

15. Conidia formed in an acervulus........................................................................ 16

15a. Conidia not formed in an acervulus ................................................................ 17

16. Conidia hyaline, one-celled, elliptical, cylindrical or sickle shaped, dark setae within and around the acervulus present or absent (Fig. 11) ........................................................................................*Colletotrichum* (14)

16a. Conidia with several celled, central cells dark, end cells hyaline, hyaline appendages from the apical cell (Fig. 12) ............. *Pestalotiopsis* (15)

17. Conidia formed in sporodochia or naked conidiophore ........................................... 18

17a. Conidia formed on naked conidiophore, no sporodochia are formed ............. 22

18. Sporodochia disc, cup or cushion-shaped ............................................................. 19

18a. Sporodochia not disk, cup, or cushion-shaped but thin, irregular, and consisting mainly disaggregated conidiophores .......................................... 20

19. Sporodochia cushion-shaped to irregular, at first hyaline then with a black or olive-black layer of conidia, conidiophores tightly packed, irregularly branched, conidia fusiform, pale-green (Fig. 13) ....................*Dendrodochium* (16)

19a. Sporodochia disk- or cup-shaped, with or without setae, conidial mass at first green then black, conidiophore branched, terminated by an apical whorl of phialides, conidia hyaline, one-celled, olive-brown (Fig. 14) ................................................................. *Myrothecium* (17)

20. Sporodochia are formed by clustered chlamydosporous cells releasing orange pigment, conidiogenous cells which arise singly or in sporodochial cells either directly or on globose subtending cells, directly ampulliform sporogenous cells producing cluster of conidia (Fig. 41) ......................................................................................... *Microdochium* (18)

20a. Slimy sporodochia formed by the aggregation of simple or branched conidiophores ........................................................................ 21

21. Conidia, large, globose, septa in various planes, dark coloured (Fig. 16) .........................................................................................*Epicoccum* (19)

21a. Macroconidia hyaline, 3 to many celled, fusoid generally curved. Basal cell somewhat resembling a foot conidiophores simple or branched aggregated into slimy sporodochia (Fig. 15) ............ *Fusarium* (20)

21b. Irregularly branched free or sporodochial terminating in phialides conidia elliptical to cylindrical -2-celled, hyaline (Fig. 58) ....................................................... *Cylindrocarpon* (21)

22. Conidiophore form a distinct synnemata (Fig. 42) ..........................*Metarrhizium* (22)

22a. Conidiophore free and borne on individual hyphae ...................................... 23

23. Conidiophores basauxic ........................................................................................ 24

23a. Conidiophores acroauxic ................................................................................. 26

24. Conidia form in chains which are oriented in sideways to the long axis colony pink (Fig. 22) .......................................................*Trichothecium* (23)

24a. Conidia formed in singly, terminal or lateral single celled, sterigmata present ........ 25

25. Stroma absent (Fig. 44) ........ *Arthririum* (24)

25a. Stroma present (Fig. 59) ........ *Pteroconium* (25)

26. Conidiophore simple (i.e. unbranched) ........ 27

26a. Conidiophore branched ........ 39

27. Conidia hyaline or light coloured ........ 28

27a. Conidia dark coloured ........ 33

28. Conidia 1-celled ........ 29

28a. Conidia 2-celled ........ 32

29. Conidiophore mostly with straight parallel walls throughout or tapering gradually towards the apex ........ 30

29a. Conidiophore not with straight parallel walls throughout but modified by swelling or across ........ 31

30. Conidia aggregated in a slimy mass at the tip of the conidiophore, globose, oval to short cylindric (Fig. 17) ........ *Acremonium* (26)

30a. Conidia arising from a convoluted mass of compact hyphae (stroma) in the young ovriries of certain grasses, conidia hyaline, ovoid on simple conidiophores in slime. Conidial state of *Claviceps* (Fig. 50) ........ *Sphacelia* (27)

31. Conidiophores (conidiogenous cell) somewhat swollen at the base, borne directly on the hyphae, tapering to a long slender tube, conidia in chains (Fig. 19) ........ *Paecilomyces* (28)

31a. Conidiophore terminated by a swelling (vesicle). Sterigmata arising in one or two series from upper part of the surface of the vesicle (Fig. 20) ........ *Aspergillus* (29)

31b. Conidiophore (conidiogenous cell) borne directly or on the hyphae, flask shaped or globose, terminated by a threat like zigzag apex. Conidia very small, ovoid to spherical. Colonies white (Fig. 35) ........ *Beauveria*(30)

32. Conidia with long thin rod like appendages arising from the apical cell (Fig. 21) ........ *Clavariopsis* (31)

32a. Conidia without appendages, formed in chains, the latter composed of conidia oriented sideways to the long axis of the conidiophore (Fig. 22) ........ *Trichothecium* (23)

33. Conidia one celled, brown to black, subglobose or ovoid, conidiophore inflated at the base tapering towards apex (Fig. 23) ........ *Nigrospora* (32)

33a. Conidia 2 to many celled ........ 34

34. Conidia only with transverse septa ........ 35

34a. Conidia with septa in two or more planes ........ 38

35. Conidia acropleurogenous, polyblastic, 1-3 septa, geniculate or denticulate (Fig. 55) ........ *Pyricularia* (33)

35a. Conidia polyblastic, acropleurogenous, conidiophores form synnemata, 0-septate (Fig. 18) ............................................*Dematophora* (34)

35b. Conidia acrogenous ................................................................................ 36

36. Conidia cylindrical, smooth, pigmented in light shades of yellow, green or brown or combinations of three, germinating laterally perfect stage in *Pyrenophora* (Fig. 24)................................................................. *Drechslera* (35)

36a. Conidia not cylindric, perfect state not in *Pyrenophora* ................................ 37

37. Conidia fusiform or ellipsoid, straight or curved. Germination only from end cells. Conidia pigmented in various shades of brown (Fig. 25)..................................................................................*Bipolaris* (36)

37a. Conidia fusiform, dark coloured with light coloured end cells. Central cells somewhat larger (Fig. 26)................................. *Curvularia* (37)

38. Conidia ovoid, obclavate, solitary or in chains with a beak, pale yellow to dark olive –brown conidia with both transverse and vertical septa (Fig. 27) .........................................................................................*Alternaria* (38)

38 a. Conidia oval, dark-brown, usually with 3-transverse and two longitudinal septa, roughened, holoblastic with protruding hilum (Fig. 28) ........................................................................... *Pithomyces* (39)

38b. Conidia large, globose to pyriform brown to black, each conidium formed by the swelling of the tip of the conidium (Fig. 16) ... *Epicoccum* (40)

39. Conidia formed in chains ........................................................................... 40

39a. Conidia not in chains ................................................................................ 42

40. Conidiophore, brush like, composed of a stipe terminated by a few or many sterile branches and conidiogenous cells or phialides only conidia ovoid and long chains of dry conidia (Fig. 29) .............................. *Penicillium* (41)

40a. Conidiogenous cells swollen at the base tapering to a long slender, widely divergent tube, conidia hyaline in chains (Fig. 19)...*Paecilomyces* (28)

40b. Conidiophore not brush like ..................................................................... 41

41. Conidia developing blastically from the conidiophore branches or mostly from other conidia to form a complex of branched conidial chains, conidia dark, elliptical or lemon shaped 1-2 celled (Fig. 2).............. *Cladosporium* (42)

41a. Conidiophore dark with few, very short apical branches, curved down warded at the apex, conidia dark, globose, rough developing successively in chains from globose conidiogenous cells (Fig. 31) ........................................................................... *Periconia* (45)

42. Ultimate branches of the conidiophore in whorls other branching pattern may be present ................................................................................... 43

42a. Conidiophore branching not in whorls ....................................................... 44

43. Conidia one celled, ovoid, aggregating in slimy balls at the slender conidiogenous cells, irregular branching sometimes present (Fig. 32).............................................................................*Verticillium* (46)

43a. Conidiophore slender, verticillately branched, conidia hyaline, 1-celled-large spherical, chlamydospores abundant (Fig. 53)...... *Sepedonium* 45

43b. Conidiophore profusely and irregularly branched, forming tree like complex, conidiogenous cells bottle shaped disposed irregularly, opposite or alternate conidia mostly subglobose to short obvoid culture green (Fig. 33) ........................................ *Trichoderma* (46)

44. Conidia phialidic ........................................ 45

44a. Conidia tertic ........................................ 46

45. Conidiophore branched irregular or simple. Each stipe or branch terminated by a cluster of club-shaped lightly pigmented phialides. Conidia one-celled formed in slimy mass around the phialides (Fig. 34) .......... *Stachybotrys* (47)

45a. Conidia hyaline to light coloured. Conidiophore penicillately branched, conidia light coloured or hyaline. Conidia aggregate at the tip of phialides in slimy masses. Culture may be rose yellow, orange or green (Fig. 36) ........................................ *Gliocladium* (48)

46. Conidiogenous cells monotretic macronematous forming arched loops (Fig. 54) ........................................ *Piricauda* (49)

46a. Conidiogenous cells tertic conidial chains branched repeatedly (Fig. 51) ........................................ *Hyalodendron* (50)

47. Ascocarp present ........................................ 51

47a. Ascocarp lacking, asci globose, arising in clusters on the mycelium, ascospores hyaline, ellipsoidal, smooth, 8-ascospores (Fig. 37) ........................................ *Byssochlamys* (51)

48. Ascocarp cleistothecium ........................................ 49

48a. Ascocarp perithecium ........................................ 52

48b. Ascocarp apothecium ........................................ 58

49. Cleistothecium with a distinct peridium ........................................ 50

49a. Cleistothecium with a distinct peridium of many layered asci scattered and globose, *Acremonium* anamorph (Fig. 62).... *Nigrosabulum* (52)

50. Ascospores scattered ........................................ 51

50a. Asci arranged in a layer no ostiole, ascospore dark fusiform (Fig. 43) ........................................ *Thielavia* (53)

51. Cleistothecium glaberous, asci catenate and part of peridium, *Polycephalum* anamorph (Fig. 63) ........................................ *Dichotomomyces* (54)

51a. Cleistothecium glaberous, evanicent, asci scattered, *Penicillium* anamorph (Fig. 49) ........................................ *Talaromyces* (55)

52. Perithecium free or on stroma ........................................ 53

52a. Perithecium immersed in stroma ........................................ 57

53. Ascospores one celled ........................................ 54

53a. Ascospores 2 to may celled ........................................ 56

54. Perithecial wall glaberous ........................................ 55

54a. Perithecial wall curved with characteristic hairs (Fig. 39) ...... *Chaetomium* (56)

55. Ascospores small, hyaline, hat – shaped. Asci globose, borne irregularly within the perithecium, asci evanescent, ascospores collect in slime at the tip of long perithecial neck, *Chalara* anamorph (Fig. 38)........ *Ceratocystis* (57)

55a. Ascospores 1-celled, produced in a perithecium borne in a felt like subiculum of dark hyphae, perithecial wall many layered, unitunicate stalked (Fig. 60) .................................................. *Rosellinia* (58)

56. Ascospores 2 – celled, elliptical to fusiform, hyaline smooth, or striate, asci cylindrical to clavate, perithecia smooth or rough, thin walled (Fig. 46) .......................................................................... *Monographella* (59)

56a. Ascospores hyaline, elliptical to fusiform. 3 or more celled ascospores elliptical to clavate asci, perithecia superficial (Fig. 40)... *Gibberella* (60)

57. Stroma globose, terminating a stout or slender stipe which arises from a hard sclerotium. Asci very narrow with a thick apex transversed with a pore. Ascospores needlelike (Fig. 30) ................. *Claviceps* (61)

57a. Ascomata in a stroma perithecia ostiolate neck, lateral paraphyses, asci unitunicate, non–amyloid apex, filiform septae ascospores (Fig. 61)...............................................................................*Epichloae*(62)

58. Sarcomata in a stroma, ostiolate, erumpt neck, 2-layered perithecial wall, unitunicate asci, 2-celled, cylindric fusoid ascospores, *Phomopsis* anamorph (Fig. 64) ............................................................... *Diaparthe* (63)

59. Apohtcium disk like, hymenium yellow – brown. Asci narrowly clavate single celled paraphyses, ascospores hyaline, 1 – celled ..................*Gloeotinia* (64)

59a. Apothecia arising from a sclerotium, small, tan colored, cup shaped, asci cylindrical pore blued with iodine, spores hyaline, elliptical, septate, paraphyses (Fig. 48) ................................................. *Sclerotinia* (65)

## iii) Figures of Mycotoxigenic Fungi

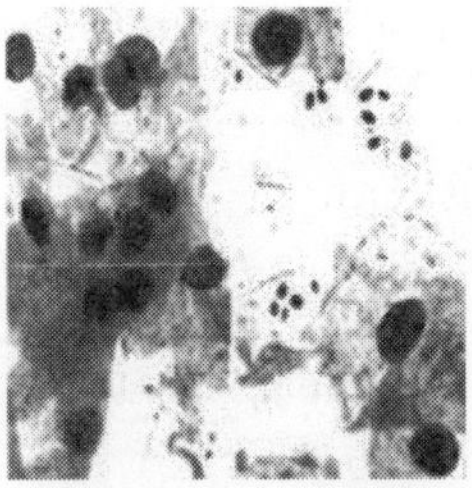

**Fig. 1.** *Torulopsis*

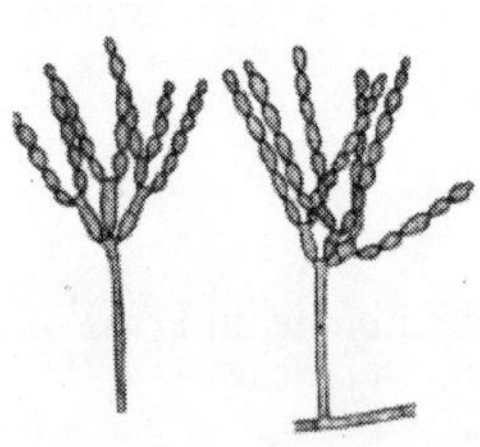

**Fig. 2.** *Cladosporium*

**Fig. 3.** *Sclerotium*

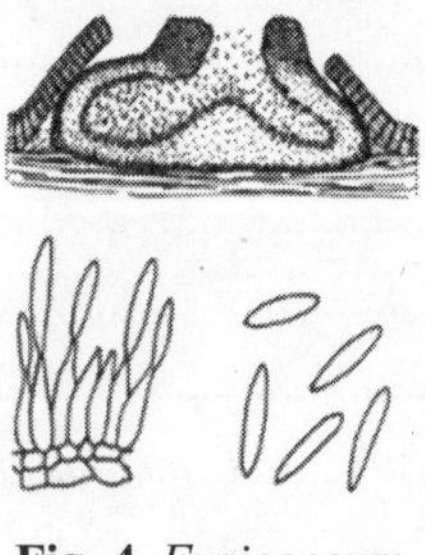

**Fig. 4.** *Fusicoccum*

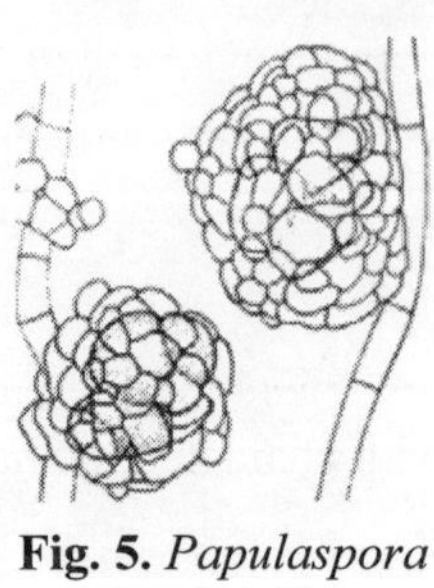

**Fig. 5.** *Papulaspora*

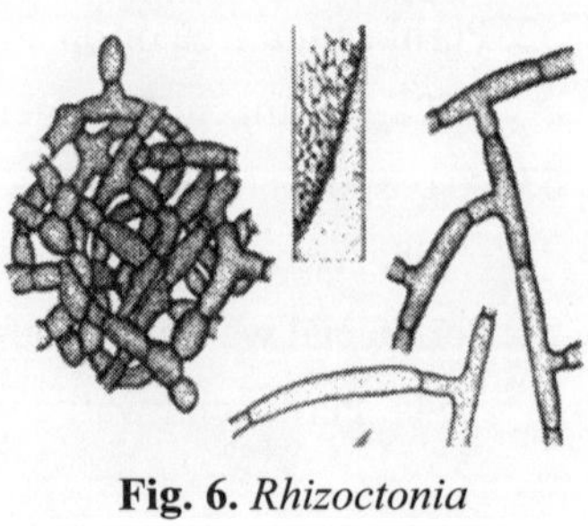

**Fig. 6.** *Rhizoctonia*

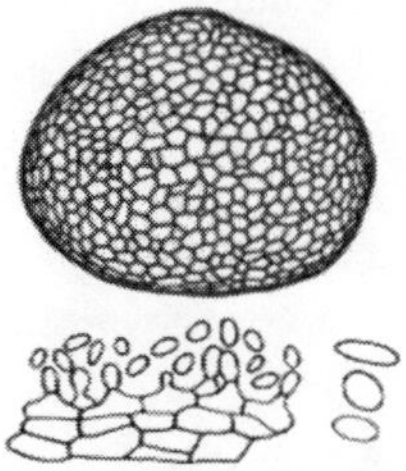

**Fig. 7.** *Phoma*

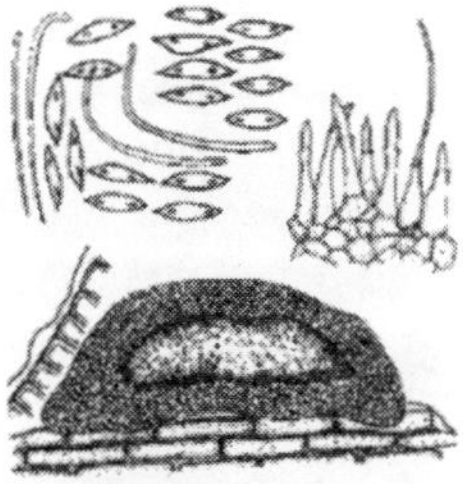

**Fig. 8.** *Phomopsis*

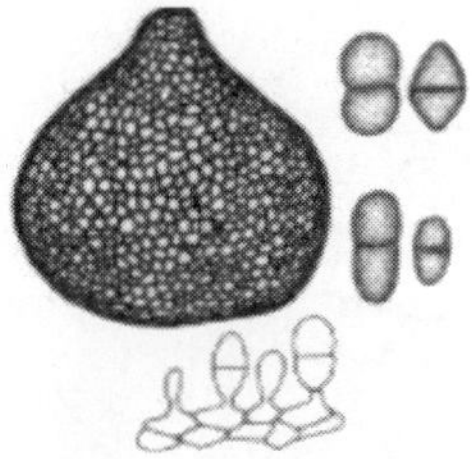

**Fig. 9.** *Diplodia*

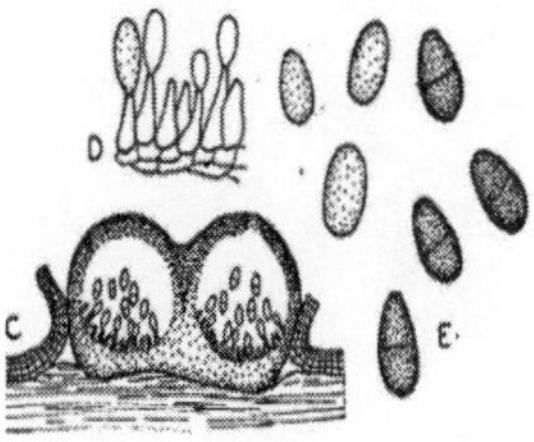

**Fig. 10.** *Lasidiplodia*

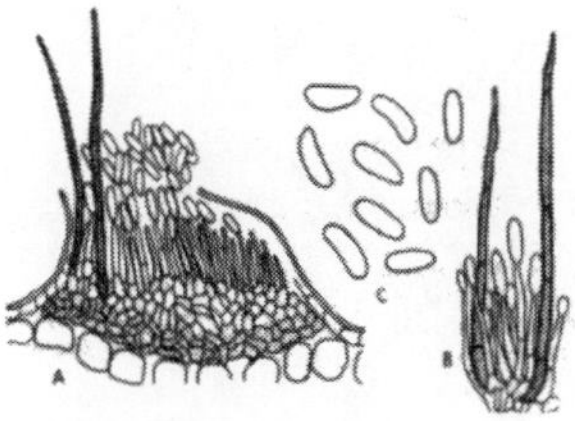

**Fig. 11.** *Colletotrichum*

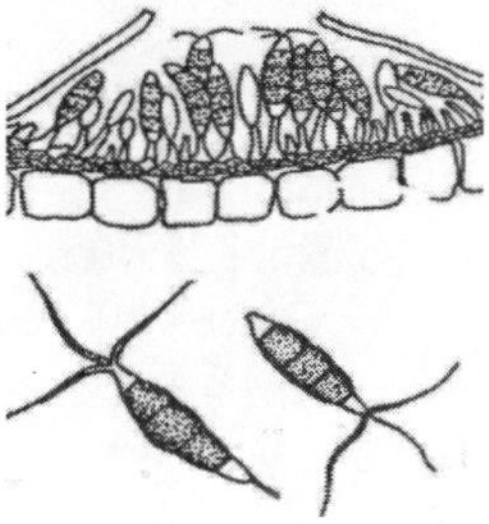

**Fig. 12.** *Pestalotiaopsis*

**Fig. 13.** *Dendrodochium*

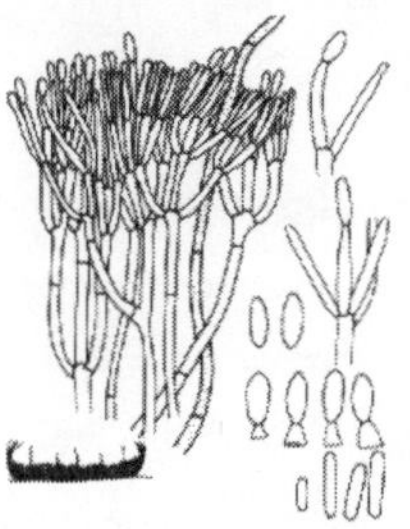

**Fig. 14.** *Myrothecium*

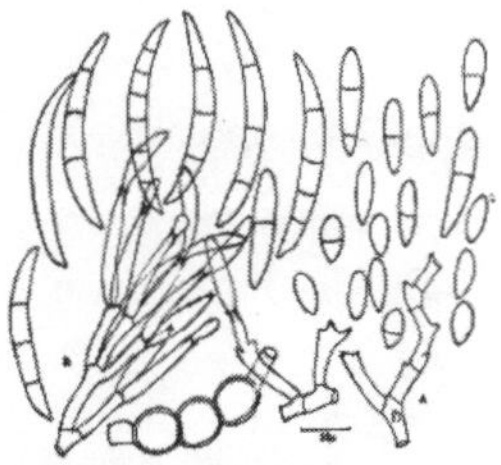

**Fig. 15.** *Fusarium*

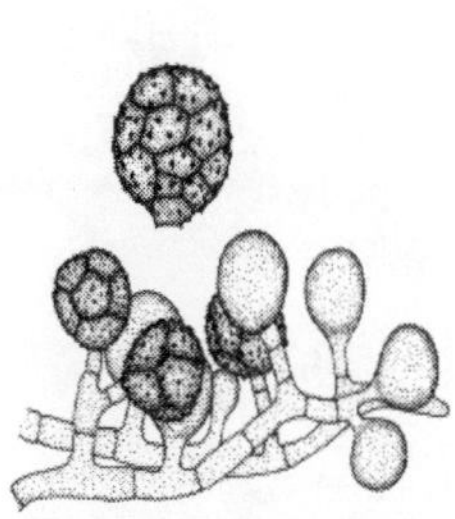

**Fig. 16.** *Epicoccum*

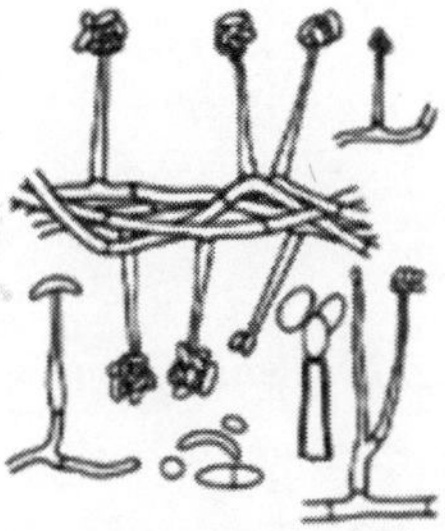

**Fig. 17.** *Acremonium*

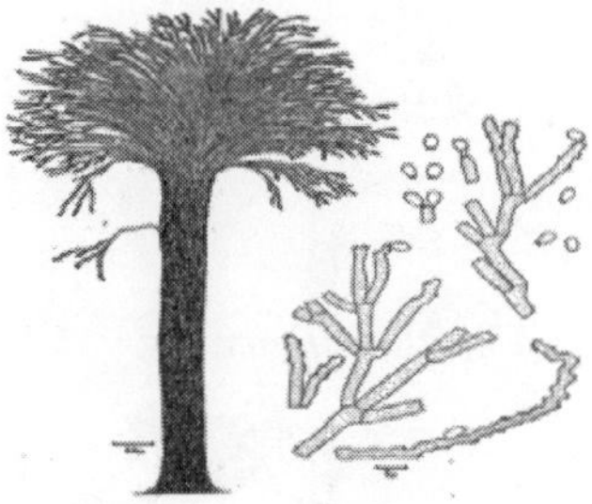

**Fig. 18.** *Dematophora*

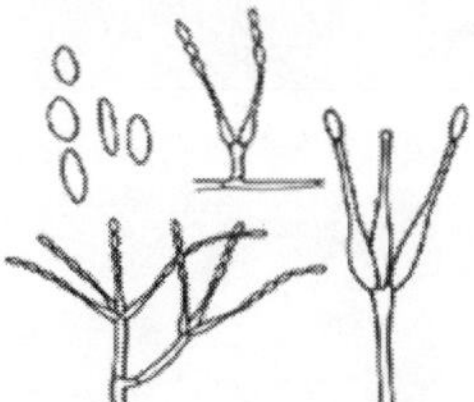

**Fig. 19.** *Paecilomyces*

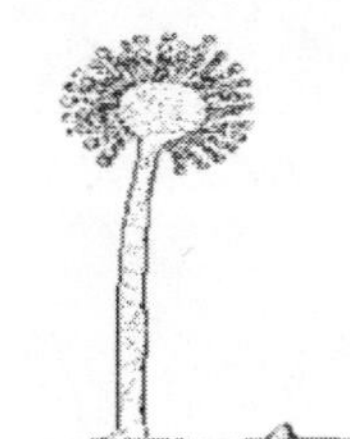

**Fig. 20.** *Aspergillus*

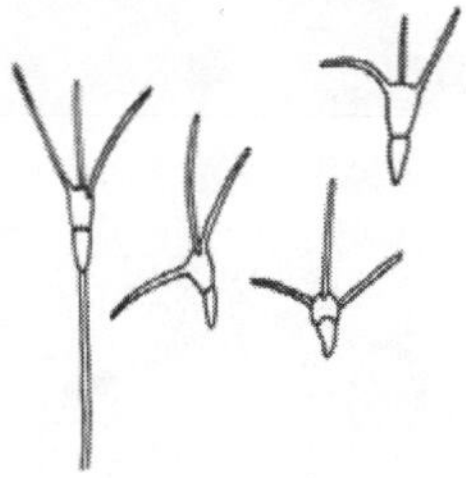

**Fig. 21.** *Clavariopsis*

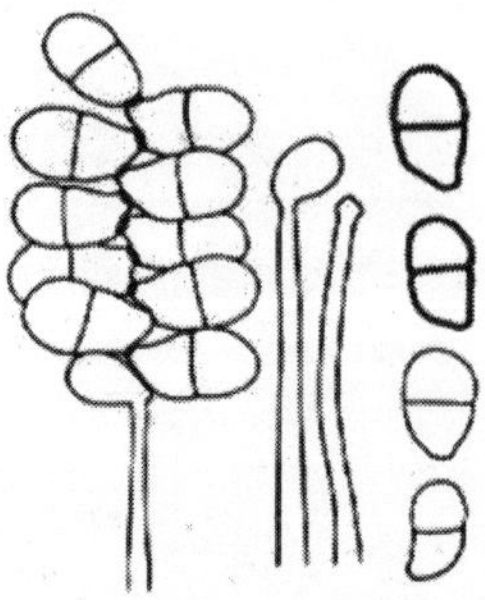

**Fig. 22.** *Trichothecium*

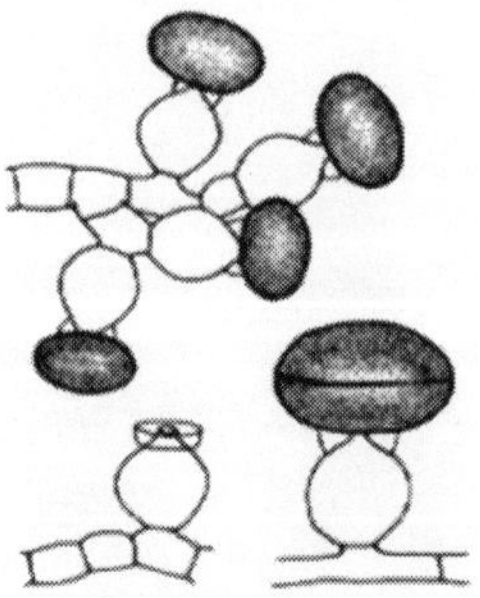

**Fig. 23.** *Nigrospora*

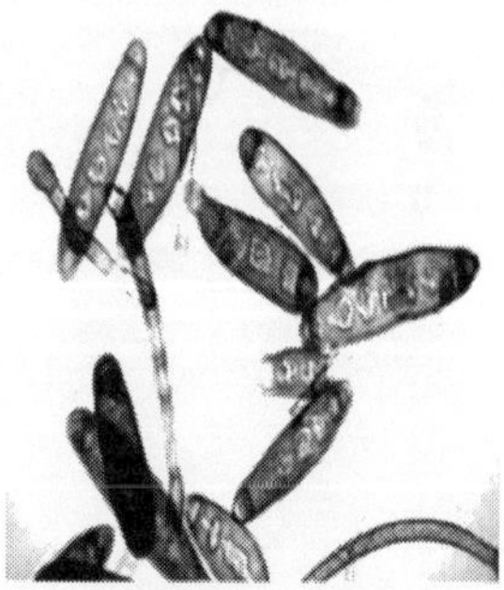

**Fig. 24.** *Drechslera*

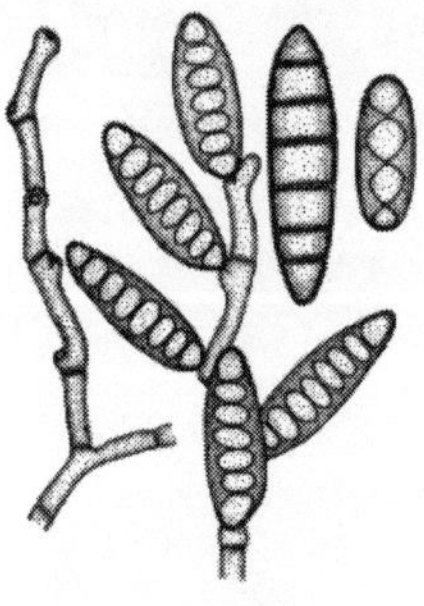

**Fig. 25.** Bipolaris

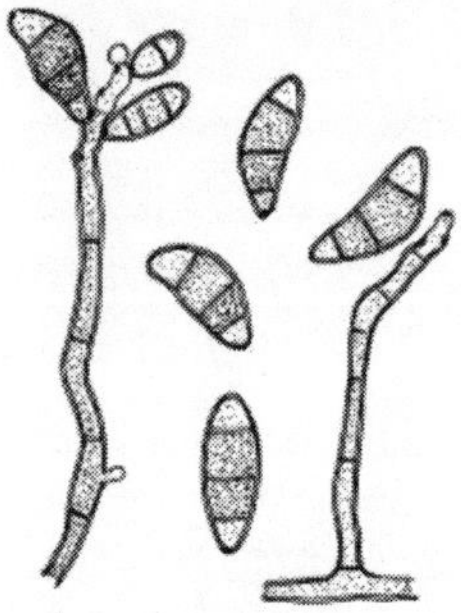

**Fig. 26.** *Curvularia*

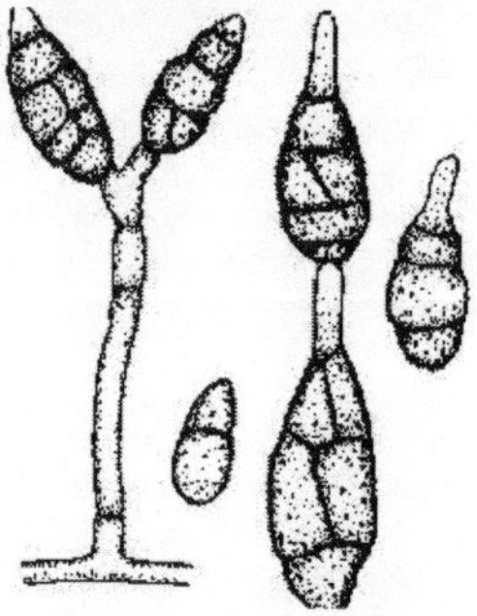

**Fig. 27.** *Alternaria*

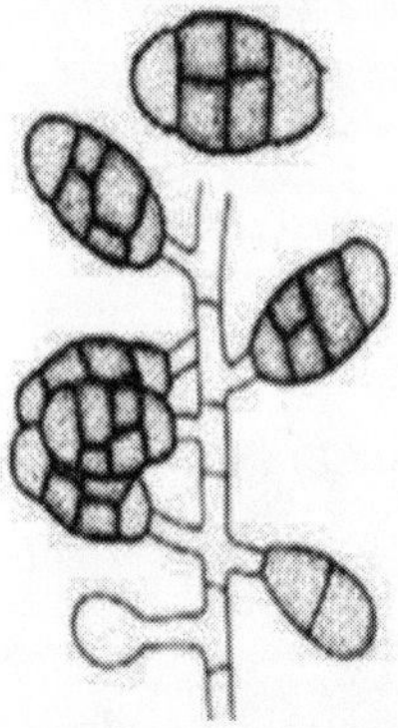

**Fig. 28.** *Pithomyces*

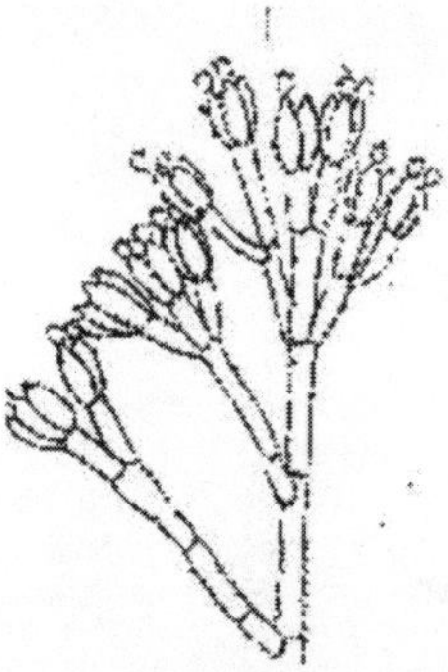

**Fig. 29.** *Penicillium*

**Fig. 30.** *Claviceps*

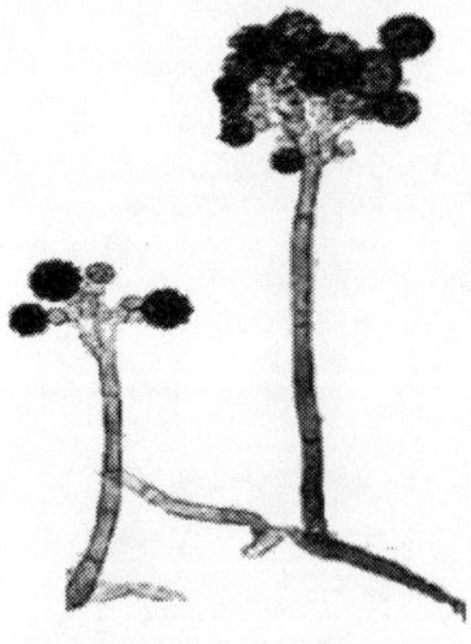

**Fig. 31.** *Periconia*

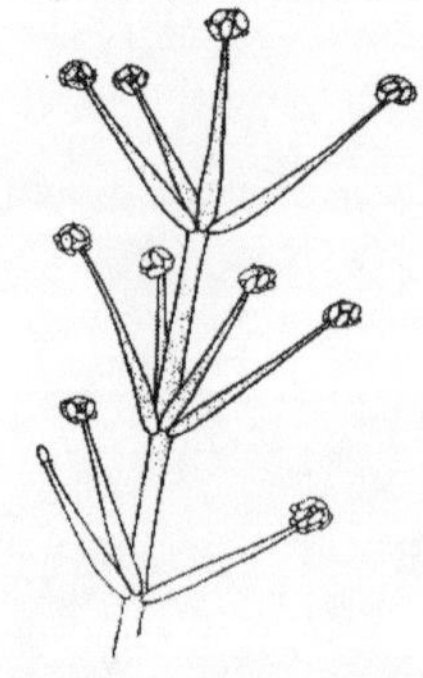

**Fig. 32.** *Verticillium*

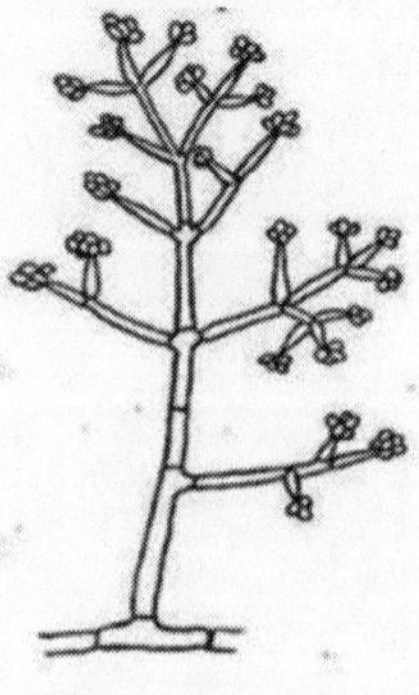

**Fig. 33.** *Trichoderma*

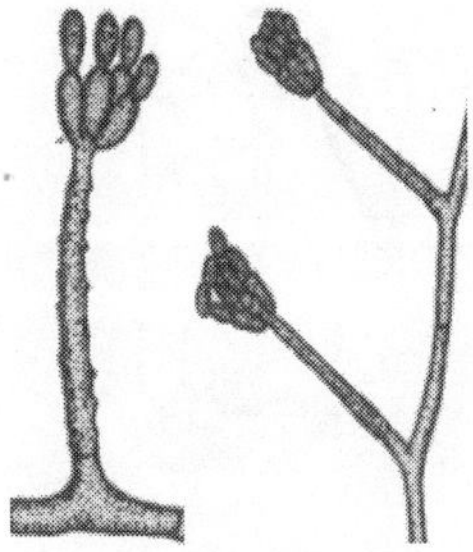

**Fig. 34.** *Stachybotrys*

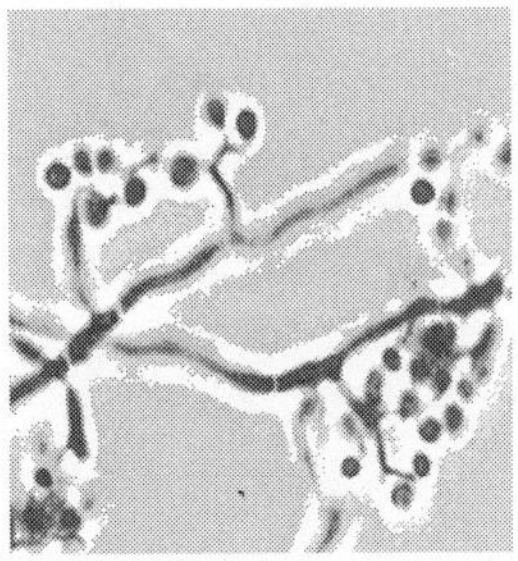

**Fig. 35.** *Beauveria*

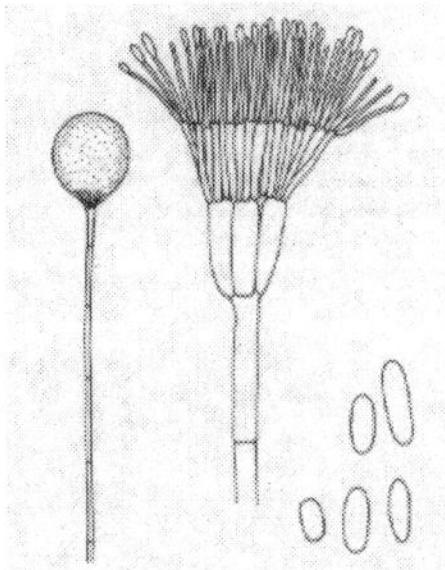

**Fig. 36.** *Gliocladium*

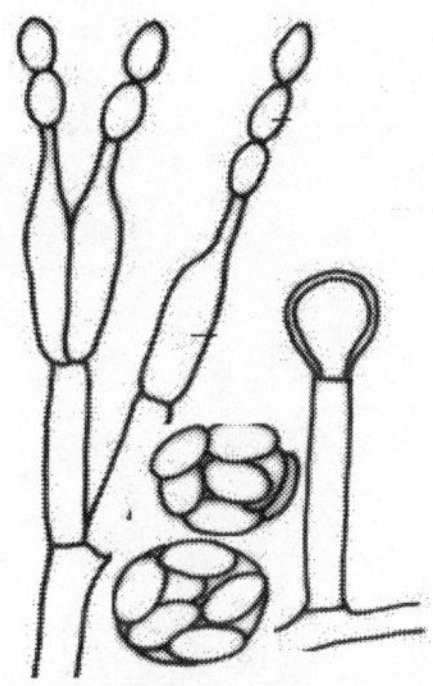

**Fig. 37.** Byssochlamys

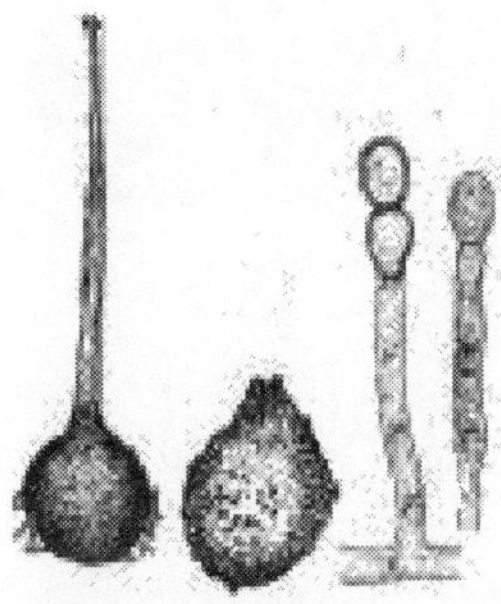

**Fig. 38.** *Ceratocystis*

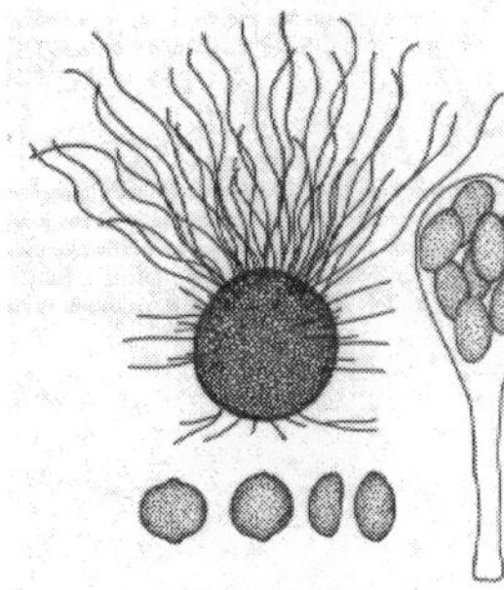

**Fig. 39.** *Chaetomium*

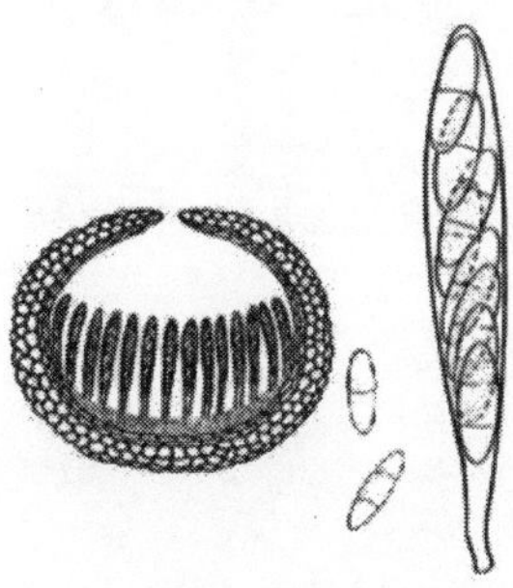

**Fig. 40.** *Gibberella*

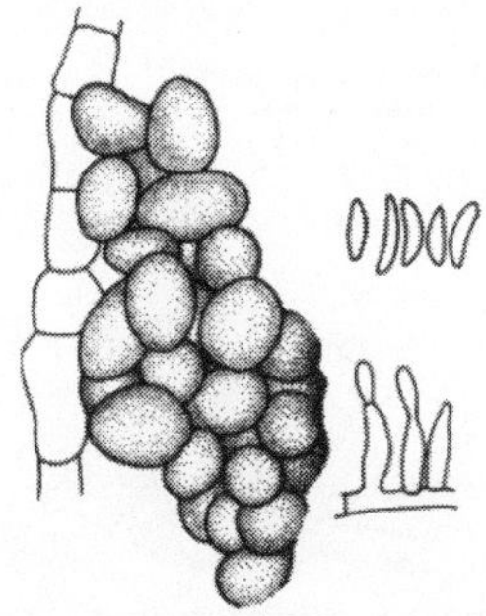

**Fig. 41.** *Microdochium*

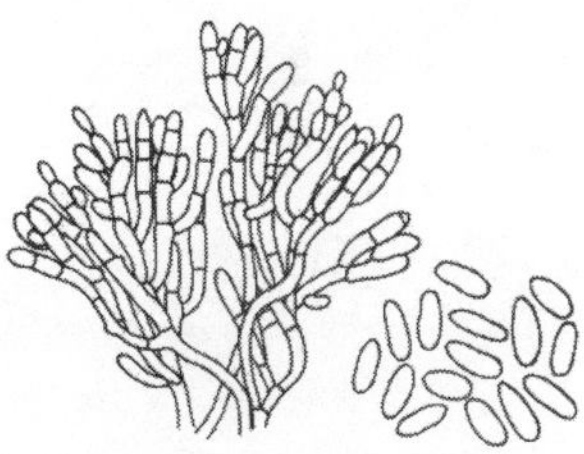

**Fig. 42.** *Metarrhizium*

**Fig. 43.** *Thielavia*

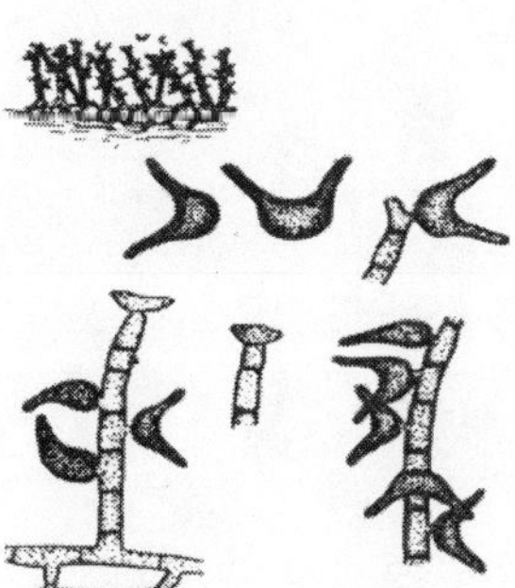

**Fig. 44.** *Arthrinium*

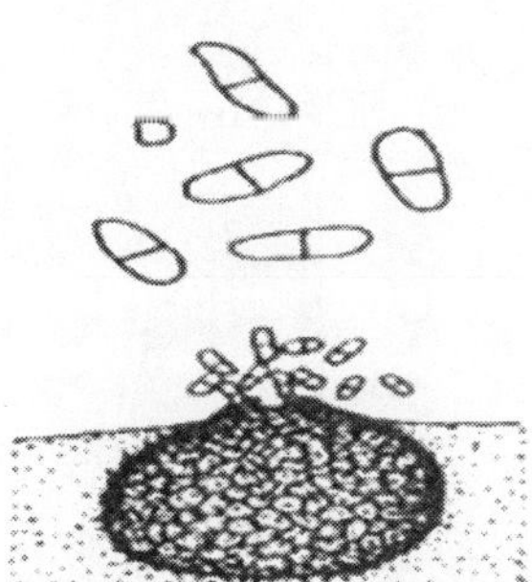

**Fig. 45.** *Ascochyta*

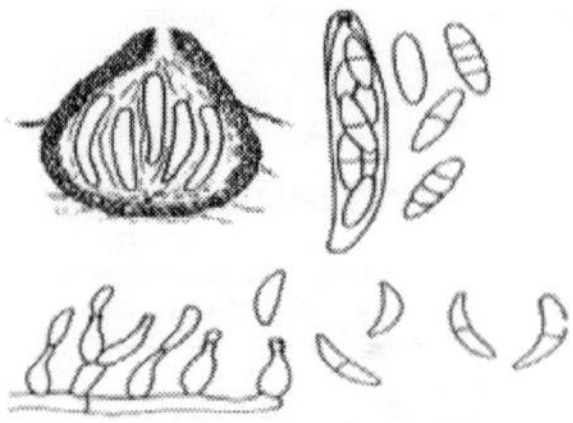

**Fig. 46.** *Monographella*

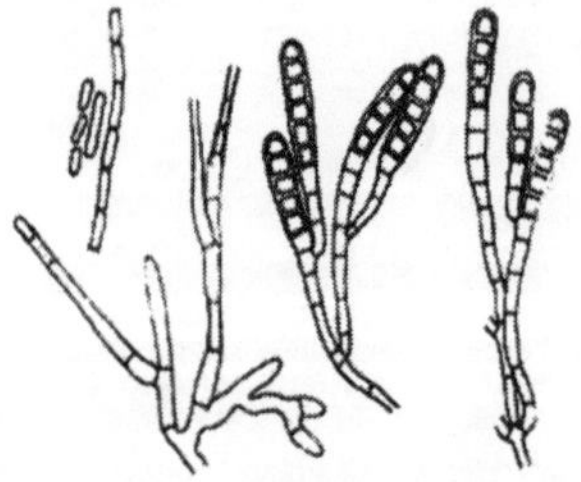

**Fig. 47.** *Chalara*

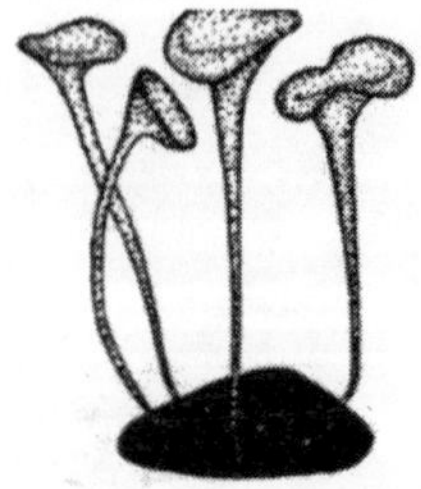

**Fig. 48.** *Sclerotinia*

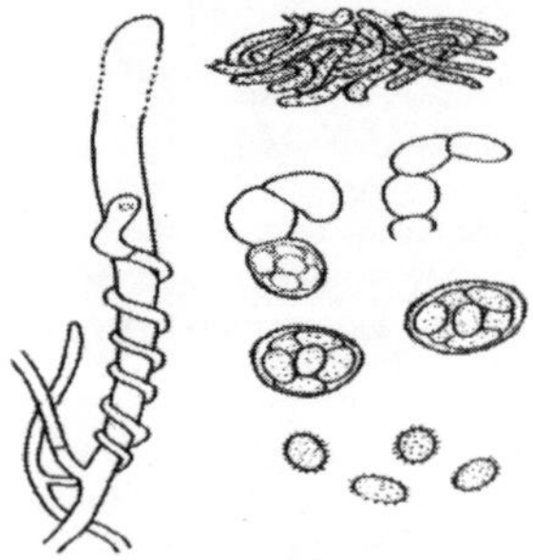

**Fig. 49.** *Talaromyces*

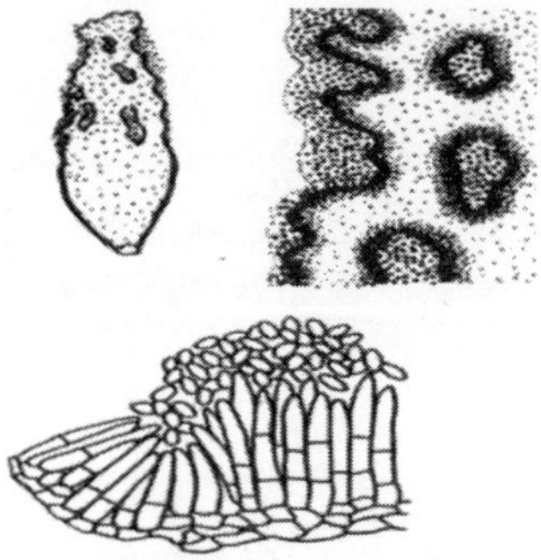

**Fig. 50.** *Sphacelia*

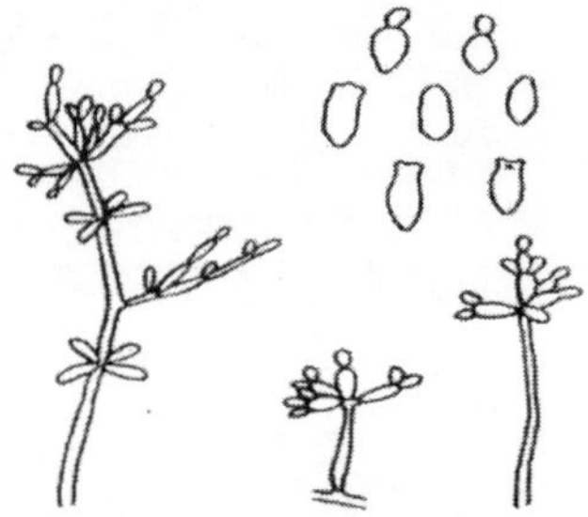

**Fig. 51.** *Hylodendron*

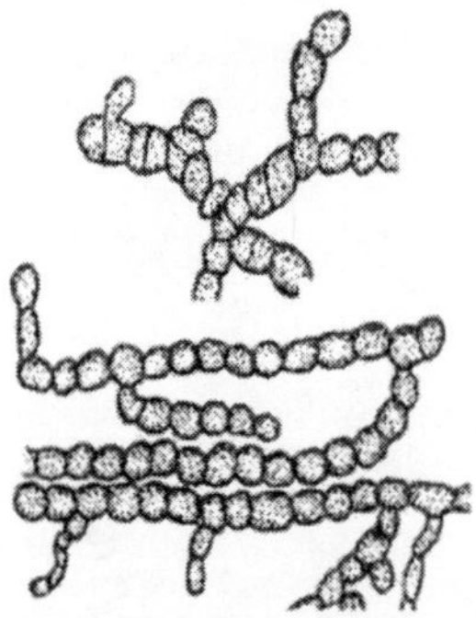

**Fig. 52.** *Hormiscium*

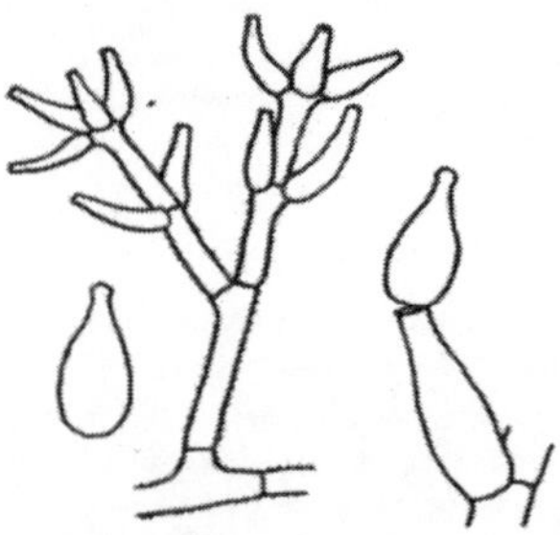

**Fig. 53.** *Sepedonium*

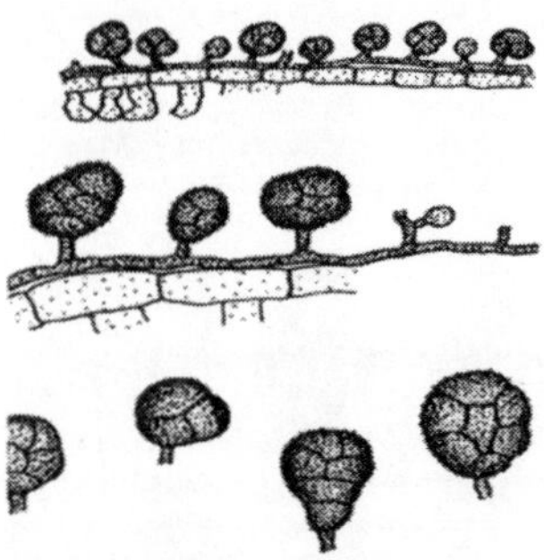

**Fig. 54.** *Piricauda*

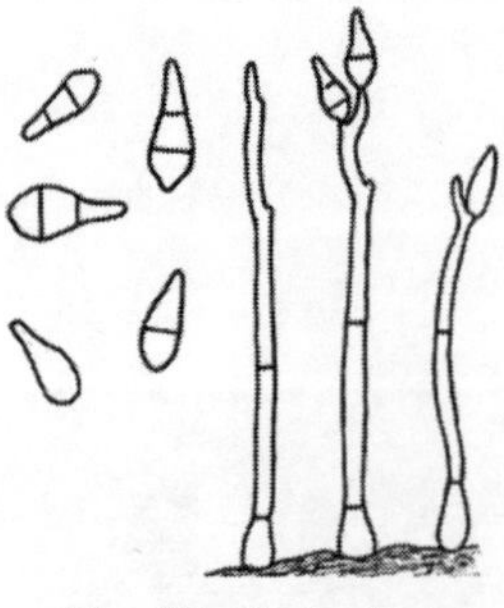

**Fig. 55.** *Pyricularia*

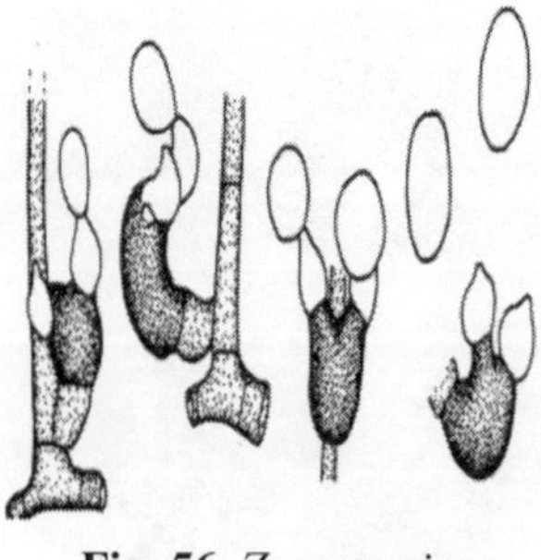

**Fig. 56.** *Zygosporium*

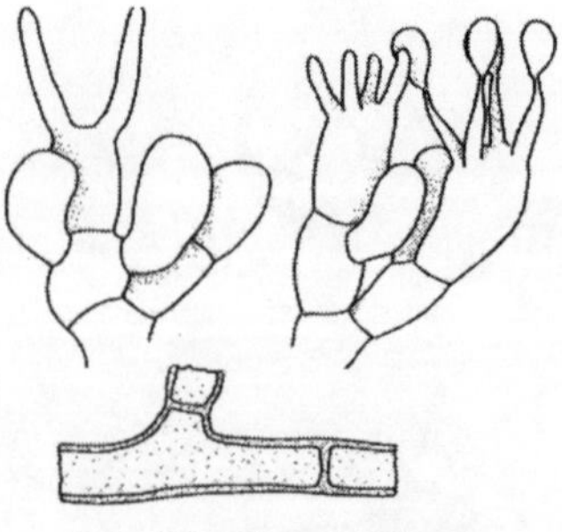

**Fig. 57.** *Thanatephorus*

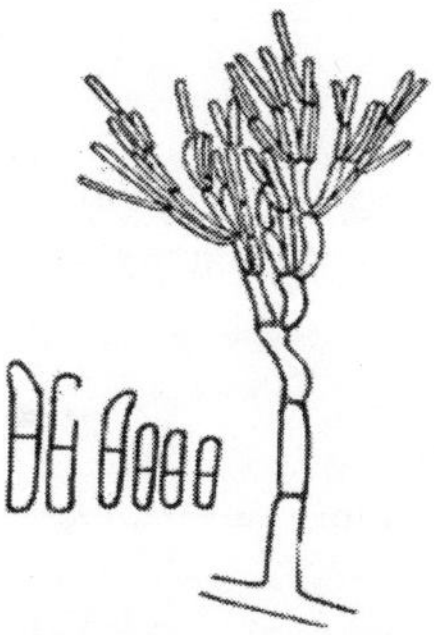

**Fig. 58.** *Cylindrocarpon*

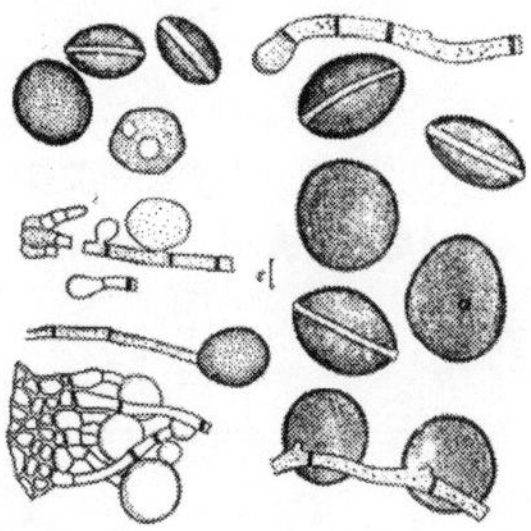

**Fig. 59.** *Ptereoconium*

**Fig. 60.** *Rosellinia*

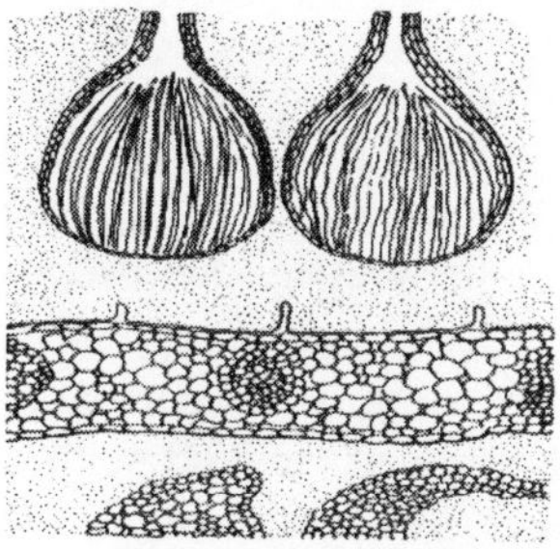

**Fig. 61.** *Epichloe*

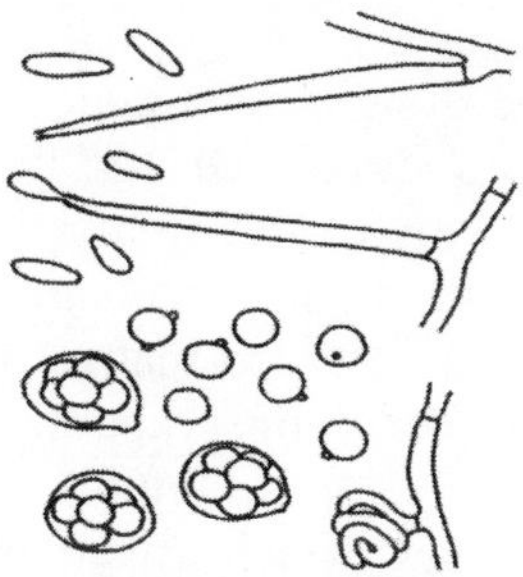

**Fig. 62.** *Nigrosabulum*

**Fig. 63.** *Dichotomomyces*

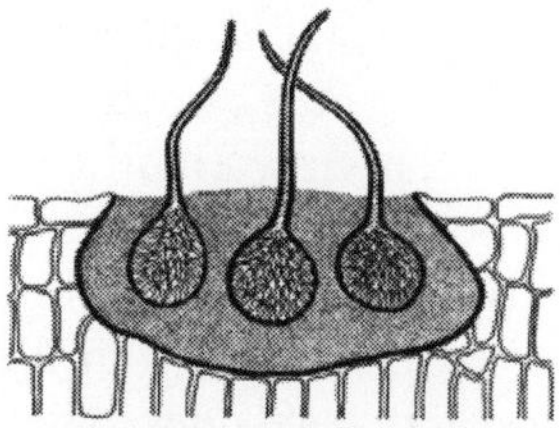

**Fig. 64.** *Diaparthe*

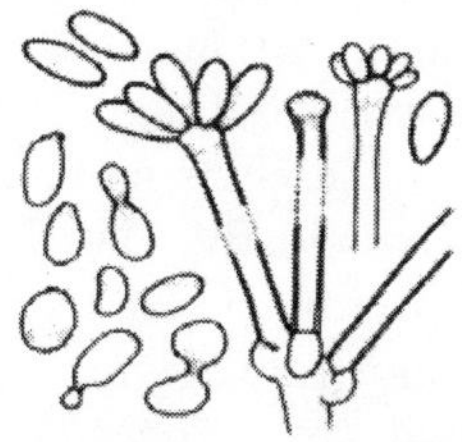

**Fig. 65.** *Filobasidium*

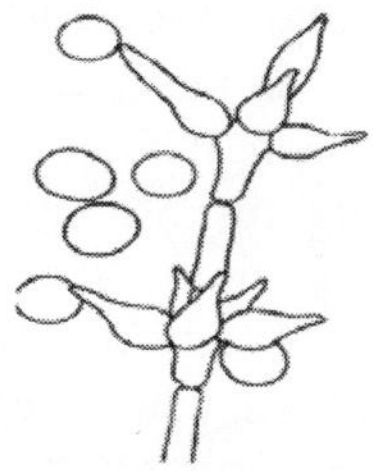

**Fig. 66.** *Verticimonosporium*

*Chapter 2*

# Isolation and Detection of Mycotoxigenic Fungi

The detection of mycotoxigenic fungi, in stored products, requires the use of specialized media and techniques developed for the purpose by internationally collaborating food mycologists. Many earlier studies have suggested that very wide range of fungi are associated with particular commodities. However, the use of modern enumeration techniques has revealed that most fungi are present merely as contaminants and that the range of fungi capable of invasion of a particular commodity or genuinely associated is quite limited. Most fungi associated with a particular commodity are 'field' fungi, which invade grains or nuts before or at the harvest. All have High Water Activity ($a_w$) requirements and cease their growth soon after the crop is harvested. The majority of these fungi play little role in spoilage and are non-toxigenic. The principal exceptions are *Fusarium* and *Alternaria* species. After harvest and during drying, 'storage' fungi take over. However, the distinctions are increasingly blurred, as some storage fungi, including *Aspergillus flavus* and related species, are now known to invade nuts and maize before harvest. Significant mycotoxigenic fungi in the tropics are mostly related to those occurring in temperate zones. In the tropics, *Eurotium* species and other Aspergilli are dominant, while *Penicillium* species play only a minor role. Significant mycotoxin producers in the tropics are relatively few. Growth of *Aspergillus flavus* and the consequent production of aflatoxins is the dominant problem.

Mostly three aspects are associated with mycotoxigenic fungi: (1) the conditions governing the growth of fungi in stored products; (2) The detection and identification of fungal growth in stored products and (3) in broad terms, the major fungi of significance in stored products.

Eight factors are reported to govern the growth of mycotoxigenic fungi in foods: (1) water activity ($a_w$), (2) hydrogen ion concentration, (3) temperature, (4) gas tensions, specifically of oxygen and carbon dioxide, (5) consistency, i.e. liquid or solid state, (6) nutrient status, (7) specific solute effects and (8) preservatives (Pitt and Hocking 1985). In stored products, several of these factors are either fixed or stable. Stored products are solids, of more or less neutral pH, and of adequate nutrient status for fungal growth. Compounds such as NaCl, which produce specific solute effects, are usually absent. Thus, the principal factors governing fungal growth are water activity, temperature, gas tension and preservatives (including fungicides and fumigants).

## 2.1. Mycological Examination of Foods

Methodology in food mycology originally developed for food bacteriology and plant pathology, where the media and methods developed suitable for bacteria and plant pathogenic fungi, are unsuitable to the xerophilic and mesophilic fungi important in stored products. Improved methodology was needed. At a specialist workshop on 'Methods for the Mycological Examination of Foods' new approaches to methodology were examined and standardized which are discussed below.

**i) Dilution plating**. Dilution plating techniques which are effective for enumerating bacteria are still widely used for this purpose. Workshop was also held in the Netherlands in 1990 for the stored products. They suggested continued development of these methods and use of collaborative studies to prove the value of the methods being advocated An International Commission on Food mycology (ICFM) evaluated different methods in food mycology as well as the moves towards standardization of methods which will allow effective comparison of studies carried out in different laboratories. Some of the methods recommended are described here:

## 2.2. Sample preparation

**i) Direct plating:** For particulate foods, including stored products, direct rather than dilution plating of samples is considered to be effective. In direct plating, grains are placed on suitable agar plates after surface sterilization in a 10% dilution of household chlorine bleach for two minutes. Surface sterilization is essential to remove surface contaminants and allow development of fungi which are actually growing, or have already grown in the grains and allow establishing the real association of particular fungus in commodities.

**ii) Protocol:** Surface sterilize a minimum of, 50 food particles (grains, nuts, etc.) in 0.4% chlorine (10% dilution of commercial household bleach) for two minutes and stir intermittently with the help of forceps. Pour off the chlorine and discard. Rinse once in sterile water, this is optional as rinsing has not been shown to be necessary.

Using sterile forceps place 6-10 pieces (depending on particle size) per plate on suitable agar medium. Incubate upright for 5 days at 25-30°C. Incubation time may vary with temperature and medium used. Count number of infected grains and express as percentage invaded. Attempt to identify the fungi growing on the grains and surrounding agar, either *in situ* or after isolation. Suitable identification media may be used for this purpose.

**iii) Dilution Plating**: Suitable medium for fungal enumeration in foods may be selected from those developed in Food Bacteriology or Plant Pathology. Such medium is usually with $a_w$, is suitable to enumeration of fungi associated with growing plants. They were of little value for detecting or enumerating the xerophilic fungi associated with stored products. A major problem in the enumeration of fungi from stored products is the presence of common rapidly spreading fungi - species of *Mucor*, *Rhizopus*, and other *Zygomycetes* which quickly overgrown Petri dishes, rendering counting and isolation of important fungi difficult or impossible. This can be avoided by employing a weak medium such as Potato Dextrose Agar (PDA), which limits the growth of such fungi. However, simple PDA not suited to the growth of storage fungi and need some modifications.

**iv) Media:** If high $a_w$ media like MEA, FDA, OGY, or DRBC are used alone, they will often not detect the most common xerophilic fungi (*Eurotium* species and members of the *A. restrictus* series) (Hocking 1981). Malt salt agar was the first medium used for isolation of storage fungi in grains. Although this medium is better than the high $a_w$ media mentioned above, it is not recommended. Salt is generally considered inhibitory to many fungi (with the exception of halophiles and it reacts with agar to form a soft, granular gel which makes spread-plating difficult). Salts reacts with some other ingredients, such as malt extract, giving rise to precipitates that can interfere with counting of fungal colonies.

## 2.3. Methods and Media for Isolation of Xerophilic Fungi

Since grains and nuts are usually contaminated with soil fungi, many fungi can be isolated from these stored commodities. Most of these fungi occur only as surface contaminants, and are not capable of growing in the commodity and spoiling it. Stored commodities may have reduced $a_w$, and the $a_w$ should be less than 0.60 for them to be microbiologically stable. However, stored commodities have moisture often more than this, or there are damp pockets due to moisture migration and xerophilic fungi are able to grow. It is important to use appropriate media and methods for isolating xerophilic fungi from these low moisture foods. If high $a_w$ media are used, then many of the important fungi, such as the *Aspergillus restrictus* series and some *Eurotium* species will not be able to grow and will be missed.

i) **Methods:** There are two methods commonly used for isolating fungi from foods. The first is dilution plating. This method is familiar to most food microbiologists, and the method for isolating fungi from foods is similar to that used for bacteria, with two basic differences. After homogenization and dilution, samples are spread plated, not pour-plated, with 0.1 ml of dilution onto the surface of the agar and then spread with a glass spreader. This allows fungal propagules equal access to oxygen and a better chance of uniform colony development. The plates are then incubated upright rather than inverted, so that fungal spores are not shed onto the lid of the petri dish.

With reduced $a_w$, commodities, particularly dried foods and meats, the sample may need to be rehydrated in the diluents for up to 1 hour, before homogenization. A suitable diluent for most samples is 0.1% peptone. Rehydration of grain samples may also aid the homogenization process. Samples can be rehydrated at room temperature. They do not need to be refrigerated during this process, as fungal spores take many hours to germinate, and the counts will not increase during this period. Gradual rehydration can aid resuscitation of yeast cells and fungal conidia, shortening germination times, increasing subsequent growth rates and also increasing number of propagules recovered.

ii) **The other method** that can be used to assess the mycological quality of stored commodities is direct plating (Mislivec and Bruce, 1977). This involves placing discrete pieces, grains, or kernels directly onto suitable agar, usually after surface disinfection with 0.35-0.4% sodium hypochlorite (Andrews, 1986). This is done to remove any contaminating spores from the external surface. Direct plating permits detection of fungal hyphae that have actually penetrated and grown in the commodity; that is, it detects the fungi that are actually growing in the grain or nut, not just the spores clinging to the outside. Direct plating provides an estimate of the extent of infection in a commodity, usually expressed as a percentage. Usually at least 50 and preferably 100 grains or kernels are plated out onto several petri dishes (about 10-15 per plate, depending on the size), and this makes calculating the percentages very easy. Direct plating is often the best way to study the degree of contamination of a commodity with a particular fungus or group of fungi, when used in conjunction with a selective or differential agar, e.g. *Aspergillus flavus* and *parasiticus* agar for aflatoxigenic fungi, DG18 for *Eurotium* species, and Y50G for fastidious xerophiles. There are two other useful, though more specialized, media for isolation of particular groups of xerophilic fungi.

Malt extract Yeast extract 50% Glucose agar (MY50G) are used for isolation of extremely xerophilic fungi from high sugar foods such as confectionery and dried fruits. It is most useful for direct plating of samples, and is used only for dilution plating for detecting xerophilic yeasts in syrups and similar products. Reduced $a_w$ diluents (e.g. 20% glucose) should be used to avoid osmotic shock when enumerating xerophilic yeasts from these concentrated products. MY50G is ideal for isolation of *Xeromyces bisporus*, xerophilic *Chrysosporium* species, *Eremascus* species, and *Eurotium halophilicum*. Because it contains no spreading inhibitors, it is not suitable as an enumeration medium if common *Eurotium* species are present, as they will quickly outgrow the slower growing xerophiles. For isolation of extreme xerophiles in the presence of *Eurotium* species, it may be necessary to use a medium of even lower $a_w$, such as MY70GF, $a_w$ (0.76), which contains 70% w/w glucose and fructose in equal proportions.

Malt extract Yeast extract 5% salt, 12% glucose agar (MY5-12) were developed for isolation of halophilic fungi from salted sea food products (Pitt and Hocking, 1985). This medium is optimal for isolating *Polypaecilum pisce* and *Basipetospora halophila*, two halophilic fungi which commonly grow on dried sea foods in the tropics. *B. halophila* is the more xerophilic of the two species, and actually grows better if 10% rather than 5% salt is incorporated into the medium. While these fungi do not have an absolute requirement for salt, they grow much faster if some salt is present (Andrews and Pitt, 1986).

There are a number of other media for xerophilic fungi (Beuchat and Hocking, 1990), many of which are modifications of the media already mentioned. Some are older modifications that are no longer in use in mainstream food mycology, because they are either unnecessarily complicated or are not as effective as those in current use. For general purpose, enumeration of xerophilic fungi from stored commodities and intermediate and low moisture foods, DG18 is recommended by the International Commission on Food Mycology (ICFM).

**i) Dichloran Rose Bengal Chloramphenicol Agar (DRBC)**

A breakthrough in medium formulation came with the introduction of dichloran. It inhibits the growth of spreading fungi. Dichloran Rose Bengal Chloramphenicol Agar medium (DRBC) was developed by King *et al.* (1979) which is effective for enumeration and isolation of fungi from foods. It has a high $a_w$, but sufficient nutrients to permit growth of many normal storage fungi including species of *Penicillium*, *Aspergillus* and common genera associated with developing grains such as species of *Fusarium*, *Alternaria* and *Curvularia*. It is recommended by ICFM as a general-purpose isolation medium. It is also highly effective for detection and enumeration of the major mycotoxin-producing fungi, including *Aspergillus flavus*, *A. ochraceus* and some *Fusarium* species. It also permits the detection and enumeration of slow growing species like *Aspergillus penicillioides* and *A. candidus* in the presence of high number of *Eurotium* colonies. The composition of DRBC is as follows:

| | |
|---|---|
| Glucose | 10.0 g |
| Peptone (Bacteriological) | 5.0 g |
| $KH_2PO_4$ | 1.0g |
| $MgSO_4$.7H20 | 0.5g |
| Rose bengal | 25mg (5% in water, 0.5ml) |
| Dichloran | 2mg (0.2% in ethanol, 1.0 ml) |
| Chloramphenicol | 0.1g |
| Agar | 15.0 g |
| Distilled water | 1000 ml |

Sterilize the medium at 121°C for 15 minutes in an autoclave. Store prepared medium in dark to prevent slow decomposition of rosebengal. The stock solutions of rose bengal and dichloran need no sterilization and are stable for very long periods.

**ii) Dichloran 18% glycerol agar (DG 18)**

For isolation of xerophilic fungi which cause deterioration of stored products, requires medium of much lower water activity ($a_w$) than DRBC. Dichloran 18% glycerol agar was developed by Hocking and Pitt (1980) specifically for the isolation and enumeration of xerophiles. DG18 provides good growth of fungi such as *Aspergillus, Penicillioides*, *Wallenzia sebi*, and *Eurotium* species all of which may escape the observations on PDA or DRBC, but are of great significance in stored products. The composition of DG18 of medium is

| | |
|---|---|
| Glucose | 10.0 g |
| Peptone (Bacteriological) | 5.0 g |
| $KH_2PO_4$ | 1.0g |
| $MgSO_4$.7H20 | 0.5g |
| Glycerol, A.R. | 220 g |
| Dicholoran | 2mg (0.2% in ethanol, 1.0 ml) |
| Chloramphenicol | 0.1g |
| Agar | 15.0 g |
| Distilled water | 1000 ml |

Take all ingredients and agar in about 800 ml distilled water and steam to dissolve agar. Then make up the volume to 1 lit with distilled water. Add glycerol (which gives a final concentration of 18% w/w). Sterilize by autoclaving at 121°C for 15 minutes. Note that glycerol is 18% (weight in weight, not weight in volume). The final $a_w$ of this medium is 0.955.

**iii) Aspergillus flavus and Parasiticus Agar (AFPA)**

More selective media are also of value for detecting specific fungi in stored products. Aspergillus flavus and parasiticus agar (AFPA) is recommended by Pitt *et al.* (1983) for the detection and enumeration of *A. flavus* and other potentially aflatoxigenic fungi in nuts, maize, spices and other commodities. When incubated at 30°C for 48 to 60 hours, colonies of *A. flavus* and *A. parasiticus* are distinguished by bright orange-yellow reverse colour. The advantage of AFPA is that a short incubation period of 48 hours is sufficient and specific for these fungi and it requires no skill in interpreting the results. In consequence, it can be a simple, routine guide for aflatoxin contamination (Pitt, 1984). Few other fungi produce colouration similar to *A. flavus* and closely related species on AFPA. Only *A. niger* can be a source of error, as it grows as rapidly as *A. flavus* and sometimes produces a yellow, but not orange reverse colour. After 48 hours, *A. niger* colonies begins production of their diagnostic black or dark-brown heads, providing a ready distinction from *A. flavus*. After prolonged incubation of 4-5 days, *A. ochraceus* colonies may also produce a yellow reverse, but this species grows slowly at 30°C and the colour reaction does not appear within 48 hours. The composition of AFPA is:

| | |
|---|---|
| Peptone (Bacteriological) | 10.0 g |
| Yeast extract | 20.0g |
| Ferric ammonium citrate | 0.5g |
| Dichloran | 2mg (0.2% in ethanol, 1.0 ml) |
| Chloramphenicol | 0.1g |
| Agar | 15.0g |
| Distilled water | 1000 ml |
| pH | 6.2 |

Sterilize in an autoclave at 121°C for 15 minutes.

**iv) Dichloran Peptone Chloramphenicol Agar (DCPA)**

For enumeration of *Fusarium* species from stored products, Dichloran Peptone Chloramphenicol Agar (DCPA), developed by Andrews and Pitt (1986), is of great value. DCPA is also effective medium for isolation of other genera of field fungi such as *Alternaria, Drechslera* and *Curvularia* which usually sporulate well on it. In addition, DCPA is very useful for the identification of *Fusarium* species because it induces the abundant formation of macroconidia. However, it is not suitable for maintaining or storing of cultures for more than two weeks as ammonia may be produce in aging cultures. The composition for DCPA is given below:

| | |
|---|---|
| Peptone | 15.0 g |
| $KH_2PO_4$ | 1.0g |
| $MgSO_4.7H_20$ | 0.5g |
| Dichloran | 2mg (0.2% in ethanol, 1.0 ml) |
| Chloramphenicol | 0.1g |
| Agar | 15.0g |
| Distilled water | 1000 ml |

Sterilise the medium in an autoclave at 121°C for 15 minutes. Different media recommended for direct plating of stored products are presented in Table 2.1.

**Table 2.1.** Media for direct Plating

| Medium | Purpose | Incubation |
|---|---|---|
| DRBC | General; suitable for most Aspergilli, 5 days for Penicillia and mycotoxigenic fungi | 5 days at 25°C |
| DG18 | Xerophiles, best for *Eurotium* spp., *Wallemia, Penicillioides* | 5-7 days at 25°C |
| AFPA | *A. flavus* and other aflatoxin producers | 2-3 days at 30°C |
| DCPA | *Fusarium, Alternaria, Drechslera* and other dematiaceous hyphomycetes | 5 days at 25°C |

In a collaborative study with the Department of Agriculture, Bangkok, Thailand, the South East Asian detection Center for Tropical Biology (BIOTROP), Bogor, Indonesia, and the National Post harvest Institute for Research and Extension (NAPHIRE), Munoz, Philippines, revealed the detection of limited number of fungi which have been consistently associated with South East Asian as commodities during storage are precised in Table 2.2.

**Table 2.2.** Fungi detected on different commodities in South East Asian Commodities

| Commodity | Field fungi | Spoilage fungi | Mycotoxigenic fungi |
|---|---|---|---|
| Cashews | *Cladosporium cladosporioides* | *Chaetomium* spp | *Aspergillus flavus* |
| Cassava | *Lasiodiplodia theobromae*<br>*Nigrospora oryzae, Phoma spp* | | |
| Copra | *N. oryzae* | *Chaetomium* spp<br>*A. tamarii*<br>*A. niger* | *A. flavus* |
| Maize | *N. oryzae*<br>*Curvularia pallescens*<br>*C. lunata, C. clavata*<br>*Lasiodiplodia theobromae*<br>*Bipolaris maydis,* | *A. niger, A. flavus*<br>*Penicillium citrinum*<br>*P.funiculosum,*<br>*A. wentii* | *A. flavus*<br>*F. moniliforme*<br>*F.semitectum* |

| Commodity | Field fungi | Spoilage fungi | Mycotoxigenic fungi |
|---|---|---|---|
| | *Arthrinium phaeospermum, Rhizopus oryzae, Phoma herbarum, Rhizoctonia solani* | | |
| Peanuts | *Cladosporium cladosporioides* *Lasiodiplodia theobromae* *Pestalotiopsis guepinii* | *Aspergillus niger* *P. pinophilum* *Chaetomium* spp. | *A. flavus* |
| Rice | *Bipolaris maydis, Fusarium semitectum, Cl. cladosporioides* *N. oryzae, Curvularia lunata* *C. genticulatus, C. oryzae* *C. pallescens, Phoma* sp. *Colletotrichum* sp. | | *Alternaria padwickii* *A. alternata* *A. longissima* |
| Sorghum | *B. maydis, Cl. cladosporioides* *F.semitectum, N. oryzae* *C. lunata, C. pallescens* *Phoma* sp., *Setosphaeria rostrata* | *A.niger , Eurotium chevalieri, E. rubrutn,* *Chaetomium* sp. | *A. flavus* *F. moniliforme* *P. citrinum* *Alternaria longissima* *A. alternata* |
| Soya bean | *Arthrinium phaeospermum* *Lasiodiplodia theobromae* *F. semitectum, Cl. cladosporioides* *N. oryzae, Curvularia lunata,* *C. Pallescens, Phoma* spp. *Epicoccum nigrum, Pestalotiopsis guepinii* | *A. niger* *A. wentii, A. restrictus* *A. penicillioides* *Eurotium rubrum* *Eupen. cinnamopurpureum* *Chaetomium* sp. | *A. flavus* *F. moniliforme* *P. citrinum* *A. alternata* *Eupenicillium* |

Table 2.2 reveals that majority of the fungi isolated from the commodities examined were field fungi. Such fungi commonly require high $a_w$ levels for growth. In many cases growth must have occurred before harvest. This implies a close relationship with the growing grain. As a result of the use of this technique, it is increasingly clear that close association between particular fungus and developing grain is more widespread than earlier suspected and is not confined to true pathogens. Most of the fungi encountered by this way are of doubtful significance as they cause little deterioration and are not mycotoxigenic. *Fusarium* and *Alternaria* are exceptions.

Spoilage fungi were represented by *Aspergillus* and *Eurotium* species, which are capable of growing and causing deterioration of products stored at moisture level which is little above the safe levels. Several *Chaetomium* species were also isolated, but their role in stored products is less certain. Only three species of *Penicillium* (*P. citrinutn, P. funiculosum* and *P. pinophilum*) where detected. *A. flavus* was very common in maize and peanuts and also quite frequent in copra and sorghum. *Fusarium moniliforme* was found at high levels in maize.

## 2.4. Taxonomy of Food Spoilage Fungi

Fungal classification has traditionally been based on the appearance (morphology) of colonies and microscopic structures. Because this information is qualitative, decisions on the relationships among species (taxonomy) was often based on personal opinion. Moreover, correct procedures for naming fungi (nomenclature) were often ignored. Main problem is that the literature is diffuse and requires extensive background knowledge to locate information. Clarifying and stabilizing both nomenclature and taxonomy of food spoilage and mycotoxigenic fungi has become imperative. Improvements to our understanding must inevitably involve the taxonomy of three genera: *Penicillium, Aspergillus* and *Fusarium.*

Many advances have been made in recent years, especially in *Penicillium.* The difficulties in naming fungus which possess both sexual (teleomorphic) and asexual (anamorphic) states have been discussed at length. The principles of priority, typification and type specimens are increasingly being applied to both teleomorphs and anamorphs, bringing order to nomenclature. New taxonomic approaches are providing criteria independent of traditional methods. The most important advances have come from studies on secondary metabolites and mycotoxins, electrophoretic patterns of certain isoenzymes and DNA Restriction Fragment Length Polymorphism (RFLP) studies.

As a result of these initiatives, the taxonomy of most *Penicillium* subgenera has become firmly established and distinctions among aflatoxin producing *Aspergillus* species have been clarified. Using techniques developed from *Penicillium* studies, the taxonomy of the ascomycetes genus *Monascus* has been revised. Two text books on food spoilage fungi, **'Fungi and Food Spoilage'** by Pitt and Hocking (1985) which provides keys and descriptions to 200 commonly occurring food-borne fungi. Another book **'An Introduction to food-borne Fungi'** by Samson and van-Reenen Hoekstra (1988) is helpful in the identification 100 species, primarily based on microscopic morphology. They bring together the common food-borne fungi in single volume, greatly facilitating identification.

*Fusarium,* however, remains difficult. Few of the techniques applied to *Aspergillus* and *Penicillium* have been effective and specialists in the taxonomy of *Fusarium* remain divided over fundamental issues. International collaboration rather than activity by individual taxonomists, is increasingly seen as the way of the future. The recently established International Commission on *Penicillium* and *Aspergillus* will play a key role in continued development and stabilization of names in these genera. Improved agreement on *Fusarium* taxonomy can be expected as the result of formation of a broadly based international sub-commission.

A further problem for the scientist interested in fungi in foods has been that the literature is diffuse, requiring extensive background knowledge to select the books containing effective taxonomies of the species of interest. Improvements to our understanding of the taxonomy of food-borne fungi are inevitably tied up with the taxonomy of the three major food spoilage and mycotoxigenic genera: *Penicillium, Aspergillus,* and *Fusarium.* Simple, accurate, and reproducible classification of *Penicillium, Aspergillus,* and *Fusarium* species is highly desirable. Identification of species in these genera should be unequivocal, accurate, simple and immutable.

The taxonomy of *Penicillium* has always been difficult as the genus has many species that are often difficult to differentiate. The first workable taxonomy was developed by Raper and Thom (1949). *Penicillium* taxonomy in the 1960s and 1970s become uncertain, close to chaotic, with inaccurate identifications widespread in the literature. Consequent confusion was about which species was present in foods or produced mycotoxins.

*Aspergillus* taxonomy was stabilized by the publication of a monograph by Raper and Fennell (1965). Like *Penicillium* monograph, the taxonomic base of *Aspergillus* was narrow and the work suffered from severe nomenclatural problems and failed to adhere the International Code of Botanical Nomenclature (ICBN).

**i) Priority**: Raper and his coworkers sometimes failed to accept names which had priority as required by the ICBN, on the grounds that are unacceptable now. Problems of priority in *Penicillium* were largely corrected by Samson *et al.* (1976) and Pitt (1979), but at the expense of confusion by name changes. Problems over priority in *Aspergillus* and *Fusarium* are equally severe and have not been addressed adequately. It is clear that some important fungi, such as *P. chrysogenum*, *A. niger* and *A. nidulans*, are threatened by older species names and hence are being proposed as *nomina conservanda* (Frisvad *et al.,* 1990). Controversy over priority for names of some *Fusarium* species continues unabated (Nirenberg 1990).

**ii) Typification**: Raper and Thorn (1949), Raper and Fennell (1965) and Nelson *et al.* (1983) have failed to specify dried types for any of the species of *Penicillium, Aspergillus* and *Fusarium.* Even living type of material has not been identified. Descriptions and sometimes existing type material have been ignored.

**iii) Teleomorph names:** Raper and Thom maintained that teleomorph species with anamorph in *Aspergillus or Penicillium* should be named in the anamorph genera and made new combinations as necessary (Raper, 1957). For example, *Eurotium amstelodami* was renamed as *Aspergillus amstelodami* (Thom and Church, 1926) and maintained this way in later taxonomies. However, the ICBN and many nomenclaturalists' opined, that teleomorph name has priority over anamorph names. A second problem concerns with species described in an anamorph genus which may be included the teleomorph. For example for *Penicillium javanicum* van Beyma. Raper and Thom (1949) accepted such species being correctly placed in *Penicillium*, contrary to the ICBN.

Hawksworth and Sutton (1974) feel that the ICBN current was ambiguous in naming of species such as *Penicillium javanicum.* As a consequence, the International Botanical Congress held in Sydney altered articles in new ICBN (Voss *et al.*, 1983) to make naming of such species unequivocal. The price was that some well established names, especially in *Aspergillus* and *Penicillium*, had to be changed. The nomenclatural problem could be overcome, but at the expense of introducing new and unfamiliar names. Pitt (1979) made the necessary changes in *Penicillium.* For example, *Penicillium javanicum* Beyma is now known correctly as *Eupenicillium javanicum* (van Beyma) Stolk and Scott (1967) when the teleomorph, or the whole fungus, under consideration, is *Penicillium indonesiae* Pitt when the anamorph is being considered. Samson and Gams (1985) carried out the necessary alterations in *Aspergillus*: *Aspergillus amstelodami* (Mangin) Thom and Church now correctly called as *Eurotiurn amstelodami* Mangin when teleomorph is present and *Aspergillus hollandicus* Samson and W. Gams for the *Aspergillus* state alone.

**iv) Intrageneric taxa**: Thom and his coworkers arranged clusters of species in *Aspergillus* and *Penicillium* into 'Groups', a subgeneric identification without nomenclatural status under the ICBN. Pitt (1979) replaced the 'Groups' structure in *Penicillium* by a nomenclaturally correct sectional structure. Gams *et al.* (1985) also made similar changes in *Aspergillus.* The use of teleomorph names in *Fusarium* remains enigma. Teleomorphs are seen in culture for only a few food-borne species. Hence, most taxonomists prefer to continue identifying *Fusarium* species by its anamorph.

**v) New Approaches-cultural methods, media and incubation conditions:** New selective medium and incubation conditions were developed to improve the basis for *Penicillium* identification. The use of gross physiological characters growth at 37°C, and a discriminatory medium containing nitrite as selective agent for differentiating species are the some of the suggestions made for correct identification. A standard classification of *Penicillium* developed by employing 2.5% glycerol nitrate on (G25N, 0.935 $a_w$) as adaptation to low water activity.

Pitt *et al.* (1983) also introduced Aspergillus flavus and parasiticus agar (AFPA) for the recognition of *Aspergillus flavus* and *A. parasiticus*. Frisvad (1983) developed Dichloran Yeast Extract crone agar (DRYS) for distinguishing *Penicillium verrucosum* and *P. viridicatum* from other related species. Frisvad (1985a) used Creatine Sucrose Agar (CSA) for separation of species in subgenus *Penicillium* based on preferential utilization of protein or carbohydrate substrates. A selective medium for isolation and enumeration *Fusarium* species was developed by Abildgren *et al.* (1987). Czapek Iprodione Dichloran Agar (CIDA) was reported to be selective for wide range of common food-borne fungi. Hawksworth and Pitt (1973) revised the taxonomy of *Monascus*, a genus important in foods. They reduced 16 species and varieties to 3 species, on the basis of colony diameter and ascospore morphology. Gerlach and Nirenberg (1982) published an atlas of *Fusarium* species based on microscopic morphology. Nelson *et al.* (1983) produced a new taxonomy of *Fusarium* based on growth of isolates in culture under carefully defined conditions, allowing effective identification of fresh isolates from food-borne sources. Secondary metabolites and isoenzymes electrophoresis pattern was also employed in the identification of fungi.

Until recently, the powerful genetic tools of DNA hybridization and analysis of DNA and RNA sequences have been little used in the systematics of these genera (Kurtzman *et al.,* 1986). On the basis of studies of DNA homology among *A. flavus* and related species, several well known species have been reduced to subspecies. Genetic studies of fungi have great potential for aiding identification and understanding of phylogenetic relationship among genera and subgeneric taxa (Taylor *et al.,* 1990). However, genetic studies are only one of the many tools for defining species.

**vi) International Commission** : Due to sustained efforts of international working groups, International Commission on *Penicillium* and *Aspergillus* was formed which inturn formed a sub-commission on *Penicillium* and *Aspergillus* systematics (SPAS). This was upgraded to full commission in 14 MS (International

Commission on *Penicillium* and *Aspergillus* (ICPA) consisting of 12 members from seven countries combining both morphologists and taxonomists and those skilled in new techniques. They plan to carryout collaborative taxonomic studies, revisions of names, culture collection names, development of taxonomic techniques and other activities.

A new approach to the stabilisation of names of fungal species has emerged. A strong international group which directed drawing up lists of 'Names in Current Use, and the establishment of protected status for them. In the next few years, this approach promises to bring much-needed stability to names of *Penicillium* and *Aspergillus* species. The production of a similar list for *Fusarium* species is equally important.

Species of about 32 dematiaceous Hypomycetes producing mycotoxins and metabolites such as *Alternaria* (*Pleospora*), *Bipolaris* (*Cochliobolus*), *Botrytis* (*Botryotinia*), *Cercospora* (*Mycosphaerella*), *Cercosporidium* (*Mycosphaerella*), *Cladosporium, Corynespora, Curvularia* (*Cochliobolus*), *Dematophora* (*Rosellinia*), *Drechslera* (*Pyrenophora*), *Epicoccum, Exserohilum* (*Setosphaeria*), *Fulvia; Gmphium, Helminthosporium, Memnoniella, Mycocetztrospora, Nigrospora, Nodulisporium, Periconia; Phaeoramularia, Phytnatotrichiopsis, Pithomyces* (*Leptosphaerulina*), *Pseudocercospora* (*Mycosphaerella*), *Pyricularia* (*Magnaporthe*), *Rhynchosporium, Sirosporium, Spilocaea* (*Venturia*), *Stachybotrys, Stemphylium* (*Pleospora*), *Thedgonia* and *Zygosporium* have been recognised. The majority of these hypomycetous anamorphs having bitunicate teleomorphic state belonging to Ascomycetes.

Spoilage of stored commodities due to growth of xerophilic fungi, are controlled by many factors including $a_w$ and interactions with pH (Fig. 2.1). If the $a_w$ of foods or commodities is below 0.90 or the pH is less than 4.5, then few bacteria will grow. If the $a_w$ of a food or commodity is below 0.85, then spoilage will be caused almost exclusively by moulds and yeasts. Foods of reduced $a_w$ are often referred to as intermediate moisture foods, and their $a_w$, is generally between 0.85 and 0.60. Commodities with below $a_w$ 0.60 are microbiologically stable as no microorganism can grow. Intermediate moisture foods can be separated into several groups on the basis of the types of fungi that are able to grow in them and cause spoilage (Hocking, 1991).

They are

1. Stored commodities — cereals, oil seeds, nuts, and spices
2. High sugar foods — confectionery, dried fruits and jams
3. Dried meats and meat products and
4. Dried seafood products.

The mycoflora of stored commodities and of dried meats is fairly similar. Usually, *Eurotium* species are dominant, with the *Aspergillus restrictus* series which is also very common.

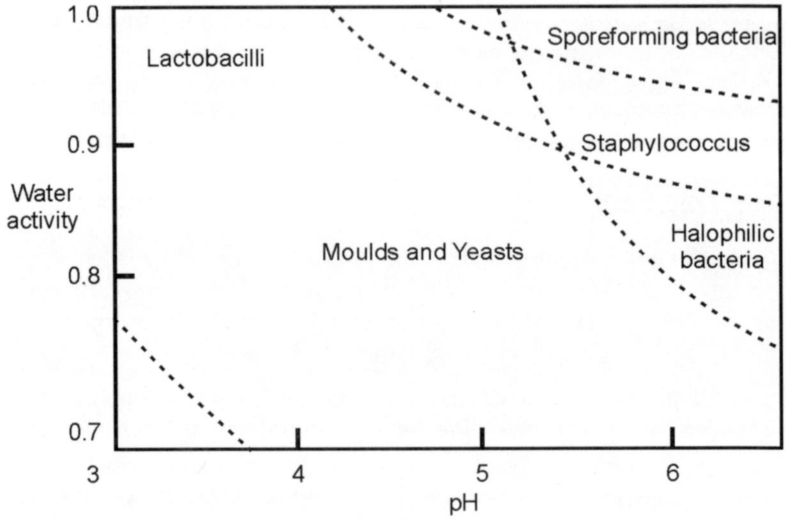


**Fig. 2.1.** Diagrammatic representation of the effects of interactions between pH and $a_w$ on microbial growth. In foods of high $a_w$ and neutral pH, bacteria will be the dominant spoilage organisms. If the $a_w$, is decreased, yeasts and moulds become increasingly important as spoilage agents

## 2.5. Culture Collections and their Importance in Mycotoxin Research

Culture collections are important germplasm resources. They conserve representatives of all kinds of microorganisms which can be made available to other for propagation and use. Their value for fundamental and applied research on fungi and mycotoxins in stored products is not only to provide isolates for biochemical research but also reference culture for identification and educational purposes. The international commission on *Penicillium* and *Aspergillus* (ICPA) is currently developing a set of *Penicillium* and *Aspergillus* isolates which can be used for the purpose.

Culture collection maintains isolates reported to produce mycotoxins. However, there are many misidentification of either fungus or mycotoxin. In several cases, isolates producing specific mycotoxin or originating from interesting sources were not deposited in major culture collections.

Most major culture collections are located in the developed countries and their existence is secured by continuous support of governmental or industrial bodies or by income gained from services to scientific, technological and commercial customers. Though there is need for culture collection centres, many centres in developing countries many constraints make their establishment difficult.

There are many collection centers throughout the world maintaining strains of microorganisms (Table 2.3). Demand for cultures, isolates with particular properties, for various related services, has greatly increased in the last decade. Several investigators dealing with taxonomic, biochemical and especially mycotoxicological aspects have stressed the importance of depositing strains in well recognized collections in which isolates can be kept under optimal conditions (Hawksworth and Kirsop, 1988).

**Table 2.3.** Some culture collection centres maintaining fungal isolates

| | |
|---|---|
| **Australasia** | |
| | Australasia national Reference Laboratory in Medical Mycology (AMMRL), St Leonards, NSW Australia |
| | Victoria University of Wellington, New Zealand |
| | CSIRO Division of Food Research (FRR), North Ryde, NSW, Australia |
| | Material Research Laboratory (MRL), Ascot Vale, Vic., Australia |
| | International Collection of Micro-organisms from plants (PDDCC), Auckland, New Zealand |
| | Plant pathology Branch Culture Collection (QA) Department of Primary Industries, Indooroopilly, Qld, Australia |
| **Europe** | |
| | Aquatic Phycomycete Culture Collection (AP), Reading, UK |
| | Centraalbureau voor Schimmelcultures (CBS), Baarn, The Netherlands |
| | Culture collection of Basidiomycetes (CCBAS), Prague 4, Czechoslovakia |
| | Culture collection of Fungi (CCF), Prague 2, Czechoslovakia |
| | Czechoslovak collection of Microorganisms (CCM) |
| | Collection de Champignons et d' Actinomycetes Pathogens de la CNCM, Paris, France |
| | CAB International Mycological Institute (IMI), Kew, UK |
| | IPCO-Collection Fungal Pathotypes (IPO), Wageningen, The Netherlands |
| | Laboratorie de Souches de Cryptogamie du Musem National Historie Naturelle (LCP), Paris, France |
| | Mycotheque de I' Universoitie 'Catholique de Louvain –la-Neuve, Belgium |
| | Fedrich –Schiller-Universite Jena, (MW)Weimar, Germany |
| | National Collection of Pathologenic Fungi (NCPF), London, UK |
| | National Collection of Wood Rotting fungi (NCWRF-FPRL), Garston, Watford, UK |
| | Portsmouth Polytechnic Culture Collection (PPCC), Portsmouth, UK |
| **North America** | |
| | Antrtic Regional Laboratory (ARL), Halitax, Nova Scotia, Canada |
| | American Type Culture Collection (ATCC), Rockville, Maryland, USA |
| | Canadian Collection of Fungus Culture (CCFC), Ottawa, Canada |
| | Mycological Culture Collection (DLR), Albany, USA |
| | Forinet Collection of Wood-inhabiting Fungi (EFPL), Ottawa, Canada |
| | Fungal Genetics Stock Centre (FGSC), Kansas, USA |
| | Agricultural Research Service Culture Collection (NRRL), Peoria, USA |
| | Ontario Ministry of Health Fungal Collection (TPHL), Toronto, Ontario, Canada |
| | University of Alberta Microfungus collection and Herbaria (UAMH), Alberta, Canada |
| | Upjohn Culture Collection (UCR), Michigan, USA |
| **South America** | |
| | Departamento de Micologia, Fundacao Instituto Osvaldo Cruz OCF, Rio de Janeiro, (RJ), Brazil |

Many microbiological laboratories maintain small scale culture collections for research use. However, in many laboratories facilities are lacking for optimal preservation and important preserved isolates may be lost during or after research project. It is, therefore, recommended that important isolates be deposited in major culture collections. In these centers cultures are professionally maintained and generally employ more than one preservation technique. Moreover, data on isolates and their properties are stored in databases and available to users through catalogues or requests through on-line access. In some cases, where a strain has potential commercial value or has an industrial application, services are provided for safety-deposit. Where patents are involved, the strain must be deposited in a culture collection centers with IDA (International Depository Authority) status.

Basically, two types of storage methods are adapted in culture collections: (1) Those in which metabolism is maintained (although often reduced) (2) those in which metabolism no longer occurs (Kirsnop and Snell, 1984). With the exception of liquid drying and drying at ultra-low temperatures (<-135°C), some of the methods currently used for various fungi with comments on costs (material and labour), longevity and stability are listed in Table 2.4.

**Table 2.4**. A comparison of methods of culture preservation

| **Method of Preservation** | **Cost Longevity** | | **Longevity** | **Genetic stability** | **General comments** |
|---|---|---|---|---|---|
| | **Material** | **Labour** | | | |
| Regular sub culturing on agar storage at room temperature | Low | High | 1-6 months | Variable depends on culture in case of curation | Keep stock culture in case of contamination of working culture |
| Storage in refrigerator | Medium | High | 6-12 months | Variable | |
| Under oil | Low | Low/ medium | 1-32 years | Poor | |
| In water | Low | Low/ medium | 2-5 years | Moderate | |
| Deep freeze | Medium | Low/ medium | 4-5 years | Moderate | Must not allow thawing of material whilst sub-culturing |
| Drying in soil | Medium | Low/ medium | 2-5 years | Moderate to low | |
| Silica gel | Low | Medium | 5-11 years | Good | Recent work reports that over |
| Freeze drying | High | Initially Medium | 4-40 years | Good | drying can cause DNA damage |
| Freezing Liquid nitrogen | High | Low | Infinite: 14 Years to date at CMI | Good | |

### 2.5.1. Methods of Metabolic Activity Maintained Fungal Cultures

**i) Storage on agar slants.** This is the most common method requiring periodic transfer at intervals varying from a few weeks to 6 months at room temperature or at lower temperature (e.g. 4°C) to one year. Though, the method is laborious it does not require expensive equipment. However, by this method the chance of degeneration (irreversible changes in properties) and contamination are relatively high.

**ii) Storage on agar slants under mineral oil** (Ellis, 1979). In this method metabolism is reduced to about 10% or less, which is considered to be useful for storage periods up to 10 years.

**iii) Storage at -20°to -40°C** is acceptable only for a short period (<2 years) (Alexander *et al.,* 1980), but can be useful during a research project. Freezing of cultures at —70° or —80°C can be preserved up to 5 years. However, thermophilic isolates may not survive or may remain viable for shorter periods.

**iv) Drying in soil or with silica gel** (Onions, 1983). These methods are not generally used, but may give good results for 5-10 years. They are relatively cheap and are extremely useful for maintaining *Penicillium* and *Aspergillus* cultures when other preservation methods are not available. They are not suitable for long-term storage of *Fusarium*.

### 2.5.2. Methods of Metabolically inactive Fungal Cultures.

**i) Freeze drying.** In this method of preservation of organisms the residual moisture content is reduced to less than 3% (and more than 1.5% to avoid permanent damage) by freezing and subsequently sublimating the ice, followed by a period of secondary drying. It is useful for many sporulating fungi, but fungi with species with large, septate and coloured conidia are more sensitive. The parameters like cryoprotectant, freezing rate, pre-culturing conditions will influence this method of preservation.

**ii) Liquid drying.** This method is used for storage of bacteria and yeasts and is often used for preservation of bulk samples.

**iii) Ultra deep freezing** (freezers operating at— 135°C or lower or in liquid nitrogen). Fungi are stored in a cryoprotectant after prefreezing at a specified rate. The method can be applied to many fungi. For the resuscitation of the fungus, thawing must be rapid.

### 2.6. Importance of Culture Collections for Research on Mycotoxgenic Fungi of Stored Products

Due to identification problems many reports in the literature are based on doubtful or incorrect names. Brayford (1990) discussed the problems in *Fusarium* where taxonomy of the genus is hampered by inaccurate identification of strains, either through misapplication of names or erroneous determinations, particularly in physiological and molecular studies which can provide important criteria for the systematics, researchers are often taxonomically unskilled and serious mistakes can be made, rendering the data unreliable. Further, many mycotoxins have been named after isolates which were misidentified or carried incorrect names. Frisvad (1989) provided several such examples of incorrect identifying of mycotoxins (Table 2.5).

**Table 2.5.** Examples of misidentified mycotoxins and isolates reported in the literature

| Mycotoxin | Incorrect identification | Correct taxon |
|---|---|---|
| Cyclopiazonic acid | *Penicillium cyclopium* | *P. commune* |
| | *Aspergillus versicolar* | *P. camemberti, A. oryzae* |
| Verruculogen | *P. verruculosum* | *P. brasilianum* |
| Viridicatum toxin | *P. viridicatum* | *P. aethiopicum* |
| Ochratoxin A | *P. cyclopium* | *P. verrucosum* |
| Nivalenol | *Fusarium solani* | *F. sporotrichioides* |
| Diacetoxyscirpenol | *F. nivale* | *F. sporotrichioides* |

To correct these names, the relevant isolates need to be available for examination, but often they were never deposited in a culture collection. Nevertheless, Frisvad (1989), Marassas *et al.* (1984), Wei and Jong (1986) and van Reenen-Hoekstra *et al.* (1990) have investigated misidentified species of *Penicillium, Aspergillus* and *Fusarium*. The International Commission on *Penicillium* and *Aspergillus* (ICPA) is also considering this problem and has plans to investigate, in a collaborative study, important toxigenic isolates of *Penicillium* and *Aspergillus* in major service collection. Moreover, ICPA is currently preparing a set of reference cultures of *Penicillium* and *Aspergillus* with morphological and biochemical descriptions for use by applied researchers who need to be able to identify members of these important groups of fungi.

## 2.7. Culture collection and Developing Countries

Most culture collections are in the developed countries but there is a need for such resources in developing countries. In many Third World countries small research collections exist but too often their sustainability is limited because of financial constraints. Important isolates obtained from research projects are often lost or not kept for long term preservation.

The Netherlands and Indonesia jointly established a culture collection centre which trains culture collection staff, develop expertise in identification of contaminating fungi and mycotoxin in food stuff of tropical countries. The world fungal culture collection (WFCC) established a network of microbiological research centres (MIRCENs) to organise conferences, training courses and fellowship programmes these are discussed.

The World Federation of Culture Collections has resolved these problems in 1975, by establishing a network of microbiological resources centers (MIRCENs) to organize conferences, training courses, and fellowship programs. These centers are designed: (1) to provide the infrastructure for building a world network incorporating regional and inter-regional functional units geared to the management, distribution, and utilization of the microbial gene pool; (2) to strengthen efforts related to the conservation of microorganisms with emphasis on *Rhizobium* gene pools in developing countries with an agrarian base; (3) to foster the development of new, inexpensive technologies that are appropriate to the particular regions (4) to promote the application of microbiology for strengthening of rural economies and 5) to serve as focal points for the training of man power and dissemination of microbiological knowledge.

*Chapter 3*

# Description of Mycotoxigenic Fungi

## 1. *Acremonium* Link ex Fr

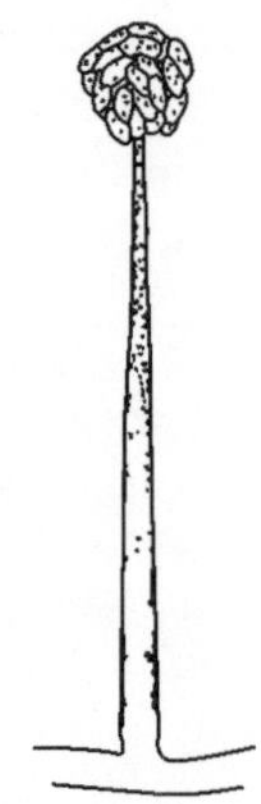

It is common and found in soil, hay, compost, cellulose, bamboo and other plant debris. The colonies generally slow growing but can be moderate to rapid growth rate. Hyphae hyaline, smooth to lightly downy or fluffy and can be powdery. White, yellow, orange, pink, gray to grayish-green even reddish-brown to black with a species having a reverse side of the plate violet colonies can be flat with a slight raised center. The texture of the macro-colony is velvety to slimy to granular to powdery. The reverse of the colony is beige, brown, voilet, pink to orange. Phialides are generally solitary long to not so long with a septation at the base. It can have a faint collarette at the spore formation. Spores are 2–8 μm round, ovoidal, sub-spherical, verruculose, ellipsoidal, to oblong, ob-ovoidal, tear shaped, spindle shaped, cylindrical, falcate, reniform to falciform, 1-2-celled even 3-celled in chains or slimy clusters or balled up. Some species have chlamydospores.

### Key to species

Phialides simple, arising generally from aerial hyphae, rarely from conidiophores, hyaline, smooth, conidia, fusiform to ellipsoidal ...................... - *implicatum*

Phialides simple, arising from submerged or slightly fasciculated aerial hyphae conidia cylindrical ........................................................................... - *strictum*

### (i) *A. crotrocinigenum*

**Synonymys** = *Cephalosporium*

= *Hyalopus*

= *Gliomastix*

**Colony:** Hyaline, slow growing

**Conidiogenous cell**: Phialidic, mostly simple, awl-shaped, erect, arising from substratum or from fasciculate aerial hyphae. If compound conidiophores occur then branching is various, confined to lower part.

**Conidia:** One-celled, hyaline or pigmented in mass in droplets or dry chains

**Mycotoxins:** Crotocin, crotocol, oosporein, paxillin

**Diagnostic features**: Simple oval shaped phialides, arise from fasciculate aerial mycelium, one-celled, hyaline conidia are produced in droplets.

**Teleomorphs:** *Nectra*, *Emericellopsis*, *Mycoarchis*

### 2. *Alternaria* Nees Ex Fr

Colonies effuse usually grey, dark-blackish-brown or black. Conidiophores macronematous, mononematous, simple or irregularly and loosely branched, pale-brown or brown, solitary or in conidiogenous cell, integrate, terminal, becoming, intercalary, polytretic, sympodial, sometimes monotretic, cicatrized. Conidia catenate or solitary, dry, typically ovoid or obclavate, often rostrate, pale to mid-olivaceous brown or brown, smooth or verrucose with trans-verrucose and frequently also oblique and long longitudinal-septa. Species of *Alternaria* produce Alternariol (AoH), Alternariol mono ethyl ether (AME), Alternuene (ALT), isoaltneuene, Altertoxins I, II, III, ATX-I, ATX-II, ATX III and Tenuazonic acid (TeA) (Bottalico and Logrieco, 1998), which belongs to three structural classes:

(1) Dibenzopyrone derivatives-AOH, AME, ALT,

(2) Pyrelene derivatives ATX-I, II, III, ALT I, II & III

(3) Tetramic acid derivatives -TeA

Mycotoxins produced by different species of *Alternaria* are précised in table 3.1.

**Table 3.1.** The production of important mycotoxins by *Alternaria* species

| Species | Mycotoxin | Reference |
|---|---|---|
| *A. alternata* (Fries) Keisster | AOH, AME, ALT, TEA and ATX-I, -II, -III | Andersen and Thrane (1996) |
| *A. brassicae* (Berk.) Sacc | AOH, AME | Bottalico and Logrieco (1998) |
| *A. capsici-anui* Săvul. & Sandu | AOH, AME, TeA | Bottalico and Logrieco (1998) |
| *A. cassiae* Jurair & A. Khan | ATX-I, -II | Hradil *et al.* (1989) |
| *A. citri* Ell. & Pierce | AOH, AME, TeA | Kinoshita *et al.* (1972) |
| *A. cucumerina* (Ell. & Ev.) Elliott | AOH, AME | Freeman (1965) |
| *A. dauci* (Kühn) Groves & Skolko | AOH, AME | Freeman (1965); Raistrick *et al.* (1953) |
| *A. japonica* Yoshii | TeA | Kinoshita *et al.* (1972) |
| *A. kikuchiana* Tanaka | AOH, AME, TeA | Kameda *et al.* (1973) |
| *A. longipes* (Ell. & Ev.) | AME, TeA | Bottalico and Logrieco (1998) |
| *A. mali* Roberts | ATX-I, -II, -III, TeA | Kinoshita *et al.* (1972) |
| *A. oryzae* Hara | TeA | Kinoshita *et al.* (1972) |
| *A. porri* (Ell.) Cif. | AME, TeA | Bottalico and Logrieco ( 1998) |

| Species | Mycotoxin | Reference |
|---|---|---|
| *A. radicina* Meier, Drechsler & Eddy | ATX-I, -II, -III, TeA | Bottalico and Logrieco (1998); Solfrizzo *et al.* (2005) |
| *A. solani* Sorauer | AOH, AME, TeA | Bottalico and Logrieco (1998) |
| *A. tenuissima* (Kunze) Wiltshire | AOH, AME, ATX-I, -III, TeA | Bottalico and Logrieco (1998) |
| *A. tomato* (Cooke) Jones | AOH, AME, ATX-I, -II, -III, TeA | Bottalico and Logrieco (1998) |

**Key to species**

1. Conidia produced solitary .................... *A. radicina*

    1a. Conidia produced in chains .................... 2

2. Conidia solitary or in short chains .................... 2

    2a. Conidia in long chain .................... 3

3. Conidia solitary or in short chains, obclavate, beak of the length of the conidium, golden brown 20-95 mM .................... *A. tenuissima* (1)

    3a. Conidia in short chain, shape variable .................... *A. solani* (2)

4. Conidia mostly in chains often or more chains branched, Conidia irregularly shaped, beak short not exceeding 1/3 of conidial length, medium golden brown .................... *A. alternata* (3)

    4a. Conidia beaked, they are produced in long chains .................... *A. kikuchiana* (4)

    On cucumber plants .................... *A. cucumerina* (5)

**Synaptic Key**

1. Conidia beakless .................... *A. radici*

    1a. Conidia beaked .................... 2

2. Conidia on crucifers .................... *A. brassicae*

    *2a.* Conidia on cucumbers .................... *A.cucumerina*

    2b. Conidia on daucus .................... *A.dauci*

    *2c.* Conidia on Gossypium .................... *A.macrospora*

    2d. Conidia on Allium .................... *A.porri*

    2e. Conidia on solanaceace crops .................... *A.solani*

**II.** Conidia catenate –Conidia are produced in chains

i) Conidia in short chains (2-5 conidia)

*A. tenuissima* on wheat

ii) Conidia in long chains (more than five conidia).................... *A. alternata*

*A. citri* (on lemon)

*A. kikuchiana*

*A. longipes*

**i)** ***A. alternata*** **(Fries) Keisster**

**Synonymys=** *A. tenuis* Nees ex Perx

= *Torula alternate* Fr.

**Colony:** Usually black or olivaceous black- sometimes grey

**Conidiophores:** Arise singly or in small groups, macronematous, mononematous, simple or branched, geniculate, pale to mid-olivaceous or golden-brown, smooth

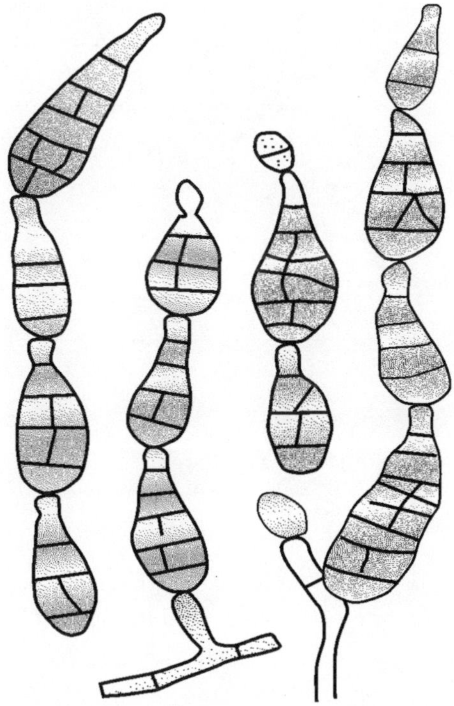

**Conidia:** Porospores, catenate, chains long, often branched, obclavate, pyriform, ovoid or ellipsoidal, short-conical or cylindrical beak, up to not more than one third of the conidia, pale to mid-golden-brown, smooth or verruculose with upto 8-transverse and several longitudinal or oblique septa, conidia muriform.

**Mycotoxins:** Alternariol, alternariol monomethylether, altermuene, tenuazoic acid and Altertoxin I, II, III.

**Diagnostic features:** Conidia irregularly shaped, break short golden brown.

**ii)** ***A. brassicae*** **(Schw) Wiltshire**

**Synonyms** = *Helminthosporium brassicicola*

= *Macrosporium eheirarithi*

= *Alternaria circinus*

**Colony:** Amphigenous, dark, olivaceous-brown to dark-blackish-brown

**Conidiophores:** Arising singly or in groups, usually simple, erect or ascending, straight or curved, geniculate, olivaceous-brown and smooth

**Conidia:** Mostly in chains of a acropleurogenous arising through small spores in the conidiophores walls, straight, cylindrical, basal-cells rounded, apical cell rectangular rescumbling a truncated core, 6-transverse septa and up to 6-longitudinal septa, smooth pale to dark-olivaceous brown.

**Mycotoxins**: Alteraniol, (AOH), Alternariol Monoethyl Ether (AME)

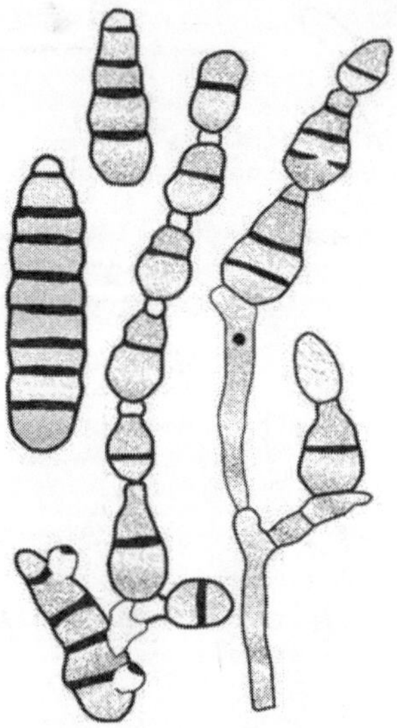

**Diagnostic features:** Conidiophores arising in groups of 2–10, pale grayish olive, with one or more distinct conidial scars. Constricted at septa

**iii)** ***A. citri*** **Ellis &Pierce**

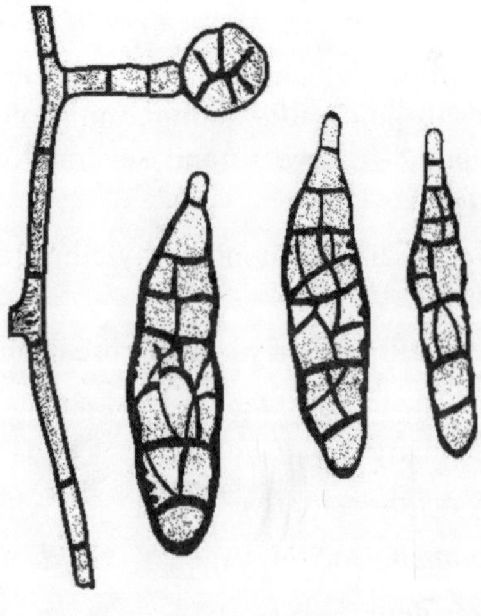

**Colony:** Grey, olivaceous brown or black

**Conidiophores:** Simple or branched, straight, septate, olivaceous-brown, terminal scar and sometimes one or two lateral ones.

**Conidia:** Solitary, simple or branched chains, straight, variously shaped but commonly obclavate, or oval, dark brown, smooth to verruculose with up to 8 transverse and numerous longitudinal septa, beaks thick and pale-brown

**Mycotoxins**: Alteraniol, (AOH), Alternariol monoethyl ether (AME), TeA.

**Diagnostic features:** Conidia solitary, slightly curved, numerous longitudinal septa.

### iv) *A. cucumerina* (Ellis & Everh)

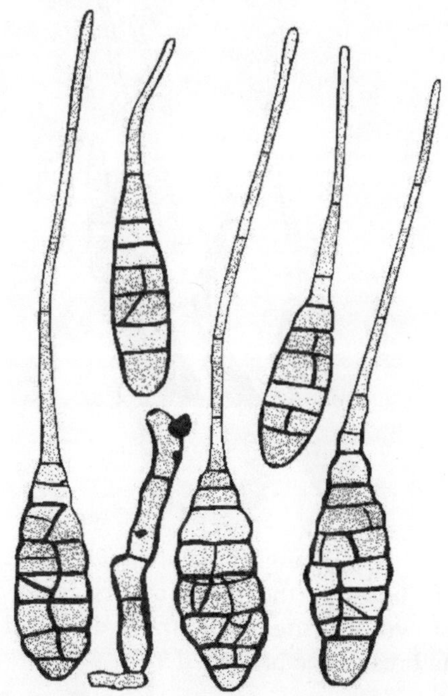

**Synonymys** = *Macrosporium cucumerinum* Ellis & Everh

**Colony:** Amphigenous

**Conidiophores:** Mononematous, macronematous, singly or in small groups, erect, straight or flexuous, geniculate, septate. Pale to mid-brown with several conidial scars.

**Conidia:** Acroporogenous, solitary or occasionally in chains of 2, obclavate, rostrate with long beak, pale to mid-golden-brown, smooth to verrucose, 6-9-transverse septa, many long or oblique septa, beak pale-brown, sepatate, not branched.

**Mycotoxins:** Erythroglacin, AOH, AME

**Diagnostic feature:** Conidia in short chains, variable in shape, 9-11 transverse septa and many longitudinal or oblique septa.

**v) *A. dauci* (Kuhn) Groves & Skolko**

**Synonyms** = *Sporidesminum exitiosum* Khun var.
= *Macrosporium carotae* (Ellis & Longrois kiihn)
= *Alternaria carotae* (Ellis & Longrois) Stevenson &Wellman

**Colony:** Olivaceous-brown or black

**Conidiophores:** Arising singly or small groups, straight, sometimes geniculate, septate, pale olivaceous-brown.

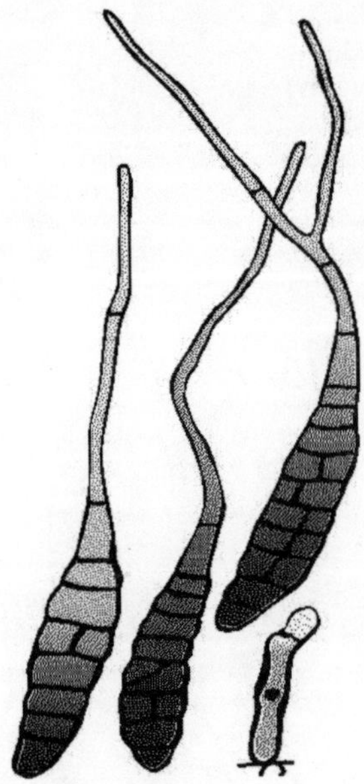

**Conidia**: Usually solitary, straight, obclavate, rostrate, beak up to 3 times the length of the body of the spore at first pale olivaceous brown often becoming brown with age 7-11 transverse and several longitudinal or oblique septa breaks one branched

**Mycotoxins:** Alteraniol (AOH), Alternariol monoethyl ether (AME).

**Diagnostic features:** Solitary conidia beak length is 3 times larger than spore body, beak once branched.

**vi) *A. longipes* Ellis & Everh**

**Synonymys** = *Macrosporium longipes* Ellis & Everh.

**Colony:** Amphigenous

**Conidiophores:** Single or in groups, erect or ascending simple or loosely branched, straight or flexuous, cylindrical septate, pale-olivaceous-brown

**Conidia:** Sometimes solitary but usually in chains obclavate, rostrate, pale to mind-pale-brown, smooth or verrucose, tapering gradually into the pale-brown beak, 5-7 transverse septa and 1 to several longitudinal septa

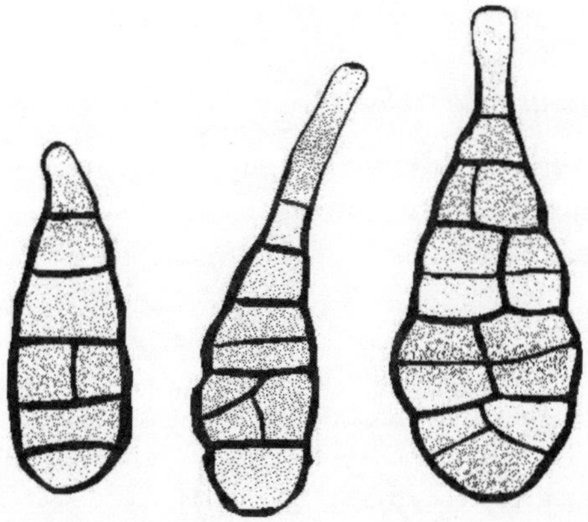

**Mycotoxins:** AME, TeA

**Diagnostic features:** Conidiophores cylindrical septate, conidia in chains obclavate, rostrate.

**vii) *A. porri* Ellis**

**Synonyms** = *Macrosporium porri* Ellis

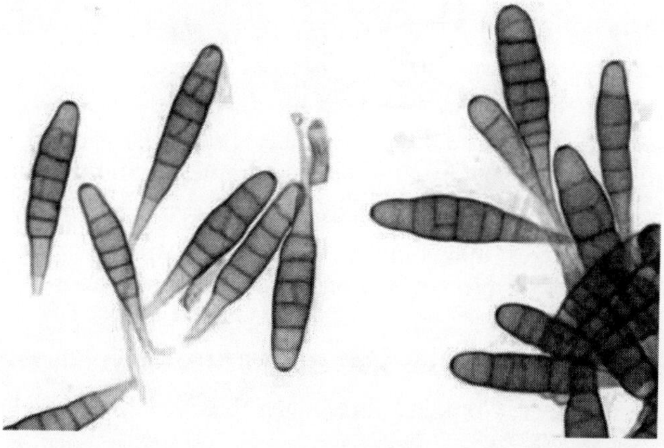

**Colony:** Colony colour varied from pinkish white through dull orange to olivaceous, smooth growth but all the isolates had compact mycelium.

**Conidiophores:** Single or in groups, straight or flexuous, sometimes geniculate, septate, pale to mid-brown with many conidial scars

**Conidia:** Solitary, straight or curved, obclavate or with the body of the conidium ellipsoidal tapering to the beak, pale to mid-golden brown, smooth or minutely verrucose, 8-12 transverse septa and 0- several longitudinal or oblique septa, beak flexuous, pale tapering

**Mycotoxins:** AME, Tenuazonic acid

**Diagnostic features:** Conidiophore straight, septate, with many conidial scars. Conidia solitary, obclavate, smooth.

**viii) *A. radicina* Meier**

**Synonymys** = *Thyrospora radicina*
= *Stemphylium radicinum*

**Colony:** Spreading, olivaceous-brown

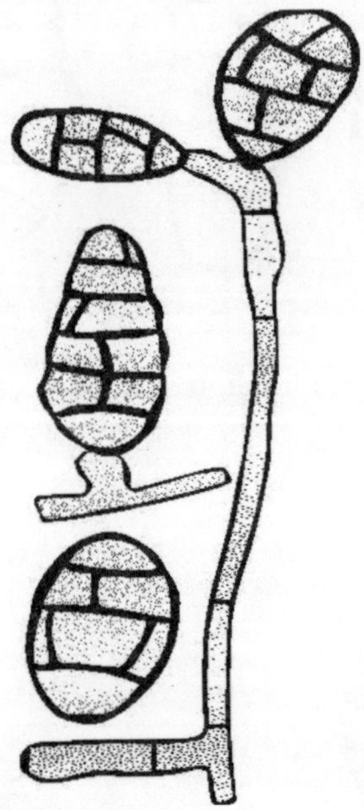

**Conidiophores:** Arising usually singly, simple, straight, cylindrical, septate, pale to mid-brown, smooth

**Conidia:** Solitary or in chains 2-3, straight, variable in shape, ellipsoidal, obclavate, dark-brown, smooth,3-7 transverse and several longitudinal septa.

**Mycotoxins:** Alteraniol, (AOH), Alternariol Monoethyl Ether (AME), TeA.

**Diagnostic features:** Conidia solitary, variable in shape, straight.

**ix) *A. solani* Sorauer**

**Synonyms**=*Macrosporium solani* Ellis & Martin

**Colony:** Effuse, usually grey, dark, blackish-brown or black

**Conidiophores:** Mononematous, macronematous, straight or flexuous, septate, pale-brown or olivaceous-brown.

**Conidia:** Solitary, porospore, obclavate tapering to a beak, pale or mid-pale-golden or livaceous-brown smooth walled, 9-11 transverse and a few longitudinal or oblique septa beak flexuous, pale sometimes branched.

**Mycotoxins:** Alternariol, monoethyl ester, alternariol, alteruence, alternarine.

**Diagnostic features:** Conidia in short chains, variable shape

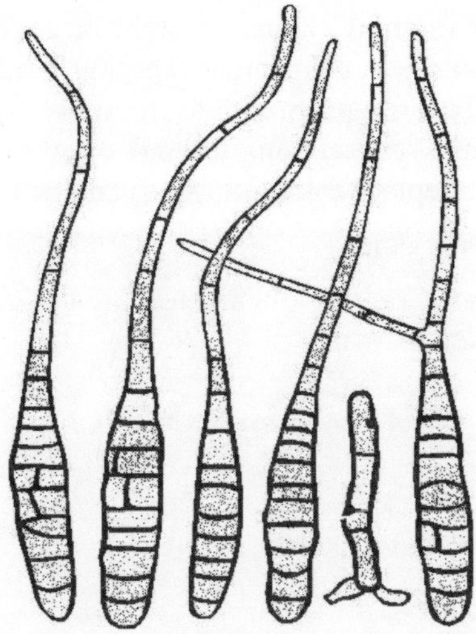

**x)** ***A. tenuissima*** **Kubize expers**

**Synonyms** = *Helminthosporium tenuissimum* Kunze
= *Macrosporium tenuissimum* Fries

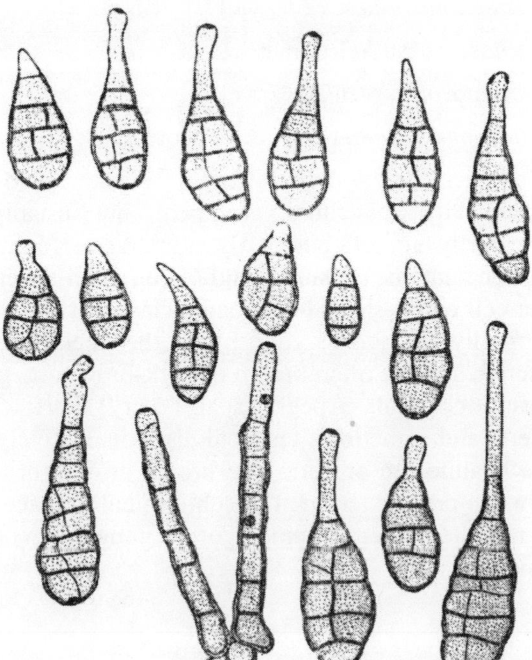

**Colony:** Effuse, usually olivaceous brown, darkish-brown

**Conidiophores:** Mononematous, macronematous straight or flexuous, septate, pale or mid- pale-brown, smooth, acronematous with distinct conidial scars.

**Conidia:** Solitary or in short chains, straight or curved, obclavate or with the body of the conidium, ellipsoidal tapering gradually to the beak, pale to mid clear-golden-brown, smooth to minutely verrucose, generally with 4-7 transverse and several longitudinal or oblique septa, slightly or not restricted at the septa, beak thick, swollen apex.

**Mycotoxins:** Tenuazoic acid

**Diagnostic features:** Conidia obclavate, break length equal to length of conidium, golden brown.

### 3. *Arthrinium* Kunze ex Fr

**Synonyms** = *Camptoum* Link
= *Coniosporium* Link
= *Sporophleum* Nees ex Link
= *Papularia* Fr.
= *Gonatosporium* corda
= *Microtypha* Speg
= *Tureenia* Hall
= *Pseudobasidium* Teng wall
= *Phaeoharziella* Lonbiere
= *Innotospora* Van Beyama
= *Racemosporium* Mme. & F. Moreau

**Colonies:** Compact or widely effused, black to dark-blackish-brown fractifractions, occasionally mycelium erumpent but usually superficial and frequently pulvinate partly superficial, partly immersed. Conidiophores basauxic, macronematous mononematous, arising singly from sub-spherical, ampulliform, conidiophore mother cell barrel-shaped or broadly clavate, simple often narrow more or less cylindrical usually colourless concept for the thick transverse septa which may be highly refractive and are often brown or dark-brown integrated, terminal and intercalary, polyblastic denticulate pegs usually very short, cylindrical truncate conidia, solitory lateral and sometimes terminal also, distinctively shaped frequently flattened and with a hyaline rim or germ slit, brown or dark-brown, smooth, single celled. Sterile cells when present terminal or subterminal in place of conidia, usually smaller, paler and not the same as conidia often containing one or more highly refractive cubical bodies.

**Key to species**

1. Conidia truncate at the base, often verruculose ........................................ *A. spegazzini* 1

 1a. Conidia not truncate at the base always smooth

2. Conidia almost round or polygonal in face view........................................................ 2

 2a. Conidia nearly all polygonal in view puccinioides conidia nearly all round or almost round in face view ............................................................. 8

3. Conidia lenticular ........................................ 3
    3a. Conidia not lenticular
    3b. Conidia 7-9 μm diameter in face view .................. *A. phaerospernum*
4. Diameter of conidia 9-10 μm in face view usually ........................ 5
    4a. Diameter of conidia 6-8 μm in face view usually ........................ 6
5. Conidiophores 2-4 μm thick *saccharicola* ........................ 4
    5a. Conidiophores 1.0- 1.5 μm thick *phaeospermum* ........................ 5
6. Conidiophores 1.0-1.5 μm, septa numerous brown *sacchari* ........................ 6
    6a. Conidiophores 0.5 μm thick, septa few hyaline ............. *Appiospora montagnei*

### i) *A. caricicola* Kunze ex Ficinus & Schubert

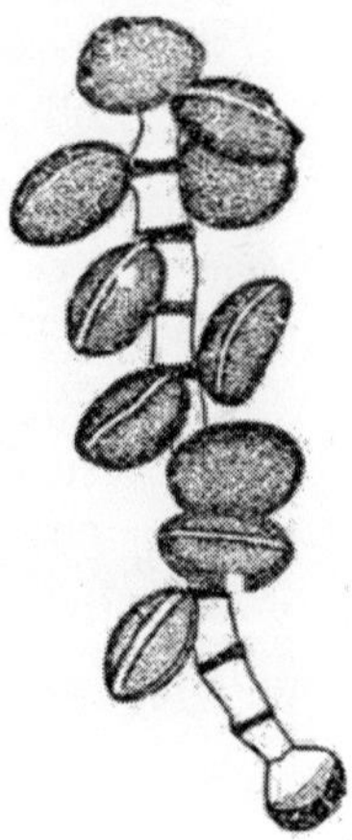

**Colony:** Compact or widely effused, black or dark-blackish-brown, occasionally erumpent but usually superficial and frequently pulvinate

**Mycelium:** Partly superficial, partly immersed

**Stroma:** None

**Setae and hyphopodia:** Absent

**Conidiophores:** Basauxic, macronematous, mononematous, arising singly from sub-spherical ampulliform, barrel shaped or broadly clavate, conidiophores on other cells simple often narrow more or less cylindrical, usually colorless except for thick transverse septa which may be highly refractive and are often brown or dark-brown.

**Conidiogenous cells:** Integrated, terminal and intercalary, polyblastic, denticulate pegs, usually very short, cylindrical, truncate.

**Conidia:** Solitary, lateral and sometimes also terminal, distinctively shaped, frequently flattened and with a hyaline rim or germ slit, brown or dark-brown, smooth as a rule (0-sepatate).

**Sterile cells:** When present terminal or sub terminal in place of conidia usually smaller or paler not the same shape as conidia often containing 1 or more highly refractive cubical bodies

### ii) *A.phaeospermum* (Eorda) Ellis

**Colony**: Fast growing, floccose, white becoming even more widely effused by the production of large number of mycelial mats on which fresh conidiophores arise often producing a red pigment on the agar, sporulation localized spots on the aerial mycelium.

**Conidiophores:** Hyaline to pale –brown with dark septa.

**Conidium**: Lenticular, dark-brown with equatorial germs slit, smooth

**Mycotoxins:** β-nitropropionic acid

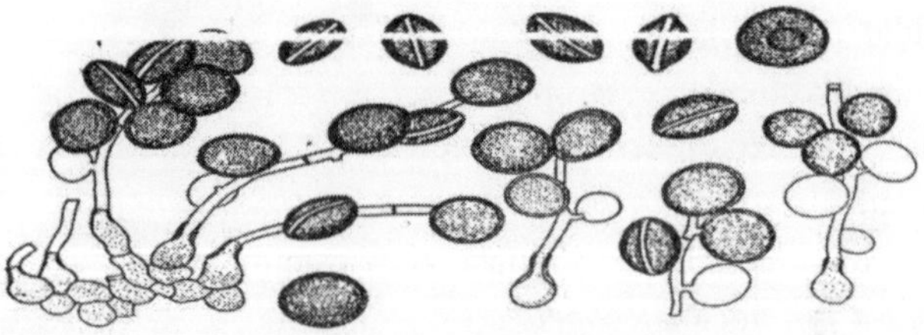

**Diagnostic features:** Conidiophores pale-brown with dark septa, conidium lenticular, dark-brown with equatorial germ slit

**iii)** ***A. saccharicola***

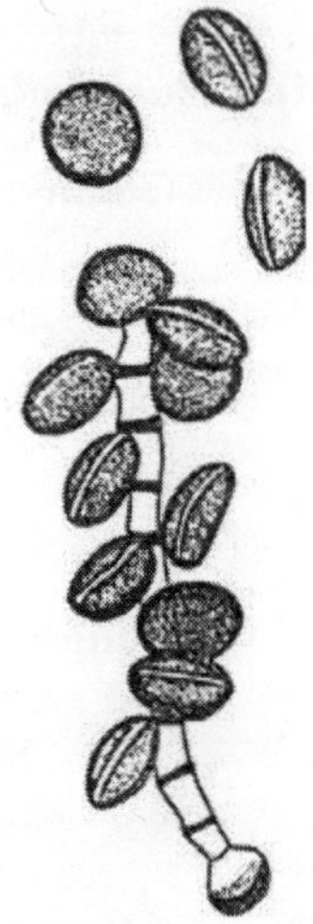

**Colony:** Flat, spreading with sparse aerial mycelium, surface iron-grey on OA, MEA and PDA umber with patches of olivaceous –grey.

**Conidiophores:** Reduced to conidiogenous cells, aggregated in clusters on hyphae, medium brown, finely verruculose, ampuliform.

**Conidia:** Brown, smooth, granular, globose to ellipsoid in surface view, lenticular in side view with pale equitorial slit at times appearing like ridge of polar pigment.

**Mycotoxins:** β-nitropropionic acid

**Diagnostic features:** Colonies flat, spreading with sparse aerial mycelium, conidiophores reduced to cluster of conidiogenous cells on hyphae, globose to ellipsoid in surface view and lenticular in side view.

**iv)** ***Arthrinium*** **state of** ***Apiospora montagnei***

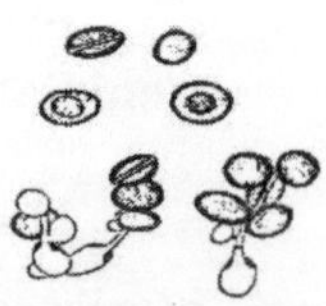

**Teleomorph:** *Apiospora montagnei*

**Colony:** Growing fast, floccose, white, becoming widely effused producing a red pigment sporulation localized spores on aerial mycelium

**Conidiophores:** Hyaline to pale-brown with a few thick hyaline septa

**Conidia:** Variable in shape, dark-brown with equatorial germ-slit, smooth, diameter of conidia between 6-8 μm in face view

**Mycotoxins:** β-nitropropionic acid

**Diagnostic features:** Conidiophore with a few thick hyaline smooth base and 6-8μm diameter in surface view

***Apiospora montagnei***

= *Sphaeia apiospora* (Dur & Mont.)

= *Hyploptteris apiospora* (Dur & Mont.) Berk

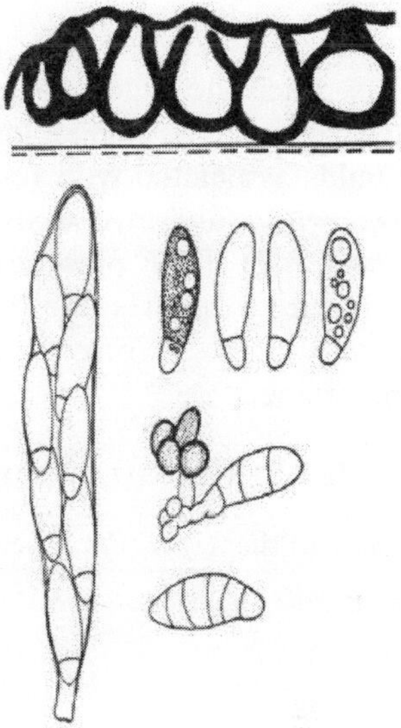

= *Papularia arundiris* (Corda) Fr.

**Ascomata**: Perithecia formed in clusters of 1-7 either in uniseriate rows or irregularly arranged crumpent, globose or ovoid, papillate ostiolate

**Ascus**: Shortly stalked, unitunicate clavate, 8-spored

**Ascospores**: Hyaline, granulose, curved in equilateral, taper to the base and have rounded ends. Initially unicellular later becoming bicellular at the time of germination cells in equal.

**Mycotoxins:** β-nitropropionic acid

**Diagnostic features**: Ascocarp globose, asci bitunicate, clavate, short stalked filiform, pesudoparaphyses. Ascospores biseriate, hyaline or yellow, curved pyriform bicelled.

## 4. *Ascochyta* Lib

### i) *A. pisicola* (Berk.) Sacc.

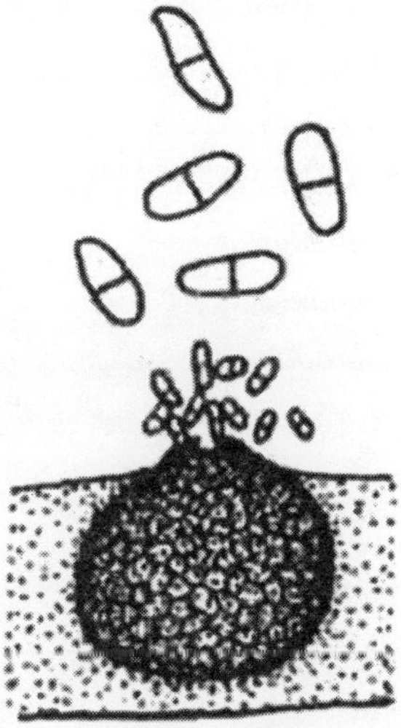

Colonies spreading, hyaline to pale-brown, pycnidia distinct globose, brown, uniloculate, ostiolate. Conidia are enterobalstic, phialides, doliform to lageniform, hyaline on the inner cells of pycnidia, hyaline, two-celled, ovoid to oblong, cylindrical to irregular.

**Synonymys** = *Ascochyta pisi* Lib. (1830)

**Colony**: Spreading, hyaline to pale- brown.

**Conidiomata:** Pycnidia, amphigenous, globose, brown, immersed, uniloculate, thin walled texture angularis, 1-3 celled thick, ostiolate, slightly papillate.

**Conidiophore:** Absent

**Conidiogenous cells**: Enterobalstic, phialidic, determinate, discrete, doliform to lageniform, hyaline, smooth, formed from the inner cells of the pycnidial wall.

**Conidia**: Hyaline, median, 1-septate, continuous or constricted, thin walled, smooth, guttulate, ovoid to oblong, cylindrical to irregular.

**Mycotoxins:** Cytochalasin-A, B, ascochalasin, ascochitine

**Diagnostic features:** Pycnidia glaberans uniloculate, enteroblastic phialidic determinate conidia producing two-celled hyaline conidia.

## 5. *Aspergillus* Mich.exFr.

Ever since the report of carcinogenic aflatoxins produced by *A. flavus*, the genus *Aspergillus* has gained importance. Extensive and intensive search for toxigenic moulds associated with foods and feeds revealed quite a large number of species of *Aspergillus* are mycotoxigenic and are important from the health point of view (Table 3.2). Their presence leads to deterioration and spoilage of many food stuffs. Exposure to their spores frequently results in allergic reactions of man often of a serious nature. Correct identification of the species is, therefore, of paramount importance.

**Table 3.2.** Mycotoxins produced by species of *Aspergillus*

| **Name of the *Aspergillus* species** | **Name of the mycotoxin** |
|---|---|
| *Aspergillus acanthosporus,* | Ochratoxin B |
| *A. aculeatus* | Secalonic Acid D, Cyclopiazonic Acid |
| *A. affinis* | Ochratoxin A |
| *A. albertensis* | Ochratoxin A |
| *A. alliaceus* | Ochratoxin A, Ochratoxin B |
| *A. amstelodami* | Echinuline, Neoechinuline |
| *A. arachidicola* | Aflatoxin B1.B2, G1, G2, M1, M2, Kojic Acid |
| *A. aureus* | Emodin /Archin/Emodol/Frandulic Acid |
| *A.auricomus* | Ochratoxin A, Ochratoxin B |
| *A. bombycis* | Aflatoxin B1. B2, G1, G2 , M1, M2 |
| *A. brevipes* | Viriditoxin |
| *A. caespitosus* | Fumitremorgin A |
| *A. calidoustus* | Austin |
| *A. candidus* | Citrinin, Kojic Acid, Xanthocillin |
| *A. carbonarius* | Ochratoxin A, Ochratoxin B |
| *A. carlsbadensis* | Brevianamide A |
| *A. carneus* | Citrinin |
| *A. chevalieri* | Xanthocillin |
| *A. citricus* | Ochratoxin A |
| *A. clavatus* | Patulin, Tryptoquivaline A |
| *A. discophorus* | Sterigmatocystin |
| *A. egyptiacus* | Verruculogen, Fumitremorgin A |
| *A. elongatus* | Fumitremorgin C |
| *A. fischeri* | Terrein |
| *A. flavipes* | Citrinin |
| *A. flavus* | Aflatoxin B1.B2, G1, G2, M1, M2, Cyclopiazonic Acid, Gliotoxin, O-methyl-Sterigmatocystin, Sterigmatocystin, Versicolorin A |

| Name of the *Aspergillus* species | Name of the mycotoxin |
|---|---|
| *A. fonsecaeus* | Ochratoxin A |
| *A. fumigatiaffinis* | Helvolic Acid |
| *A. fumigatus* | Helvolic Acid, Methyl-sulochrin, Phthioic Acid, Pyripropene A, Restrictocin, Terrein Tryptoquivaline A, Verruculogen, Viriditoxin |
| *A. fumigatus (Sartorya fumigata)* | *9-deacetoxyfumigaclavine C |
| *A. herbariorum ser. minor* | Echinuline |
| *A.lanosus* | Ochratoxin A |
| *A. lentulus* | Pyripropene A, Terrein, Cyclopiazonic Acid |
| *A. longivesica* | Patulin |
| *A. melleus* | Ochratoxin A , Penicillic acid, Viomellein, Xanthomegnin |
| *A. minisclerotigenes* | Aflatrems, Aflatoxins B1.B2, G1, G2, M1, M2, Cyclopiazonic acid, Kojic acid |
| *A. monodii* | Terrein |
| *A. multicolor* | Sterigmatocystin, |
| *A. nidulans* | Sterigmatocystin |
| *A. niger* | Asnipyrone A,B, Austin, Fumonisin B2, B4, Gliotoxin, Malformin C, Malformins, Nigerapyrone A, B, Ochratoxin A, Ochratoxin B |
| *A. nomius* | Aflatoxin B1.B2,G1,G2 ,M1,M2 |
| *A. novofumigatus* | Helvolic Acid, Terrein |
| *A. ochraceopetaliformis* | Ochratoxin A, B, C |
| *A. ochraceus* | Ochratoxin A, B, C, Penicillic acid, Secalonic Acid D, Viomellein Xanthomegnin, Aflatoxin B1. B2, G1, G2, M1, M2 |
| *A. olivicola* | Sterigmatocystin, Terrein |
| *A. oryzae* | 3-Nitropropionic Acid, Cyclopiazonic acid, Kojic Acid |
| *A. oryzae var. microsporus* | Maltoryzine |
| *A. ostianus* | Ochratoxin A, Penicillic acid |
| *A. parasiticus* | Sterigmatocystin, Aflatoxin B1.B2, G1, G2 , M1, M2, O-methyl-Sterigmatocystin |
| *A. parvisclerotigenus* | Aflatoxin B1.B2,G1,G2, M1, M2, Cyclopiazonic acid |
| *A.petrakii* | Ochratoxin A |
| *A. pseudotamarii* | Aflatoxin B1.B2,G1,G2, M1, M2, Cyclopiazonic Acid |
| *A. puniceus* | Austocystin |
| *A. rambellii* | O-methyl-Sterigmatocystin, Sterigmatocystin, Aflatoxin B1.B2,G1,G2, M1, M2 |
| *A. restrictus* | Restrictocin |
| *A. sclerotiorum* | Emodin/Archin/Emodol/Frandulic Acid, Ochratoxin A,B |

| Name of the *Aspergillus* species | Name of the mycotoxin |
|---|---|
| *A. sulphureus* | Ochratoxin A,B, Xanthomegnin |
| *A. tamarii* | Cyclopiazonic acid, Fumigaclavine A |
| *A. terreus* | Citreoviridin, Emodin/Archin/Emodol/Frandulic Acid, Gliotoxin, Patulin, Phthioic Acid, Terreic Acid, Terrein, Terretonin, Territrem A |
| *A. toxicarus* | Aflatoxin B1, B2, G1, G2, M1, M2 |
| *A. ustus* | Brevianamide A, Sterigmatocystin, Austdiol, Austin |
| *A. uvarum* | Secalonic Acid D |
| *A. versicolor* | Versicolorin A, Cyclopiazonic acid, O-methyl-Sterigmatocystin, Sterigmatocystin, Versicolorin A, |
| *A. viridinutans* | Viriditoxin |
| *A. wentii* | 3-Nitropropionic Acid, Emodin/Archin/ Emodol/ Frandulic Acid, Ochratoxin B |
| *Emericella venezuelensis* | Aflatoxin B1. B2, G1, G2, M1, M2, Sterigmatocystin |
| *E.enezuelensis* | Terrein |
| *E.striata* | Emestrin |
| *E. unguis* | Dechloronidulin |
| *E.variecolor* | Terrein |
| *E.nidulans* | Sterigmatocystin |
| *Eurotium chevalieri* | Gliotoxin, Xanthocillin |
| *Neosartorya fischeri* | Fumitremorgin A,C, Tryptoquivaline A, Verruculogen |
| *N. aureola* | Fumitremorgin C |
| *N.pseudofischeri* | Gliotoxin |

*Aspergillus* was first described almost 300 years ago and is an important genus in foods. They are extremely common in stored commodities such as grains, nuts and spices and occur more frequently in tropical and subtropical regions than in temperate climates. It is a large genus containing more than 100 species. Though there are number of teleomorphic genera which have *Aspergillus* conidial states, only two genera *Eurotium* and *Neosartorya* produce heat resistant ascospores and are mycotoxigenic and sporulate heavily in stored products.

Raper and Fennel (1965) monograph is the most widely used monograph for identification of species of *Aspergillus*. However, concepts have changed due to the addition of some more species (Samson and Pitt, 1990). Gams *et al*. (1985) proposed nomenclaturally correct classification for species of genus *Aspergillus* by grouping then into six sub-genera which are sub clubbed with sections. Klich and Pitt (1988) have also provided a key to species of *Aspergillus* which are important in foods. In addition to traditional morphological taxonomic characters, chemical characteristics such as isozyme patterns (Cruickshank and Pitt, 1990), secondary metabolites (Frisvad, 1985), ubiquinone system (Kuraishi *et al*., 1990) and molecular characteristic (Mullaney and Klich, 1990) have been used to clarify relationships within the genus *Aspergillus*.

Raper and Fennell (1965) assigned species into three groups based on conidiogenous apparatus (Table 3.3). They are

1. Those with one series of conidiogenous cells-uniseriate
2. Those with two series of conidiogenous cells plus supporting cells-biseriate
3. Those in which both conditions are found. However, recent change in the taxonomy of *Aspergillus,* the genus has been divided into six sub genera (Gams *et al*., 1985), each containing one or more sections. Sections correspond to the groups of Raper and Fennel (1965). Only those sections that have mycotoxicological importance will be discussed here.

**Table 3.3.** *Aspergillus* sections based on the structure of conidiogenous cells

| **Sections with uniseriate conidiogenous cells** | **Sections with biseriate conidiogenous cells** | **Sections with uniseriate and biseriate cells** |
|---|---|---|
| *Clavati* | *Nidulans* | *Flavi* |
| *Restricti* | *Versicolores* | *Wentii* |
| *Fumigati* | *Flavipedes* | *Nigri* |
| *Aspergillus (A. glaucus* group*)* | *Terrei* | *Candidi;Circumdati* (*A. ochraceus* group) |

For identification of species of *Aspergillus* primary emphasis was on arrangement of conidiogenous cells, secondary emphasis on shape and pigmentation of conidial heads and finally on the presence or absence of ascomata, Hulle cells (thick-walled cells of unknown function) and pigmentation of the conidiophores.

*Aspergillus,* one of the oldest genera of fungi, was described by Michaeli in 1729. He gave the name from a device used by Roman Catholic clergy to sprinkle holy water during a part of the liturgy called the asperges (Ainsworth, 1976). It is derived from Latin verb `aspergese', the ceremony of people with holy water before high mass in the Roman Catholic Church with which the priest begins the ceremony. The brush used for sprinkling holy water is aspergilla (*Aspergillus*). Among the fungi most frequently found in number exceeding levels that they reach out doors are *Aspergillus* and *Penicillium* (Ren *et al.,* 1999).

Thom and Church (1926) brought out a first manual for identifying most common and important species. It was revised by Thom and Raper (1954) and Raper and Fennel (1965). Samson and Gams (1985) proposed new scheme of classification of *Aspergillus* (Table 3.4). Classification of species of *Aspergillus* underwent changes with time was well as with persons. Samson and Gams (1985) divided *Aspergillus* into 7 subgenera and 19 sections (Table 3.4).

**Table 3.4.** Classification of genus *Aspergillus* (Samson and Gams, 1985)

| Sub genus | Section | Teleomorphy/Anamomorph |
|---|---|---|
| 1. *Aspergillus* | 1. Aspergillus<br>2. Restricti | 1. Eurotium / Dichlaena<br>*(A. glaucus)* |
| 2. Fumigati | 3. Fumigati<br>4. Cervini | 2. Neosartorya *(A. fumigatus)* |
| 3. Ornati | 5. Ornati | 3. Warcupiella<br>4. Sclerocleista *(A. ornatus)*<br>5.Hemicarpenletes<br>*(A. paradioxus)* |
| 4. Clavati | 6. Clavati | 10. Neocarpenulteles<br>(*A. clavatus*) |
| 5. Nidulantes | 7. Nidulantes<br>8. Versicolores<br>9. Usti<br>10. terrei<br>11. Flavipedes | <br>6. Emesicella *(A.nidulans)*<br><br>7. Fennellia *(A. cereus)* |
| 6. Circumdati | 12. Wentii<br>13. Flavi<br>14. Nigri<br>15. Circumdating<br>16. Candidi<br>17. Cremei<br>18. Sparsi<br>19.Ochraceorosei<br>(New section) | 11. Chaetosartorya *(A. wentii)*<br><br>8. Retromyces<br>*(A. flavus,* possible *A. niger)*<br><br>9. Neopetromyces |
| 7. Stilbothamorium | Species forming synmata<br>Aspergillus | |

Klich and Pitt (1988) brought out a laboratory manual which facilitated for precise identification of species of this genus. Kozakiewing (1989) gave an exhaustive account of species of *Aspergillus* that grows in stored products. Samson and Pitt (2000) tried to resolve taxonomic problems of this intriguing genus by integrating with modern taxonomic methods. Samson and Varga (2008) with vast experience of working on *Aspergillus* discussed elegantly the taxonomy in genomic era. He gave detailed account of molecular biology in resolving species identification of this important genus. There are about 250 known species of *Aspergillus* (Geiser *et al.*, 2008). This mould has been involved in many industrial processes including enzymes (amylases), commodity chemicals (citric acid) and food stuffs (say sauce). *Aspergillus oryzae* and *A. sojae,* called Koji moulds, are utilized in fermented rice beverages which have numerous local variations and names such as shaoshing (China), sake (Japan), takj or yakju (Korea) and other names across Asia. *Aspergillus* carries a huge potential for finding new enzymes that could be useful to convert plant biomass into fuels and other industrially useful products (Baker *et al.*, 2008).

In 1854 de Bary observed production of cleistothecium as well as *Aspergillus* head and felt both are two different phases of one fungus. Since that time, many fungi were linked between asexual and sexual stages i.e. dual naming was started for one single fungus with two reproductive stages. Since 1905, dual naming of one organism was allowed in Botanical Code of Nomenclature (which governed the naming of the plants and fungi) one for asexual stage and other for sexual stage. In an international workshop held in 2008 dual name was allowed in the absence of any other convenient system (Bennet 2010). Some species of *Aspergillus* which are with both teleomorph and anamorphic stages in environment are listed in Table 3.5.

**Table 3.5.** List of teleomorphic and anamorphic names of different species of *Aspergillus* (Geiser, 2008)

| **Teleomorphic name (sexual stage)** | **Anamomarphic name (Asexual stage)** |
|---|---|
| *Chaetosartorya* | *A. wentii* |
| *Emericella* | *A. nidulans* |
| *Eurotium* | *A. glaucus* |
| *Fennellia* | *A. terreus* |
| *Petromyces* | *A. flavus* possible *A. niger* |
| *Sclerocleista* | *A. ornatus* |
| *Warcupiella* | *A. spinulosus* |
| *Hemicarpenteles* | *A. porradexus* |
| *Neocarpenteles* | *A. clavatus* |
| *Neosartorya* | *A. fumigatus* |

Species which reproduce both asexual and sexual means are called holomorphic genera (Table 3.6). Three teleomorphic species of *Aspergillus* (*Eurotium, Emericella* and *Neosartorya*) are important from food technology point of view. *Eurotium* species generally have bright-yellow fruit bodies. They are relatively xerophilic and are associated with *Aspergillus* anamorphs without metulae, while *Neosartorya* are non-xerophilic and have white ascomata. In *Emericella* ascomata are also white but associated with thick-walled hulle cells and anamorphic stage has metulae. There are about 16 species of *Aspergillus* have no known teleomorph (Pitt and Hocking, 1985). Still some species of *Aspergillus,* mainly represented by asexual stage are known to have teleomorphic stage.

The new taxonomics are based on polyphasic approach using phenotypical characters together with multigene DNA sequences. In a polyphasic approach micro- and macro morphology, physiology, metabolites produced and molecular data are all important and in principle no particular character should be over emphasized. Extrolites profiles have proven to be specific for the taxa and this has contributed not only to a stable classification but also has been very valuable in critical revisions of the species in their taxonomy and phylogeny. These are detailed by Samson and Gams (1985).

**Table 3.6.** Holomorphic genera (Pitt *et al.*, 2000)

| | |
|---|---|
| *Chaetosartorya* | Neocarpenteles |
| *Emericella* | Warcupiella |
| *Eurotium* | Neopetromyces |
| *Fennalia* | |
| *Hemicarpenteles* | |
| *Neosartorya* | |
| *Petromyces* | |
| *Sclerocleista* | |

Modern taxonomists seek a more natural classification scheme based on evolutionary criteria. Such phylogenetic classification seems more possible now that traditional morphological markers have to be supplemented with nucleic acid sequences. In particular, DNA markers such as ITS regions or conserved genes such as calmodulin, tuberlin or RNA polymerase are used to conduct molecular phylogenetic studies. Species with few nucleotide differences are considered more closely related than those with many. Some of the important taxonomic treatments and identification manuals of *Aspergillus* are précised in Table 3.7.

**Table 3.7.** Important taxonomic treatments and identification manuals of the genus *Aspergillus*

| |
|---|
| 1926-Thom, C and M. Church. The *Aspergillus:* Williams and Wilkins Baltimore. |
| 1954-Tham, C. and K.B. Raper A manual of the *Aspergillus*: Willams and Wilkins Baltimore. |
| 1965-Raper, K.B. and D.I. Fennell. The genus *Aspergillus*: Williams and Wilkins Baltimore |
| 1985-Samson, R.A and J.I. Pitt. Advances in *Penicillium* and *Aspergillus* systematic Plenum New York |
| 1988 – Klich, M.A. and J.I. Pitt. A laboratory guide to common *Aspergillus* species and their teleomorphs North Ryde, Australia. |
| 1989-Kozakiewicz, A. *Aspergillus* species on stored products (walking food): AB international. |
| 1990- Samson, R.A. and J. I. Pitt (eds), Modern concepts in *Penicillum* and *Aspergillus* classification: Plenum press New York. |
| 2000-Samson, R.A. and J.I. Pitt (eds). Integration of modern Taxonomic methods for *Penicillium* and *Aspergillus* classification. |
| 2002-Klich, M.A. Identification of common *Aspergillus* species: Central bureau voor schimmel cultures. Utrekht |
| 2008- Samson, R.A. and J. Varga. *Aspergillus* systematics in the genomic era CBS fungal diversity centre. Utrecht. |
| 2010-Peterson, S.W. M., Machida and K. Gami *Aspergillus*: Molecular biology and Genomics, Appl. Microbial Biotechnol. 95, 339-344. Academic press .Vcaister. |
| 2012-*Aspergillus* and *Penicillium* identification using DNA sequence, bar code or MLST. |

*Aspergillus* species are only group among large number of eukaryotes now cataloged in data banks. There are currently well over 100 fungal genome projects in various stages of completion. Expressed Sequence Tag (EST), DNA micro assays are available for increasing number of *Aspergillus* genomes and their use allows targeted functional analyses.

*Aspergillus* species are common and widespread and are most successful moulds with important roles in natural ecosystems and human economy. These may also cause animal and plant diseases. The disease caused by these fungi in animals is called *Aspergillosis*. Immunosupression is generally a prerequisite for systemic *Aspergillosis* in humans. Certain species of *Aspergillus* can infect humans directly causing both localized and systemic infections. The incidence of systemic Aspergillosis, the most serious form, is on the rise and imposes an increasing medical burden upon hospitals and physicians. Several species contaminate grains and other foods with toxic metabolites that are threat to health of humans and animals. Most species are not only adapted for the degradation of complex plant polymers but they can also dine on substrates as diversed as dung, human tissues and antique parchment. An unidentified *Aspergillus* species is capable of solubalization of low rank coal (Torzilli and Isbister, 1994). *Aspergillus nidulans* is being used to elucidate para sexual cycle as well as to understand basic concepts in regulation of metabolic pathways, the cell cycle, intron splicing and hyphal polarity. Cultural and morphological characters used in classifying species of *Aspergillus* are given below:

**I. Media and cultural characters useful for identification of *Aspergillus* species are**

1. Czapek Yeast extract Agar, temp 25°C (CYA 25).
2. Czapek yeast extract Agar temp 37°C (CYA 37).
3. Capek Yeast extract Agar (with 20% sucrose) (CYA20S).
4. Malt Extract Agar (MEA), Czapek agar (CZ).

**II. Morphological characters useful in identification of *Aspergillus* species are**

(i) Conidiophores

(ii) Vesicle (Terminal part of the conidiophore)

(iii) Metulae

(iv) Phialides

(v) Conidium bearing cell (Phialides, uni-or biseriate), foot cell

(vi) Sclerotia (some species).

**III. Molecular characters are also useful in identification of *Aspergillus* species**

Morphological, physiological and molecular characters helpful in the identification of genus *Aspergillus* are presented in Table 3.8.

**Table 3.8.** Characters useful in the identification of species of *Aspergillus*

| Section | Species | Morphology | Physiology | Molecular |
|---|---|---|---|---|
| ***Crcumdati*** | *A. persii* | I | X | *BenA* |
| | A. cretensis | I | *X* | *BenA* |
| | *A.* floculosum | I | | *BenA* |
| | *A. neobridgen* | I | *g* | *BenA* |
| | *A.* pseudoelegans | I | *X* | *BenA* |
| | *A.* roseoglobosus | I | X | *BenA* |
| | *A.steyrii* | I | *gx* | *BenA* |
| | *A.* westerdikiae | I | *x* | *BenA* |
| ***Nigri*** | *A.* costaricensis | I | *gx* | *BenA* |
| | *A.* homomorphus | I | *gx* | *BenA* |
| | *A.* lactocoffeatus | I | *gx* | *BenA* |
| | *A.* pipiens | I | *gx* | *BenA* |
| | *A.sclerotionger* | I | *gx* | *multi* |
| | *A.* vadensis | I | *gx* | *multi* |
| | *A. ibiriques* | I | *gx* | *multi* |
| | *A.* brasiliensis | I | *gx* | *multi* |
| | *A.* uvarum | I | *gx* | *multi* |
| | *A.* aculeatus | I | *gx* | *multi* |
| ***Flavi*** | A. sclerotiocarbonarius | I | *gx* | *multi* |
| | *A.* bombycis | I | *gx* | *multi* |
| | *A.* pseudotamarii | I | *gx* | *multi* |
| | *A. arachidicola* | I | *gx* | *multi* |
| | *A. minesclerotigenes* | I | *gx* | *multi* |
| | *A. parvisclerotigenes* | I | *gx* | *BenA* |
| ***Ochraceorosei*** | *A.* rambelli | I | *gx* | *ITS,RAPD* |
| ***Fumigati*** | *A. lentulus* | I | | *multi* |
| | *A. fumigatus* | I | *x* | *multi* |
| | *A. novofumigatus* | I | *x* | *multi* |
| | *A. turcosus* | I | *gx* | *multi* |
| ***Neosartorya*** | *N. nishimurae* | a I | *g* | |
| | *N. otani* | a I | *g* | |
| | *N. takakii* | a | | |
| | *N. indohii* | aI | | |
| | *N. tsurutae* | A | | |
| | *N. coreana* | | | |
| | *N. laciniosa* | a | | *BenA/calm* |

| Section | Species | Morphology | Physiology | Molecular |
|---|---|---|---|---|
| | *N. assulata* | a | *gx* | *BenA/calm* |
| | *N. denticulata* | a I | *gx* | *multi* |
| | *N. galapagensis* | a I | *gx* | *multi* |
| | *N. australensis* | a I | *gx* | *multi* |
| | *N. ferenczii* | a I | *gx* | *multi* |
| | *N. papuensis* | a I | *gx* | *multi* |
| | *N. warcupii* | a I | *gx* | *multi* |
| ***Emerciella*** | *Em.qingiriakani* | a I | | |
| | *Em. venezuelensis* | a | *x* | *BenA* |
| ***Eurotium*** | *Eu. takimakanense* | a I | *g* | |

Abbreviations: a = aeciospores characters; I= anamorph characters; g = Growth rate; x = Extrolites; BenA = β-tubulin; Calm = calmoclulin; multi =Three or more molecular probes

Taxonomy is the trinity involving identification, nomenclature and classification. The genus *Aspergillus* poses taxonomic problems in all three realms.

On the other hand (Klich *et al.,* 2000) *Aspergillus* was divided into 6 subgenera and 16 sections (Table 3.9).

**Table 3.9.** Classification of *Aspergillus*

| Sub-genus | Section |
|---|---|
| Aspergillus | Aspergillus |
| | Restricti |
| Fumigate | Fumigate |
| | Cervini |
| Ornate | Ornate |
| Clavati | Clavati |
| Nidulantes | Nidulantes |
| | Versicolores |
| | Usti |
| | Terrei |
| | Flavepedes |
| Circumdati | Wentii |
| | Flavi |
| | Nigri |
| | Circumdati |
| | Candidi |

**Klich (2002)** classified species of *Aspergillus* into 3 groups based on type of conidial heads, which are given below

**Key to common groups of *Aspergillus***

**Group I:** Conidiogenous cells uniseriate

(i) Vesicle club shaped heads clavate. *A. clavatus*

(ii) Vesicle otherwise

(a) Heads loosely to definitely columnar, conidia cylindrical at first *A.restrictus*.

(b) Conidial heads compactly columnar : conidia only *A. fumigatus*, *cleistothecium A.fischeri*

**Group II:** Conidiogenous cells mixed (both uniseriate and biseriate)

1. Conidial heads globose radiate splitting with age:

(i) Brown to black -*A.niger*

(ii) White to cream- *A. candidus*

(iii) Yellow to ochraceues -*A.ochraceues*

2. Conidial heads radiate or forming poorly defined columns:

(i) Yellow-green, rough conidiophores-*A.flavus*.

(ii) Yellow-brown, smooth conidiophores -*A.wentii*

**Group III:** Conidial heads are tri seriate.

1. Conidial heads green:

(i) Heads radiate stipes uncolored: Two types of head-*A. jianus*. One type of head-*A.versicolor*.

(ii) Heads loosely columnar, cleistothecia usually present, stipes brownish- *A. nidulans*

2. Conidial heads some other colour:

(i) Heads radiate, conidiophores brown, conidia drab-olive to dull – brown – *A. ustus*.

(ii) Heads loosely columnar, conidia white to pale a vellaneous stalks yellow, brown to yellow to colorless. - *A. flavipes*.

(iii) Heads compactly columnar, conidia cinnamon to orange-brown to

(iv) pale- buff, conidiophores colorless. - *A. terreus*.

**Key to subgenera of *Aspergillus***

1. Conidiophores synnematous ........................................................ – ***Stilbothamnium***.

Conidiophores asynnematous ........................................................................ -2

2. Phialides strictly uniseriate ................................................– Subgenus ***nidulantes***.

3. Phialides strictly biseriate.............................................................. – ***circumdati***.

4. Phialides uniseriate or biseriate.

5. Conidial heads with spore mass splitting at maturity in blue-green shades vesicle strong.............................................................................. **clavate.**

6. Conidial heads radiate to columnar, variable in colour, vesicles globose to sub clavate or turbinate.
7. Conidial heads radiate too loosely to definitely columnar, osmophilic sub genus-............................................................................................**Aspergillus.**
8. Conidial heads variable not osmophilic
9. Colonies blue-green. Subgenus – ......................................................... ***Fumigate.***
10. Colonies grayish or yellowish, Green to olive-brown, White to purplish or olive. Cleistothecia are produced in some species.

**Key to sections**

**1. Subgenus-*Aspergillus***

Conidial heads radiate, bluish- green, bright-yellow, cleistothecia present in most species.

Conidial heads loosely columnar

Conidia cylindrical when young

Cleistothecia absent............................................................................................. **3. Section**

**2. Subgenus –Fumigati**

i) Conidial heads columnar – light-blue-green to very dark-green, vesicle flask shaped conidia echinulate to delicately roughened, Cleistothecia present in some species growth better at 37°C....................................... **4. Section-Fumigati**

ii) Conidial heads radiate or short columnar fawn coloured, vesicles globose to flask shaped conidia globose, smooth, thin walled show the presence of remains of connectives, light pinkish. ....................................... **5. Section – Cervin**

**3. Sub genus *Nidulantes.***

1. Conidial heads columnar, dark -yellow-green, green-bluish-green or buff, to cinnamon- brown.

   1a. Conidial heads long, compactly columnar, buff to cinnamon-brown, cleistothacia absent.......................................................................... **6. Section - *Terri***

   b. Conidial heads short, columnar, dark-yellow-green to bluish-green, clestothecia mostly present................................................ **7. Section *Nidulant***

2. Conidial heads radiate, columnar in age, white to vinaceous-brown, drab to dull-orange to gray or green.

   2a. Conidial heads drab to dull-orange or olive-grey-conidiophores brown and smooth hulle cell elongated, often strongly curved or twisted. ....................................................................................... **8. Section - *Usti***

   2b. Conidial heads white to vinaceous-brown or green; conidiophores hyaline or brown, smooth or rough; conidia smooth or rough, hulle cells often present.......................................................................... ---3

3a. Conidial heads white to vinaceous buff, conidia smooth, thin walled, cleistothecia absent, thick walled hyphal elements or elongated hulle cells present. .......... **9. Section *Flavipedes***

3b. Conidial heads in shades of green in some both green and white, conidia mostly echinulate, sclerotia globose to sub-globose, hulle cells sometimes present. .......... **10. Section *Versicolores***

**4. Subgenus Circumdate.**

Conidial heads large, radiate, vesicles strictly globose, conidiophores definitely constricted below the vesicle, sclerotia lacking. .......... ---1

Conidial heads globose, radiate and splitting in age, rarely loosely columnar, vesicles globose to subglobose, radiate and splitting in age rarely loosely columnar, vesicles globose to sub _ globose and elongate, conidiophores constricted not below the vesicle, sclerotia produced in many species. .......... 2

1. Conidial heads of one type, pale gray-green or light – brown, conidiophores usually colourless, smooth, osmophilic, cleistothecia produced in some species. .......... 11**. Section *Cremi***

1. Conidial heads of two types large heads, light -gray, green or olive-buff with conidiophores usually brown, fragmentary structures borne near or beneath the agar surface. .......... 12. **Section – Sparsi**

2. Conidial heads globose when young splitting in age, rarely loosely columnar. .......... 3

2. Conidial heads typically radiate, spore chains usually separate, sometimes forming poorly defined columns. .......... 4

3. Conidial heads persistently white or cream coloured, conidiophores smooth and colourless. .......... 13. **Section candidi.**

3. Conidial heads other than white. .......... 5

4. Conidial heads yellow-green or deep-olive-brown; conidiophores usually colourless and roughened. .......... **14. Section – *Flavi***

4. Conidial heads yellow-brown to dull-buff; conidiophores smooth or delicately roughened colourless or nearly so. .......... **15. Section *Wenti***

5. Conidial heads yellow, buff or ochraceous shades, conidiophores commonly roughened often pigmented. .......... **16. Section – *Circumdati***

5. Conidial heads in shades of black; conidiophores smooth and colourless or pigmented below the vesicle. .......... **17. Section – Nigri.**

**5. Subgenus – *Stilbothamnium***

Conidiophores synnematous 18. *Stilbothamnium* .......... **6. Sub genus – Ornati**

1. Conidial heads radiate to very loosely columnar, grayish or yellowish-green, olive-brown, not osmophilic, white to purplish or olive cleistothecia produced in some species.

1a. Conidial heads at first dark-blue-green then brownish-black, radiate, conidia globose, strongly spinulose to irregularly warty .... ***19. A. brunneo-uniseriatus.***

## Key to species of section – Flavi

1. Conidial heads dull-yellow-green to olive – brown when young changing to brownish-green or brown on CZSA, conidia conspicuously verruculose. .................... *A. tamari* ---1

1a. Conidial heads pale to intense yellow or yellow-green when young. .................... ---2

2. Conidial heads radiate or very loosely columnar, dark-yellow with age, phialides predominantly biseriate.................... *A. flavus* ----2

2a. Conidial heads radiate, colonies changing to yellow-brown with age on CZSA; conidia pyriform or elliptical when young, varying greatly in size. .................... *A. oryzae* ----3

2b. Conidial heads radiate, dark-yellow-green with age, phialides always uniseriate. .................... *A.parasiticus* ---4

2c. Conidial heads typically columnar, phialides predominantly uniseriate. .................... *A.columnaris* ---5

## Key to sections and species of *Aspergillus*

1. Conidiogenous cells strictly uniseriate .................... 2

 1a. Conidiogenous cells biseriate or uniseriate .................... 5

2. Conidial heads clavate .................... **Clavati**

 2a. Conidial structures not exceeding 4.0mm length.................... ***A. clavatus (1)***

 2b. Conidial heads not clavate.................... **3**

3. Conidial heads radiate, ascomata present.................... ***Aspergillus***

 3a. Conidial heads large, radiate to loosely columnar, borne above the surface layer of cleistothecium and enveloping hyphae, ascospores lenticular, 6 μ or less in long axis, equatorial ridges facing furrow absent or showing only as a trace .................... *4*

 3b. Equatorial ridges low or rounded, furrow broad and shallow, colonies strong- red, orange -red pigment.................... ***A. repens* (2)**

 3c. Equatorial ridges thin and flexuous, crest like ascospores resembling a pulle .................... ***A.rubrum* (3)**

 3d. Conidial heads not radiate, ascomata absent .................... ***A.chevalieri* (4)**

4. Conidial heads columnar, conidia echinulate.................... **Restricti**

 4a. Conidial heads columnar, conidia spinose.................... **Fumigati**

 4b. Cultures strictly conidial, varying from velvety to floccose.................... ***A. fumigatus*(5)**

 4c. Cultures producing cleistothecia conidial development generally limited .................... ***A. fischeri* (6)**

 4d. Vesicles often borne at an angle to the conidiophores, thin walled, vesicle, hyaline, often strongly noded conidia in pale-blue green shades .................... ***A. viridi-nutans* (7)**

5. Conidiogenous cells strictly biseriate.................... **6**

 5a. Conidiogenous cells uniseriate and biseriate .................... 10

6. Ascomata present ........................................ **Nidulantes**

6a. Ascospores present, orange-red in colour equitorial crests two in number, convex wall smooth, conidial stage dark yellow-green, crest entire ........................................ ***A. nidulans (8)***

6b. Ascomata absent........................................ **7**

7. Conidial heads columnar ........................................ **8**

7a. Conidial heads not columnar ........................................ **9**

8. Conidial heads white to brown, conidiophores brown ........................................ **Flavipedes**

8a. Conidiophores definitely pigmented in yellow to light-brown shades, conidial heads usually white to very plae-buff ........................................ ***A. flavipes*** **(9)**

8b. Conidial heads cinnamon, conidiophores not-brown ........................................ ***Terrei***

8c. Colonies velvety, conidial heads long compactly columnar in Cinnaman to orange-brown to brown shades borne on short conidiophores, sclerotium like masses of swollen, relatively heavy walved cells lacking........................................ ***A. terreus*** **(10)**

9. Conidial heads radiate, dull-brown, conidiophore brown ........................................ **Usti**

9a. Conidial heads olive-gray to drab or red-boawn, hulle cells scattered, thoughout the colony or forming irregular masses not associated with pigmented mycelium. ........................................ ***Versicolores (11)***

9b. Conidial head radiate, green conidiophores not brown ........................................

9c. Vesicles globose to somewhat elongate, fertile over most of the vesicular surface, globose to sub-globose hulle cells often present ........................................ **A. ustus** ***(12)***

9d. Conidial heads variable in colour, light yellow-green buff orange-yellow ........................................ ***A. versicolor*** **(13)**

9e. Conidial heads always blue-green when young........................................ ***A. sydowi*** **(14)**

10. Conidial heads yellow-green ........................................ **Flavi**

10a. Sterigmata single or double, heads lossely columnar Heads radiate, sterigmata uniseriate........................................ ***A. flavus (15)***

10b. Colonies shifting to light-brownish-green, conidia large elleptical to globose to subglobose, irregularly ........................................ ***A. oryzae (16)***

10c. Conidial heads not yellow-green ........................................ **11**

11. Conidial heads ginger-brown ........................................ **Wentii**

11a. Conidial heads large up to 500 μm on smooth or slightly granular stalks which may reach several mm in length ........................................ ***A. wentii (17)***

11b. Conidial heads not ginger-brown ........................................ **12**

12. Conidial heads globose, black ........................................ **Nigri**

12a. Sterigmata biseriate, conidia globose, roughened ........................................ ***A. niger (18)***

12b. Sterigmata uniseriate, conspicuously echinulate ........................................ ***A. japanicus (19)***

12c. Conidial heads globose, not black ........................................ **13**

13. Conidial heads globose, white........................................ **Candidi**

13a. Conidial heads globose, yellow to ochraceus .................................. **Circumdati**

13b. Sclerotia abundant, small, pure-yellow then brown, conidia globose to sub-globose to ovate ............................................ ***A. melleus (20)***

13c. Sclerotia large, pink to venaceous, purple when mature, ovate to cylindrical, conidia globose to sub-globose .................. ***A. ochraceous (21)***

Raper and Funnel (1965) gave key to for identification of species of *Aspergillus* based on conidiophores head and arrangement of sterigmata.

## I. Key to the common group series of *Aspergillus*

| | | | | |
|---|---|---|---|---|
| Uniseriate | Vesicles club shaped, heads clavate | | 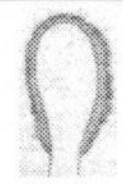 | *A. clavatus* |
| 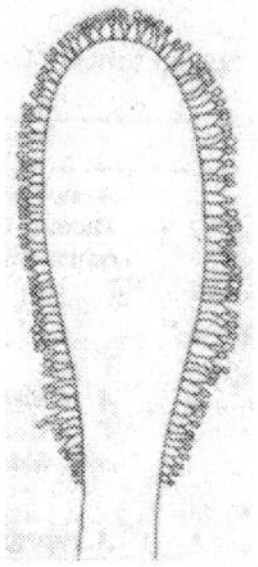 | | Heads globose, radiate, bright yellow cleistothecia | | *A. glaucus* |
| | Vesicles otherwise | Heads loosely to definitely columnar, conidia cylindrical at first |  | *A. restrictus* |
| | | Conidia heads compactly columnar | Conidia Only 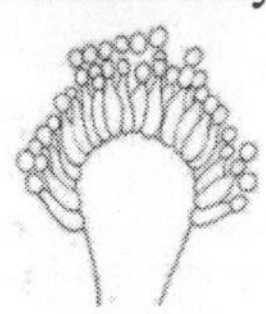 | *A. fumigatus* |
| | | | Cleistothecia | *A. fischeri* |
| **Mixed (uniseriate and biseriate)** | Conidial heads globose radiate splitting in age | Brown to black | 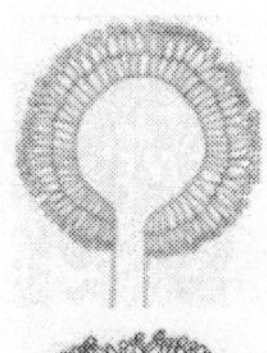 | *A. niger* |
| | | White to cream | | *A. candidus* |
| | | Yellow to ochraceous | | *A. ochraceus* |
| 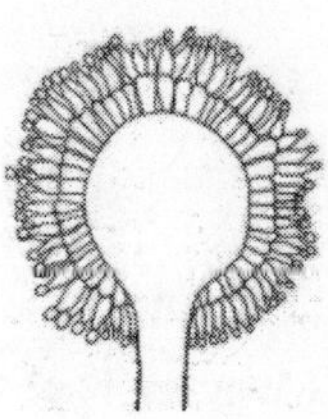 | Conidial heads radiate or forming poorly defined column | Yellow green, rough conidiophores | 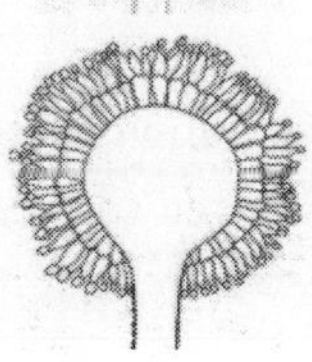 | *A. flavus* |
| | | Yellow brown, smooth conidiophores | | *A. wentii* |
| | Conidial heads large, conidiophores constricted below vesicles | | Rare | |

| | | | | |
|---|---|---|---|---|
| **Biseriate** 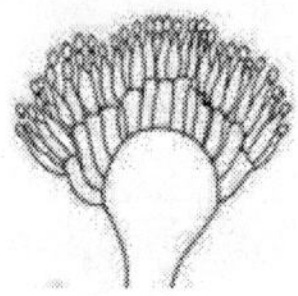 | Conidial heads green | Heads radiate, stipes uncoloured | One type of head  | *A. versicolor*<br>*A. janus* |
| | | | Two types of head | |
| | | Heads loosely columnar, cleistothecia usually present, stipes brownish | 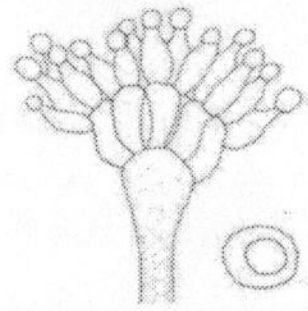 | *A. nidulans* |
| | Conidial heads some other colour | Heads radiate, conidiophores brown, conidia drab olive to dull brown | 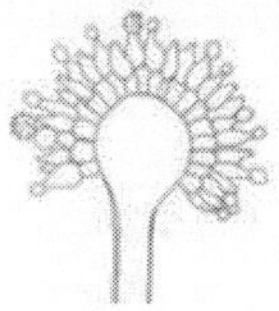 | *A. ustus* |
| | | Heads loosely columnar, conidia white to pale avellaneous, stalks yellow brown to yellow to colourless |  | *A. flavipes* |
| | | Heads compactly columnar, conidia cinnamon to orange brown to pale buff, conidiophores colourless | 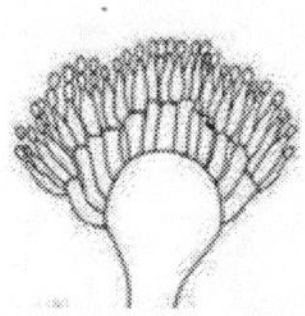 | *A. terreus* |

## II. Key to the *Aspergillus glaucus* group (Raper and Fennell, 1965)

1. Ascospores 8-10 μm in long axis ................................ *Eurotium echinulatum* Delacroix (*A. echinulatus* (Delacroix) Thom & Church)

1a. Ascospores 6 μm or less in long axis ............................................................. 2

2. Ascospores with curved surfaces smooth or nearly so ................................................ 3

2a. Ascospores with curved surfaces rough ..........*A. amstelodami* (Mangin) T. & C. (*Eurotium amstelodami* Mangin)

3. Equatorial ridges (crests) lacking; furrow nil or very shallow and inconstant .................................... *A. repens* (Coeda) Sacc. (*Eurotium repens* Corda)

3a. Ridges low and rounded; furrow broad and shallow .......................... (*Eurotium rubrum* Konig, Spieckermann & Bremer) *A. ruber* (Konig, Spieckermann & Bremer) T. & C.

3b. Ridges thin, crest-like; spore like a pulley.............. (*Eurptium chevalieri* Mangin) *A. chevalieri* (Mangin) T. & C.

**III. Key to the *Aspergillus restrictus* group**

1. Heads columnar; vesicles small, fertile on upper surface only .................................. 2

1a. Heads radiate when young, tardily loosely columnar; vesicles fertile over upper half to two thirds .............................................. *A. penicillioides*

2. Colonies growing moderately well on standard Czapek agar .................. *A. caesiellus*

2a. Colonies growing very slowly on Czapek agar .............................................. 3

3. Colonies dark olive green; spore column long, often twisted ....................*A. restrictus*

3a. Colonies light grey green; column shorter and more delicate .......................... 4

4. Conidia elliptical at first, often remaining so, 4.0-4.5 X 3.0-3.5 μm ............*A. conicus*

4a. Conidia sub-globose to pyriform, mostly 3.0-3.5 μm diam. ...............*A. gracilis*

**IV. Key to varieties of *Aspergillus terreus* Thom**

1. Colonies velvety, cinnamon to orange- brown; conidiophores short ............................ 2

1a. Colonies floccose, golden yellow; heads small; conidiophores up to 500 μm long or even lon .............*A. terreus* var. *aureus* Thom & Raper

2. Sclerotium-like masses produced on malt agar ar.................. *Africanus* Thom & Raper

2a. Sclerotium-like masses absent................................................. *A. terreus* Thom

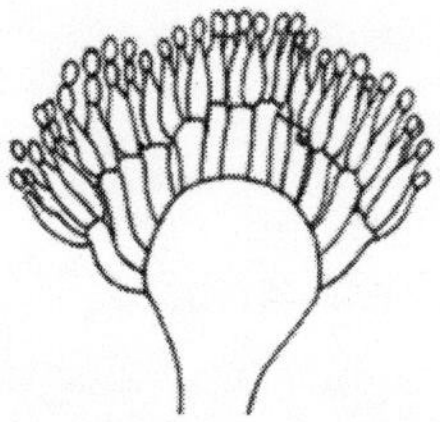

**I. General description of genus *Aspergillus* Mich. ex Fr.**

**Colony** : Restricted to fast growing.

**Colony** : White, blue, blue-green, green, yellowish-green, brown, ochraceous to black.

**Surface** : Conateur zonate, velvety, floccose, lanose or funiculose.

**Reverse of the colony:** In various shades.

**Exudates :** Present or absent.

**Conidial heads** :Radiate, globose or splitting in columns, columnar.

**Conidiophores :** Erect or sub erect, rarely synnematous

**Foot cell :** Hyaline to sub hyaline, smooth or rough, septate or aseptate.

**Vesicle :** Globose, subglobose, hemispherical, spatulate or clavate.

**Phialides :**One series (uniseriate) or two series (biseriate).

**Metulae :** The lower one of biseriate is called metulae, bear cluster of two or several phialides at its apex.

**Conidia :** Unicellular, unbranched basipetal chains.

**Sclerotia :** Are present in some species

**Ascocarp :** Cleistothecia form the basis for the classification into sub-genera.

**II. Description of species of *Aspergillus* associated with different foods**

**i) *A. aculeatus* Lizuka**

**Synonymys** = *Aspergillus japanicus* var. *aculeatus* Al-Musallam (1980)
= *A. violaceofuscus* Gasperini
= *A. yezoensis* Sasaki (1950)

**Subgenus:** Circumdati

**Section:** Nigri

**Teleomorph:** Absent

**Colony:** Heavily sporulations, dense strand of conidial heads.

**Colour:** Purple-brown or purple-black shades, yellow pigment diffusible.

**Reverse:** Pale to yellow

**Extrolites :** Lacking

**Sclerotia:** Produced by some strains, white to cream coloured.

**Vesicle:** Globose at first then splitting into relatively few compact divergent columns, fertile entire surface.

**Conidiophores:** Hyaline, slightly brownish below the vesicle.

**Phialides:** Single series.

**Conidia:** Elliptic to globose.

**Mycotoxins:** Secalonic acid D, cyclopiazonic acid.

**Diagnostic features:** Purple-brown or purple-black shades, yellow pigment diffusible globose vesicles and fertile entire surface, uniseriate phialides, elliptical to globose conidia.

**ii) *A. amstelodami* (Mangin) Thom and Church.**

**Synonymys** = *A. awamori* Nakaz Rep. amstelodami
= *A. luchvensis*
= *A. niger* var. *awamori*
= *Eurotium amstelodami* Mangin (19009)
= *E. repens* var. *amstelodami* (1920)

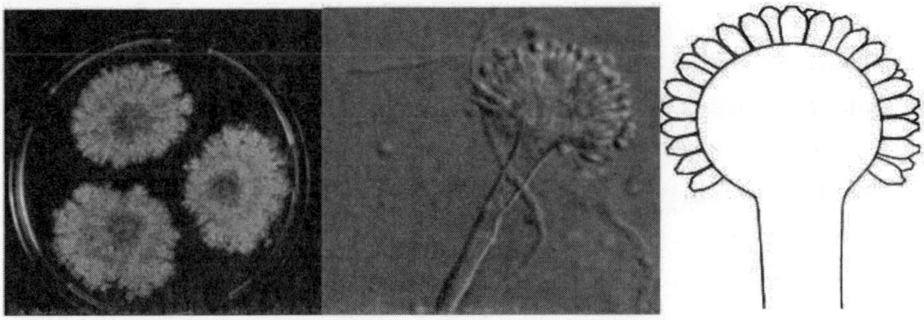

**Subgenus:** *Aspergillus*

**Section:** *Aspergillus*

**Teleomorph***:* *Eurotium amstelodami* Mongin

**Colony:** Radially furrowed

**Colour:** Dark-greenish-brown later turning dark chocolate-brown.

**Reverse:** Uncolored, irregularly mottled yellow shade becoming brown or reddish- brown

**Extrolites:** Lacking

**Vesicle:** Globose then loosely radiate splits into ill defined columns

**Conidiophores**: Thick walled.

**Phialides:** Biseriate

**Metulae:** Only one phialide per metulae

**Conidia:** Globose to subglobose, smooth to definitely roughened.

**Mycotoxins:** Echinulin, neoechinulin

**Diagnostic features:** Radially furrowed, dark chocolate- brown colonies, globose vesicles bearing biseriate phialides, globose to sub globose smooth to roughened

**iii) *A. aureoluteus* Sawsor and Gams**

**Synonymys** = *Aspergillus aurelous*

**Subgenus:** Fumigati

**Section:** Fumigati

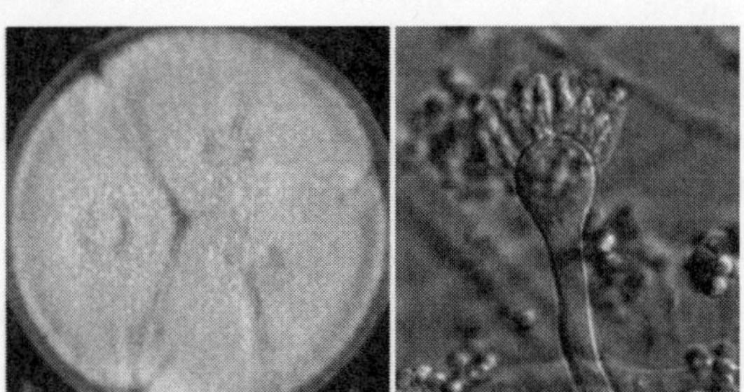

**Teleomorph:** *Neosartorya aureola* (Fennel & Raper) Malloch & Caln

**Colony:** Spreading, Radial, slightly anate

**Colour:** Apricot to light cadmium yellow

**Reverse:** Yellow ocher to ochraceous

**Conidial head:** Loosely columnar

**Vesicle:** Clavate to flask shaped.

**Conidiophores:** Color less

**Phialides:** Uniseriate

**Conidia:** Globose to subglobose, delicately echinulate.

**Ascocarp:** Homothalic, cleistothecia pale yellow, surrounded by loose walls of dark-golden yellow hyphae, lenticular ascospores with two prominent equatorial crests, conspicuously echinulate.

**Mycotoxins:** Fumigallin, triptoquivaline, tryptoquivalone, pseurolin A and visidotoxin.

**Diagnostic features:** Yellow to golden pigmentation of hyphae surrounding the cleistothecia.

**iv)** ***A. caespitosus*** **Raper & Thom**

**Subgenus:** Nidulantes

**Section:** Versicolores

**Teleomorph:** Absent

**Colony:** Plaior somewhat furrowed

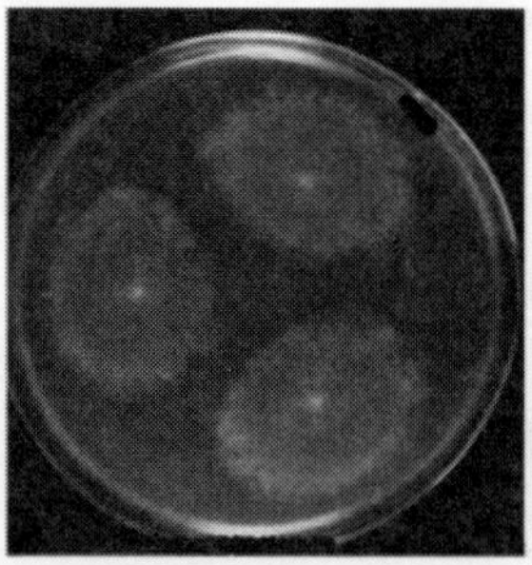

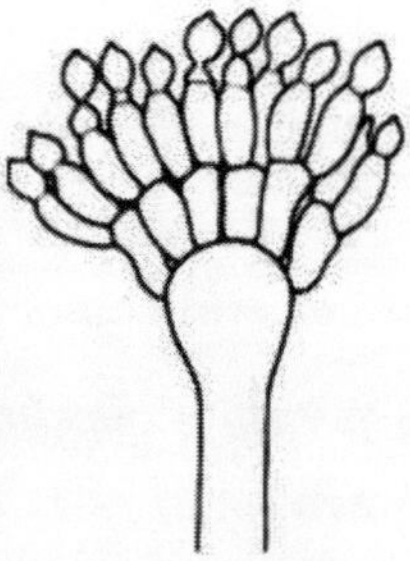

**Colour:** Dark-green.

**Reverse:** Colorless changing to dark-reddish-purplish with age.

**Hulle cell:** Thick walled, colourless at first becoming reddish-purple with age

**Odour:** Absent.

**Vesicle:** Radiate to loosely columnar.

**Conidiophores:** Thick walled and smooth to begin and becoming rough red with age turn to brown in colour.

**Phialides:** Biseriate, bottle form.

**Conidia:** Globose, spinulose, green.

**Mycotoxins:** Fumitremorgin A.

**Diagnostic features:** Dark-green coloured colonies, the cells thick walled colonies later becoming reddish-purple with age, vesicle with biseriate phialides, conidia globose spinulose and green.

**v) *A. candidus* Link**

**Synonymys=** *A. okazakii* Saito (1907)
=*A. albus* Wilh
=*A. albus* var. *thermophillus* Nakaz (1932)
=*A. tritici* Meherotra & Basu (1976)

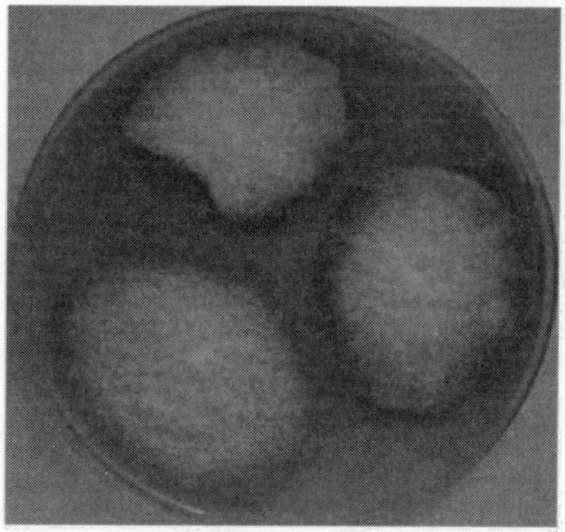

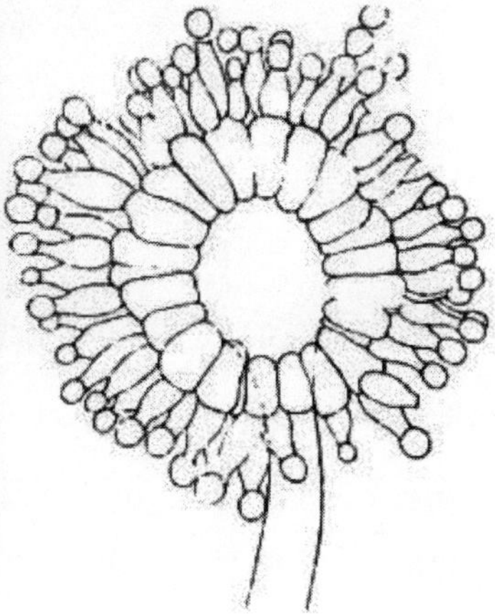

**Subgenus:** Circumdati

**Section:** Candidi

**Teleomorph:** Absent

**Colony:** Spreading, submerged

**Colour:** Persistently white or becoming cream with age.

**Reverse:** Colourless or pale gray-brown

**Vesicle:** White, globose often splitting with age

**Conidiophores:** Colourless, smooth, thick walled.

**Phialides:** Biseriate

**Metulae:** Wedge shaped

**Conidia:** Hyaline, globose to subglobose thin walled, smooth.

**Sclerotia:** If produced, white later turning reddish-purple to black.

**Mycotoxins:** Kojic acid, citrinin, xanthocillin

**Diagnostic features:** Colonies white, vesicle globose often splitting with age bearing biseriate phialides, conidia hyaline, globose to subglobose smooth walled.

**vi) *A. chevalieri* (Margin) Thom & Church.**

**Synonymys** =*Aspergillus chevalieri* Thom & Church (1926)

**Subgenus:** *Aspergillus*

**Section:** *Aspergillus*

**Teleomorph:** *Eurotium chevalieri*

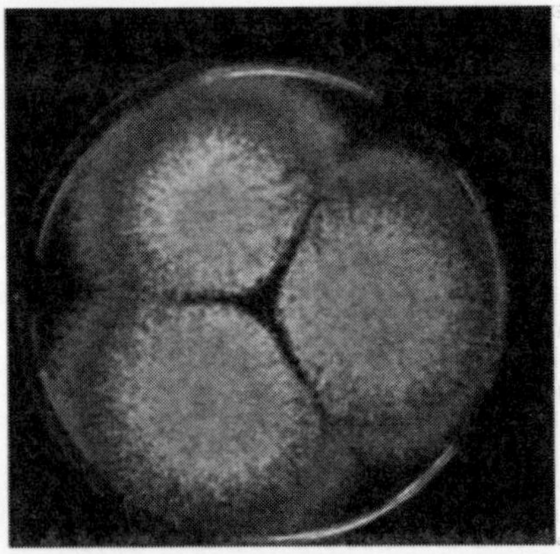

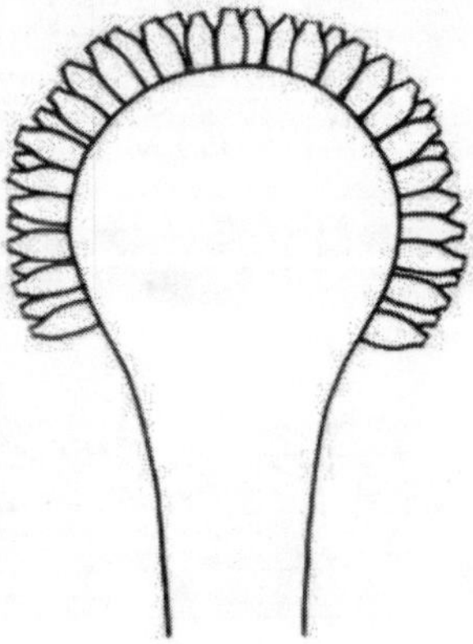

**Colony:** Spreading slowly.

**Colour:** Deep bluish-gray-green to russian-green

**Reverse:** Yellow orange to maroon.

**Conidial head:** Abundant, radiate, and scattered evenly.

**Conidiophores:** Smooth, gradually broadening upwards.

**Vesicle:** Subglobose to globose

**Phialides:** Uniseriate.

**Conidia:** Ovate to elliptical ends often flattened, spinulose.

**Ascocarp:** Cleistothecia abundant, globose to subglobose, light-yellow to brown, Ascospores lenticular, smooth walled with two thin flexuous and prominent equatorial ridges crest like resembling pulley.

**Mycotoxins:** Echinuline, gliotoxin, xanthocin

**Diagnostic features:** Deep bluish gray green to Russian green colonies, vesicles globose with uniseriate phialides, conidia ovate to elliptical and spirulose, Eurotium is its Teleomorph.

**vii) *A. clavatus* Desm**

**Synonymys** = *A. apicalis* Mehrotra & Basu 1976

= *A. pallidus* Kamyschko 1963

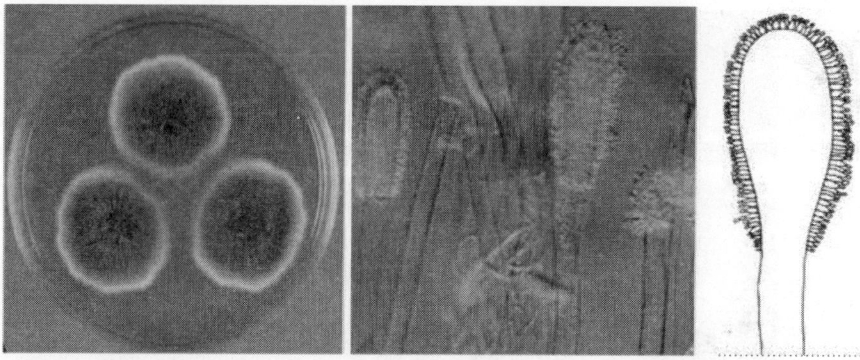

**Subgenus:** Clavati

**Section:** Clavati

**Teleomorph:** Absent

**Colony:** Spreading

**Colour of the colony:** Blue-green when young later slate-olive.

**Reverse of the colony:** Colourless at first and light-brown with the age.

**Extrolites:** Absent, strongly filed in some.

**Vesicle:** Clavate often splitting into divergent columns at maturity, fertile over the entire surface

**Conidiophores:** Thin walled, gradually enlarging at the open into a clavate vesicle.

**Phialides:** Single series, basal smaller and at the apex larger.

**Conidia:** Elliptical, heavy walled, light-green in mass.

**Mycotoxins:** Patulin, cytochalasins E & K, tryptoquivalline, tryptoquivalone, kotanins, antafumicin, ascladio, ribotoxins.

**Diagnostic features:** Conidiophore vesicle is clavate and bears single row of phialides, conidia elliptical and heavy walled light green in mass

**viii)** ***A. fischeranus***

**Synonyms** = *Aspergillus fischeri*

= *Sartorya fumigate*

**Subgenus:** Fumigate

**Section:** Fumigate

**Teleomorph:** *Neosartoya fischeri* (Wehmer) Malloch & Cain

**Colony:** Diffusive, velutinous

**Colour:** White to pale-yellow to buff

**Reverse:** Colourless to flesh coloured

**Vesicle:** Flask shaped

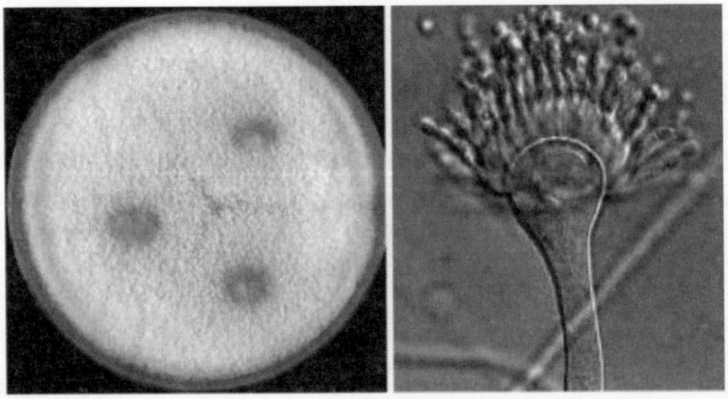

**Conidial head** – Columnar

**Conidia:** Globose to subglobose micro-tuberculate

**Ascocarp:** Cleistotheci light cream, borne singly or in small clusters within loose hyphal envelope ascospores convex, reticulate.

**Mycotoxins:** Terrein, fumitremorgins A & C, tryptoquivaline, A, trypacidin, verrucalagen, neosarotinin, helvotic acid fiscalins.

**Diagnostic features:** Reticulate ascospores ornamentation

**ix) *A. flavipes* (Baineer & Sartory) Thom & Church**

**Synonymys** = *Fernelia flavipes*

= *Sterigmatocystis flavipes* Baineer & Sartory (1911)

**Subgenus:** Nidulantes

**Section:** Flavipedes

**Teleomorph:** Absent

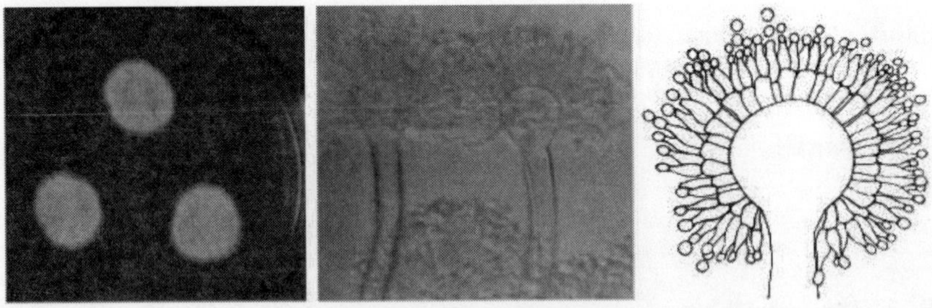

**Colony:** Growing moderately, white at first turning to bright-wheat colour later velvety

**Reverse:** Usually yellow-brown to red-brown, periphery white.

**Conidial heads:** Columnar, pale-avellaneus shades.

**Conidiophores:** Smooth or short

**Vesicle:** Subglobose to vertically elongate

**Phialides:** Biseriate

**Metulae:** Always present, curved upwards

**Conidia:** Subglobose, nearly colourless, smooth

**Hulle cells:** Sometimes present

**Mycotoxins:** Citrinin, asperthecin

**Diagnostic features:** Colonies white turning to bright-wheat colour and velvety, subglobose to vertically elongate vesicle with biseriate phialides, subglobose hyaline smooth conidia.

**x) *A. flavus* link**

**Synonymys** = *A. fasciculatus* Bat and Maia

= *A. flavus* f. magnasporus (Skaj & Yamadda)

=*A. flavus* var. *wehmeri* (Coustantin Nehira & Lucet) Blochwit.

=*A. humus* E.V. Abbot (1926)

=*A .luteus* (Tiegh) Dodge (1935)

=*A. oryzae* var. *magnosporus* Sakaj & Yamada.

=*A. wehmer*: Constantin & Ludet

**Subgenus and section:** Circumdati; Flavi

**Teleomorph :** Absent

**Colony :** Yellow when young constituting and with age to deep-gray-green.

**Colour of colony :** Yellow when young dark-yellow-green with age.

**Reverse of colony :** Colourless to pale yellow-brown

**Exolites :** In perspicuous

**Sclerotial strain :** Red-brown

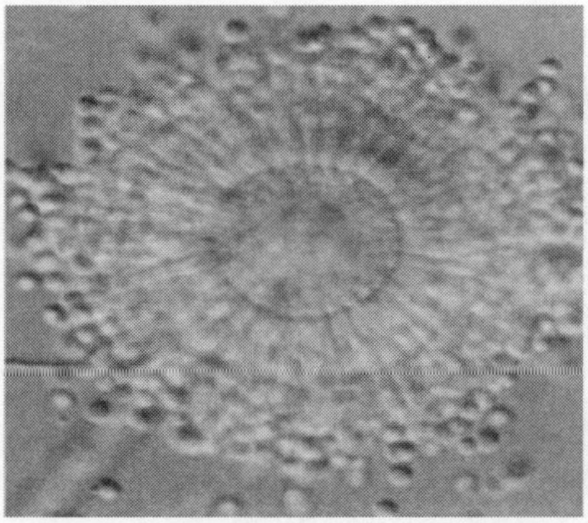

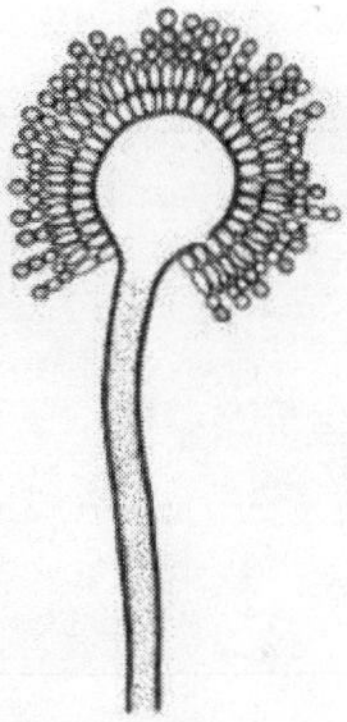

**Conidial heads :** Radiate splitting to poorly defined columns

**Vesicle :** Broadening gradually to vesicle

**Phialides :** Uniseriate or biseriate

**Conidiophores :** Long, coarsely roughered heavy walls, colourless.

**Conidia :** Globose to sub-globose, echinulate, yellow-green, sometimes elliptical.

**Mycotoxins:** Aflatoxin B1 and B2

**Diagnostic features:** Colonies yellow changing to dark yellowish green with age. Phialides uniseriate or biseriate, conidia globose to sub globose, echinulate, yellowish- green in colour characterstic features of *A. flavus* which helps in the differentiation of it from related species presented in Table 3.10.

**Table 3.10.** Differentiation of *A. flavus* and related species

| Species | Conidia | Sclerotia |
|---|---|---|
| *A. flavus* | Smooth to moderately roughered, variable in size, conidial heads dark-yellow | Large, globose |
| *A. parasiticus* | Conspicuously roughered, little variation in size, conidial dark yellow-green | Large, globose |
| *A. nomius* | Similar to those of *A.flavus* | Small, elongated (bullet shaped) |
| *A. tamari* | Conidia verruculose, conidial head brownish-green or brown | |
| *A. oryzae* | Conidia pyriform or elliptical, conidial heads yellow-brown. | |

**xi) *A. fumigati* affinis. Hang, Frisvad & Samson**

**Synonymys =** *A. arvii* Aho *et al.* (1994)

= *A. phialiseptus* Kwan-Chung (1975)

= *A.* anomalous *Mosseray* (*1934*)

= *A.cellulosae*

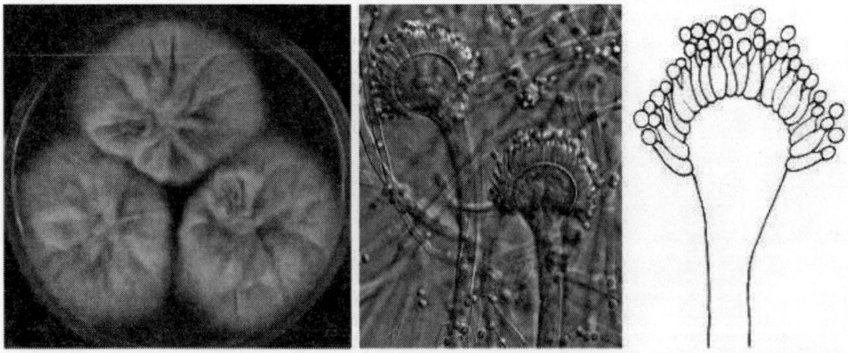

**Subgenus:** Fumigati

**Section:** Fumigati

**Teleomorph:** *Neosartorya fumigata*

**Colony:** Spreading, floccose

**Colour:** White with center dull-green

**Reverse:** Yellowish to grayish-orange

**Conidial head:** Short columnar

**Vesicle:** Globose to subglobose, smooth

**Conidia:** Globose to subglobose, smooth.

**Mycotoxins:** Fumigaclavines, helvolic acid, tryptoquivaline, tryptoquivalone.

**Diagnostic features:** Small globose subglobose vesicle grows at 10°C but not at 50°C.

**xii) *A. fumigatus* Fresenius**

**Synonymys** = *A. anamalus* Pidoplichko and Kirileuko

= *A. fumigates* var. *acaluminanis* Rai *et al.*

= *A. fumigates* var. *ellipticus* Raper & Ferinel

= *A. fumigates mut helvola* Rao *et al.*

= *A. phialiseptus* Kwon-Chung

= *A. neoellipticus* Kezakiewioz

= *A. arvi* Aho, horie Nishimua & Miyaji

= *A. fumisynnematus* Horie, Miyaji, Nizhimura Taguchi & udagawa

= *A. lentulus* Balajee & Marr

**Teleomorph:** Absent

**Subgenus section:** Fumigati

**Colour of colony:** Greyish-turquoise or dark-teurquise to dark–green to dull-green.

**Conidiation:** Abundant

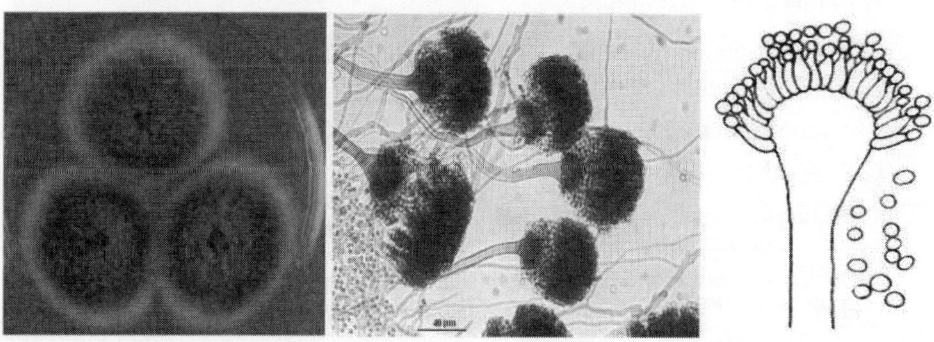

**Reverse of the colony** (CYA): creamy, yellow to orange, colony texture velutinous.

**Vesicle:** columnar, sub-globose rarely globose.

**Phialides:** Uniseriate

**Conidia:** Globose to ellipsoidal smooth to finely rough.

**Mycotoxins:** Fumigallin, furnitoxins, fumigaclavins ACC, fumitremorgins fumiquinazolines, gliotoxin, helvolic acid, pseurotins, pyripyropens, methyl-sulochrin, trypacidin, verruculogen, viriditoxin

**Diagnostic features:** Rapid growing velutinous colonies, abundant and fast conidiation, large, pyriform to semi clavate vesicle.

### xiii) *A. nidulans* (Eidam) Winter

**Synonymys** = *Sterigamatocistin nidulans* Eidam (1883)

= *Diplostephanus nidulans* Neveru (1920)

= *Aspergillus nidulellus* Samson & Gams (1986)

**Subgenus:** Nidulantes

**Section:** Nidulantes

**Teleomorph:** *Emericella nidulans* (Eidam) Vuill.

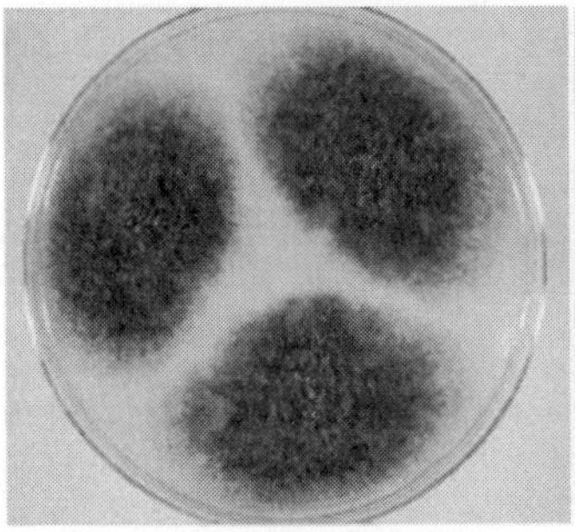
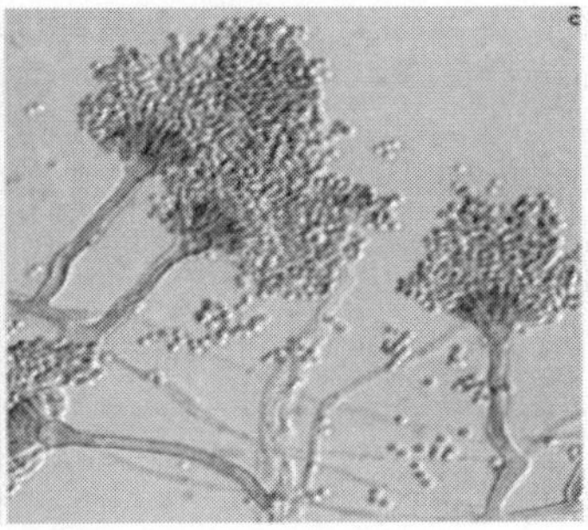
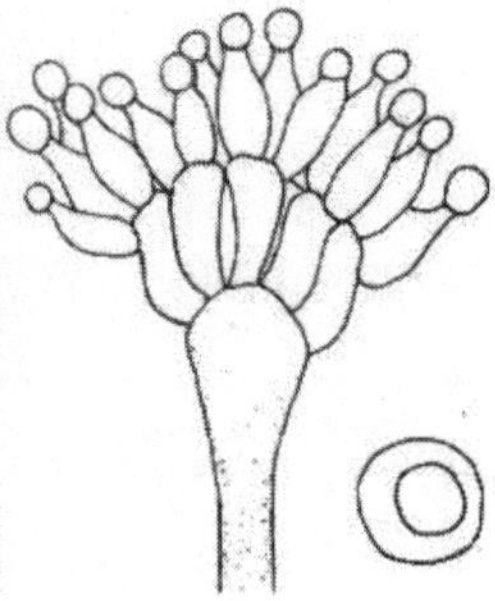

**Colony:** Typically light-green.

**Colour:** Green

**Reverse:** Olive to drab-grey

**Conidial head:** Green

**Vesicle:** Columnar

**Phialides**: Biseriate

**Conidiophore:** Short brownish and smooth walls

**Conidia:** Globose, rough walled

**Mycotoxins:** Sterigmatocystin

**Diagnostic features:** Green colonies with olive to dark green reverse, green coloured column of conidia with biseriate phialides, conidia globose. Teleomorph Emericella producing hulle cells and pink coloured spinulose ascospores.

**Table 3.11.** Illustrattion of different features of some species of *Aspergillus*

| | *A. flavus* | *A. fumigatus* | *A. nidulans* | *A. niger* | *A. terreus* | *A. versicolor* |
|---|---|---|---|---|---|---|
| Colony | Yellow-green | Blue green | Dark-green | Black | Tan- buff | Pale green |
| Conidio-phores |  | | | | | |
| Vesicles |  | | | | | |
| Sterigmata | 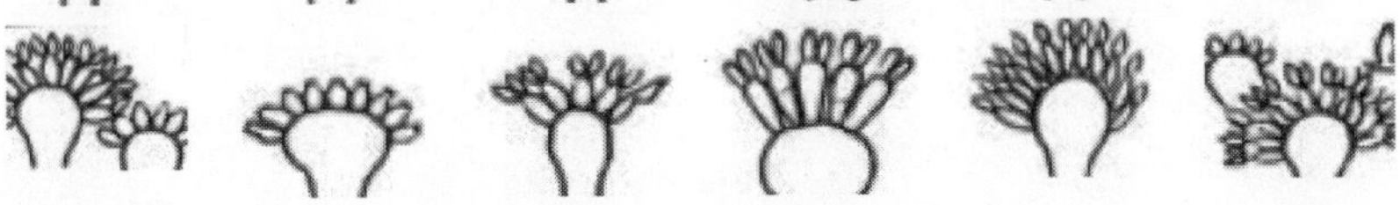 | | | | | |
| Conidia |  | | | | | |
| Other structures |  | | | | | |

**xiv) *A. niger* Tiegh.**

**Synonymys** = *Sterigmatocyotin vnigra.*

= *Sterigmatocystis nigra* Tiegh (1877)

= *Aspergillosis niger* Speg (1910)

= *P. pyri*

= *A. cinnamomeus* Schiermann (1912)

**Subgenus:** Circumdate

**Section:** Nigri

**Teleomorph:** Absent.

**Colony:** Growing moderately with abundant submerged mycelium.

**Colour:** Carbon black, sometimes deep-brownish black.

**Reverse:** Colourless to pale-yellow.

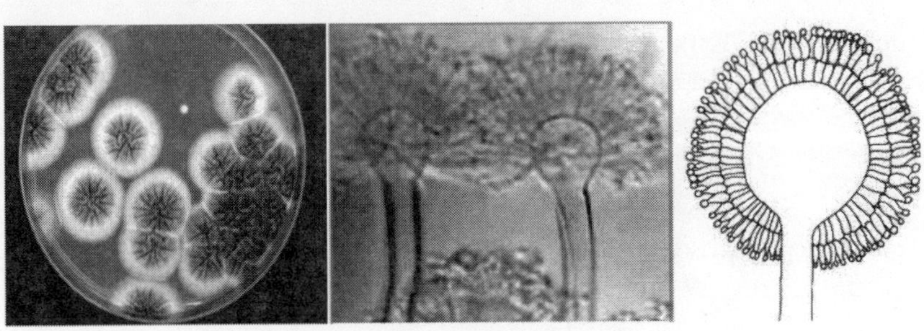

**Conidial heads:** Large and black at first, globose then radiate or splitting to well defined column.

**Conidiophores:** Smooth, thick walled, non-septate.

**Vesicles:** Globose, walls thick, biseriate phialides.

**Conidia:** Globose, spinulose, black.

**Sclerotia:** Globose to subglobose, sclerotia are produced at first cream to bufflater vinaceous- buff with age.

**Mycotoxins:** Fuminosin B2, B4, malformins, nigerapyzone A, B.

**Diagnostic features:** Carbon black colonies with abundant submerged mycelium, conidiophore non septate, vesicle globose with biseriate phialides, globose spinulose black.

### xv) *A. novofumigatus* Hong. Frisvad & Samson

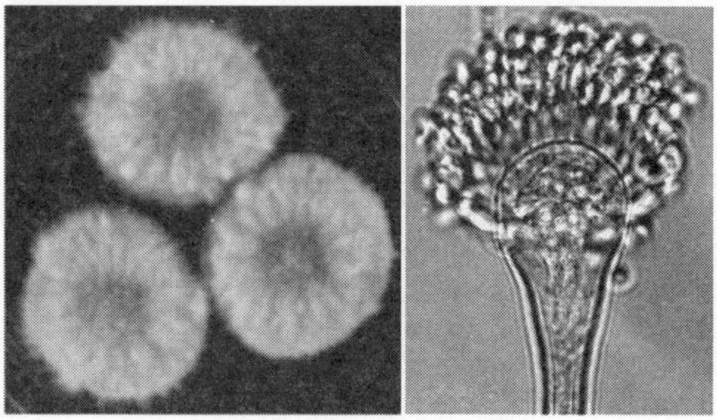

**Subgenus:** Fumigati

**Section:** Fumigati

**Teleomorph:** Absent

**Colony:** Spreading, effuse

**Colour:** Deep-green to grey-green

**Reverse:** Grayish-orange to yellowish-orange

**Texture:** Velutinous

**Vesicles:** Short, columnar, Subglobose to flask shaped large vesicle, growth at 10ºC, no growth at 50ºC.

**Phialides:** Uniseriate

**Conidiophore:** Color less

**Conidia:** Smooth, subglobose

**Mycotoxins:** Aszonalenin, cycloechine line, fiscatins, helvolic acid, neosartosin palitanin, terrain, territrem B.

**Diagnostic features:** Deep-green to grey-green colonies, vesicle large flask shaped.

### xvi) *A. ochraceous* Wills

**Synonymys** = *A. alutaceus* Berk & M.A. Curtis (1875)

**Subgenus:** Circumadate

**Section:** Circumdate

**Teleomorph:** Absent

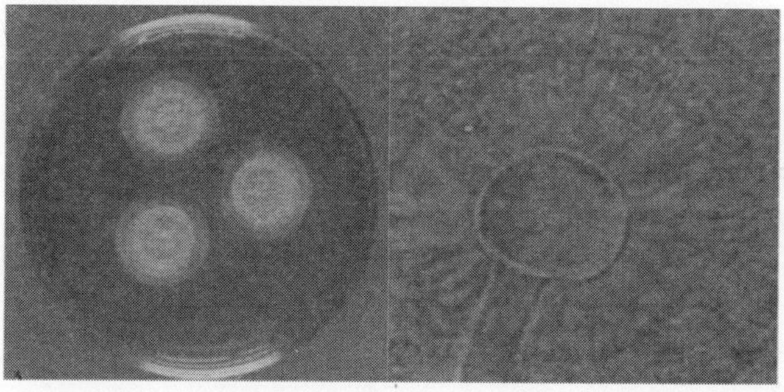

**Colony:** Growing moderately, plane with a tough basal mycelium, producing crowded conidial structures.

**Colour of the colony:** Light ochraceous-buff to warm-buff colour

**Reverse of the colony:** Yellowish, greenish-brown or reddish-purple.

**Extrolites:** Amber colour small droplets

**Vesicle:** Globose when young to 2-3 divergent columns with age, fertile over entire surface.

**Conidiophores:** Light-yellow to light-brown, walls thick and coarsely roughened.

**Metulae:** Varying in size.

**Phialides:** Biseriate

**Conidia:** Globose to subglobose

**Sclerotia:** Abundant, globose to sub-globose, white to light pink turning to vinaceous purple with age.

**Mycotoxins:** Ochratoxin A, xanthomegnin, secalonic acid D and Penicillic acid.

**Diagnostic features:** Light-ochraceous-buff to brown-buff colour colonies globose, vesicles, which is fertile, over entire surface bearing two rows of phialides produce large number of sclerotia conidia globose to subglobose.

### xvii) *A. parasiticus* Speare Hawaiian

**Synonymys** = *A. flavus* sub sp. *parasiticus* Kutzmen *et al*. (1986)

= *A. chungii* Shih (1936)

**Subgenus:** Circumdate

**Section:** Flavi

**Teleomorph:** *Pteromyces parasiticus*

**Colony:** Spreading well

**Reverse of colony:** Cream to tight-drab.

**Extrolites:** Limited

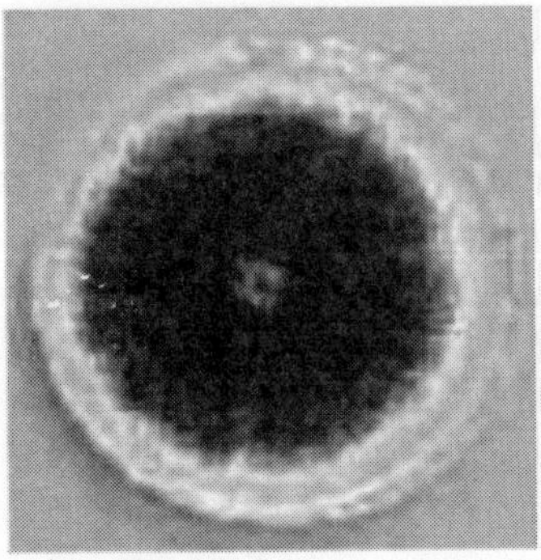

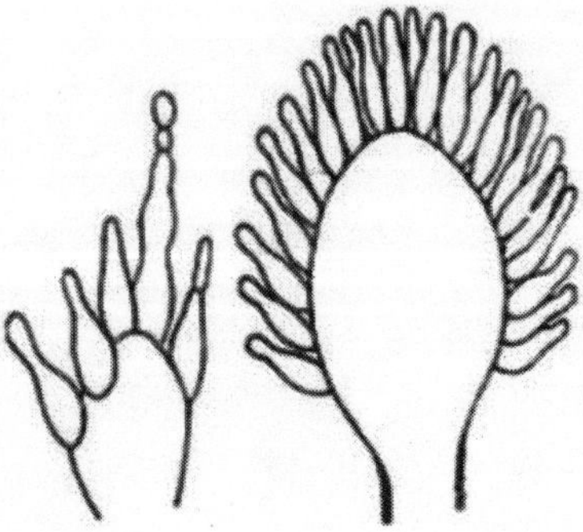

**Vesicles:** Broad

**Phialides** Uniseriate arranged half of the upper half or 2/3 area of vesicle

Colourless to pale yellow-green

**Conidia:** Globose to sub globose, coarsely echinulate, bright yellow-green in mass.

**Conidial head:** Abundant, at first parrot- green to grass- green and finally dark-dull yellow- green with age, loosely radiate.

**Sclerotia:** Not produced.

**Mycotoxins:** Aflatoxin B1, B2, G1 and G2

**Diagnostic features:** Conidial heads first parrot- green which changes to dark-yellow-green, uniseriate phialides on upper half to 2/3^rd^ of vesicle area conidia globose to sub globose, coarsely echinulate, biseriate, yellow green in mars.

**xviii) *A. repens* (Corda) Saccardo**

**Synonymys** =*Aspergillus glaucus* var. *repens* Corda (1842)

=*A. reptans* Samson & Gams (1985)

**Subgenus:** *Aspergillus*

**Section:** Aspergillus

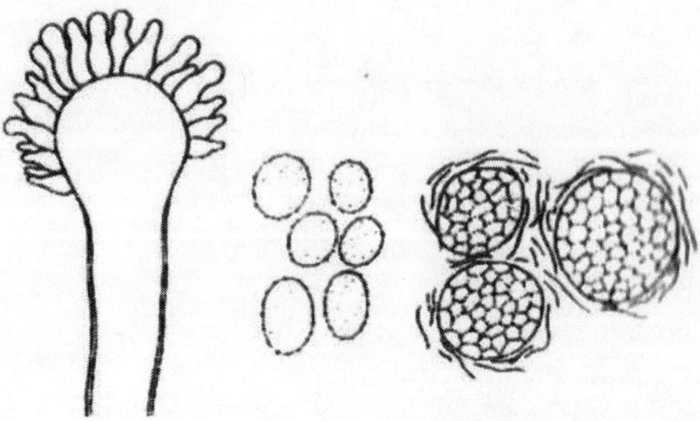

**Teleomorph:** *Eurotium repens*

**Colony:** Mixture of dirty green and yellow

**Colour:** Intimate mixture of dirty-green and yellow

**Reverse:** Dirty-brown or yellow to brownish-red (Czapek)

**Conidial heads:** Large, rough

**Vesicles:** Globose bearing uniseriate phialides

**Conidiophores:** Simple, straight

**Conidia:** Elliptical, rough

**Cleistothecia:** abundant, ascospores with trace of furrow but very indistinct and no crests.

**Mycotoxins:** Physcion, erythroglaucin, falvoglaucin, aspertin

**Diagnostic features:** Mixture of dirty green and yellow colonies. Dirty brown or yellow to brownish-red colonies on the reverse side, large and rough conidial heads, globular vesicle with uniseriate vesicles, conidia elliptical, Eurotium teleomorph with large no of ascospores.

**xix) *A. restrictus* G. Sm.**

**Synonymys** = *Penicllium fuscofulvum* Abe (1956)

**Subgenus:** *Aspergillus*

**Section:** Restri

**Teleomorph:** Absent

**Colony:** Spreads slowly, velvety when young to floccose with age.

**Reverse:** Uncoloured to greenish

**Conidial head:** Columnar.

**Conidiophores:** Smooth, shorter, broadening gradually into vesicles.

**Vesicles:** Hemispherical.

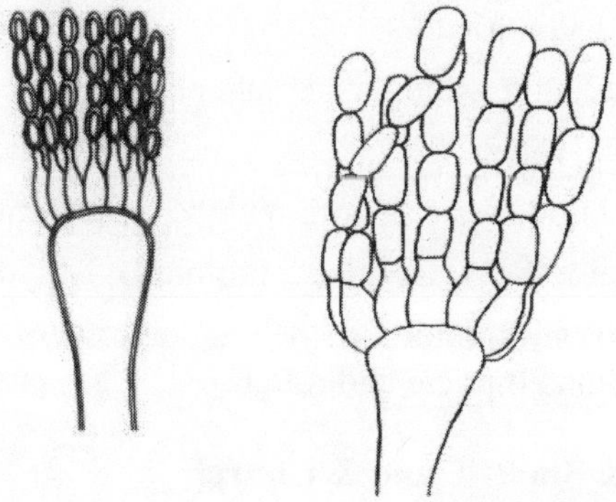

**Phialide:** Uniseriate, borne on only upper surface of the vesicle sometimes coloured.

**Conidia:** Rough, oblong cylindric to elliptic, dark in mass, adherence in long columns which remain intact in fluid mount.

**Mycotoxins:** Restictocin

**Diagnostic features:** Velvety changing to floccose with age colonies, hemispherical head, with uniseriate phialides, conidiophores oblong cylindric to elliptic, dark in mars adhere in long columns which remain intact in fluid mount.

**xx) *A. ruber* (Konig, Speckermann & Bremer) Thom & Church**

**Synonymys** = *Eurotium repens* deBary (1870)
= *E. rubrum* JOS Koning *et al.* (1901)
= *Aspergillus sejunctus* Bainer & Saratory (1911)

**Subgenus:** *Aspergillus*

**Section:** Aspergillus

**Teleomorph:** *Eurotium rubrum* Link ex Fries

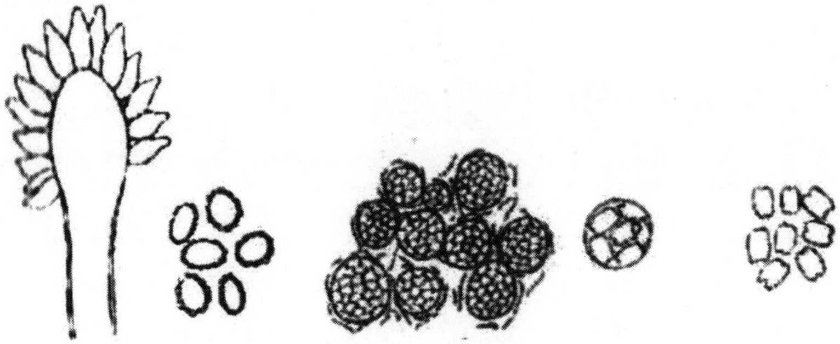

**Colony:** Rusty red, regular, uneven on morphology

**Reverse:** Deep-red to intense red-brown

**Conidial heads:** Pale blue-green turning grayish

**Conidiophores :** Cylindrical

**Vesicles:** Elliptical, thin walled uniseriate phialides

**Conidia:** Elliptical, rough

**Cleistothecia :** Abundant, ascospore with broad shallow furrow.

**Mycotoxins:** Aflatoxin B1, Cryptoechinuline G, Dihydroflavoglucin

**Diagnostic features:** Colonies rusty red, reverse of the colony deep red to intense red-brown, conidia elliptical and rough.

**xxi) *A. sydowii* (Bain & Start) Thom & Church**

**Synonymys** = *Sterigmatocystes sydowi* Bainer & Saratory (1913)
= *S. tunetana* Langeren (1924)

**Subgenus:** Nidulantes

**Section:** Versicolores

**Teleomorph:** Absent

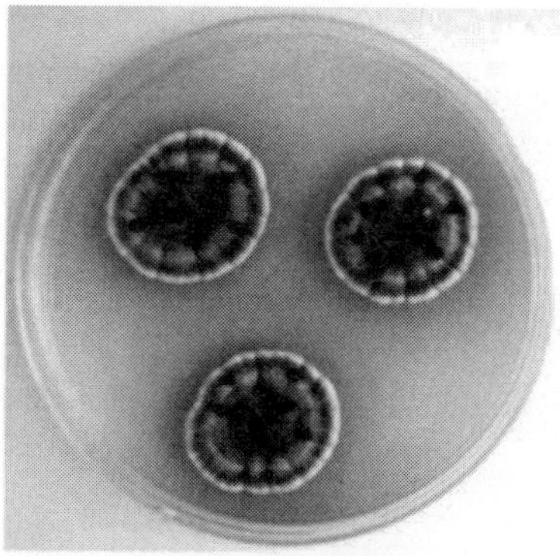

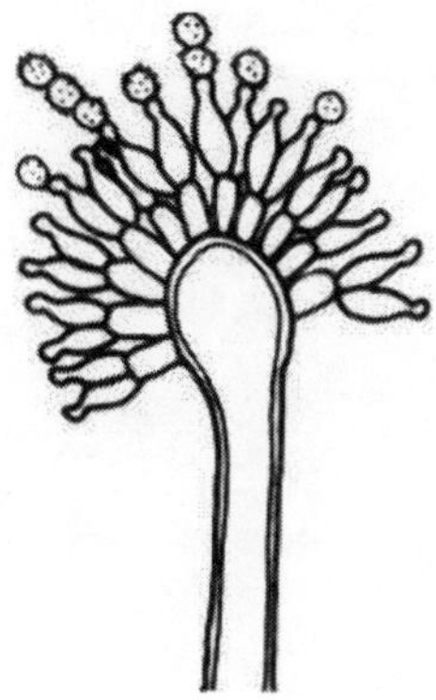

**Colony:** Velvety colonies of deep-bluish green or greenish blue color, radially sulcate, sporulating well

**Colour:** Deep-bluish, gray-green.

**Reverse:** Reddish-brown reddish brown

**Exudates:** Soluble pigment.

**Conidiophores:** Very short

**Reverse:** Tawny olive to orange -cinnamon

**Vesicles:** Subglobose

**Conidial head:** Biseriate

**Metulae:** Isolated covering most of the vesicle bearing clusters of phialides

**Phialides:** Resembling penicillate fructification abundant

**Conidia:** Globose to subglobose, spinulose.

**Mycotoxins:** Nidulotoxin

**Diagnostic features:** Deep bluish to grey green colonies, globose vesicles with biseriate phialides and metulae borne on entire surface of vesicle conidia globose to subglobose and spinulose.

**xxii) *A. tamarii* Kita**

**Synonymys** =*Aspergillus terricola* var. *bronzens* Sainel (1949)

**Subgenus:** Circumdati

**Section:** Flavi

**Teleomorph:** Absent

**Colony:** Brownish green to brown

**Colour:** Brownish-green to brown but often starting as a deep-yellow-greenish-brown.

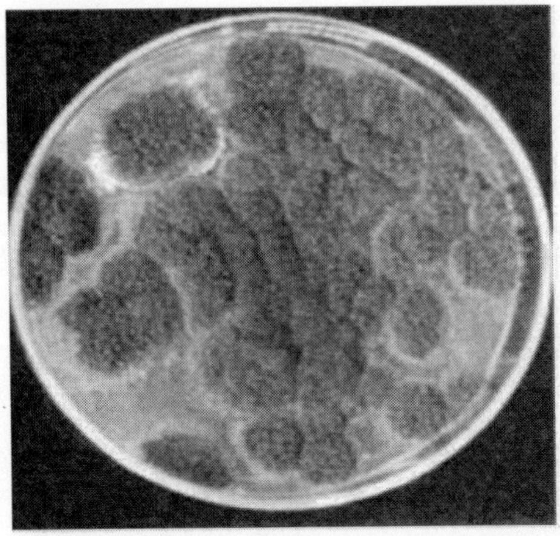

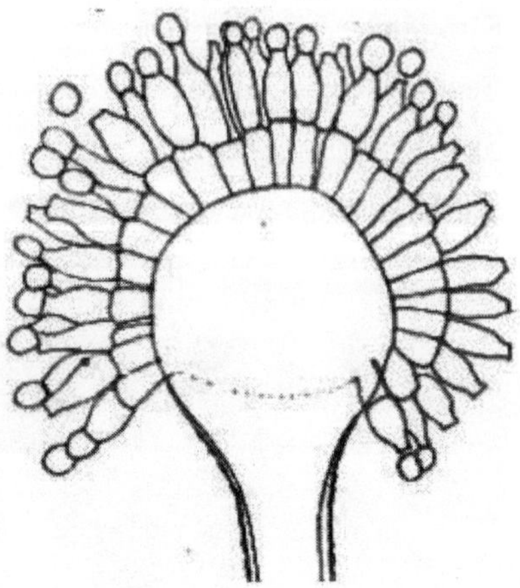

**Reverse:** Brown

**Conidial heads:** Large, radiate, single chains of conidia visible under low magnification.

**Vesicles:** Thin walled, easily crushed in moung.

**Metulae:** Usually present but absent in some heads

**Phialides:** Biseriate

**Conidiophores:** Cylindrical, borne from surface hyphae, colourless with rough and thin walls, simple, unbranched

**Conidia:** Cylindrical to pyriform when young, globose with age adhering chains coarsely roughened or tuberculate.

**Mycotoxins:** Fumigaclavine A, cyclopiazonic acid, kojic acid, cycloziazoric acid.

**Diagnostic features:** Brownish green of brown colonies, vesicles thin walled crushed easily becoming large chains of conidia, which are cylindrical pyriform adhering in long chains and tuberculate

**xxiii) *A. terreus* Thom**

**Synonymys** = *A.boedijinii* Ann. Mycol. 32.83

= *A. terreus v. boedijinii* (B1935).

**Subgenus:** Nidularites

**Section:** Terrei

**Teleomorph:** Absent

**Colony:** Spreading, plain or with radial furrows, velvety, floccose.

**Colour of the colony:** Cinnamon-buff to wood-brown

**Reverse of the colour:** Dull-brown.

**Extrolites:** Amber coloured

**Vesicles:** Long, columnar, compat diam.

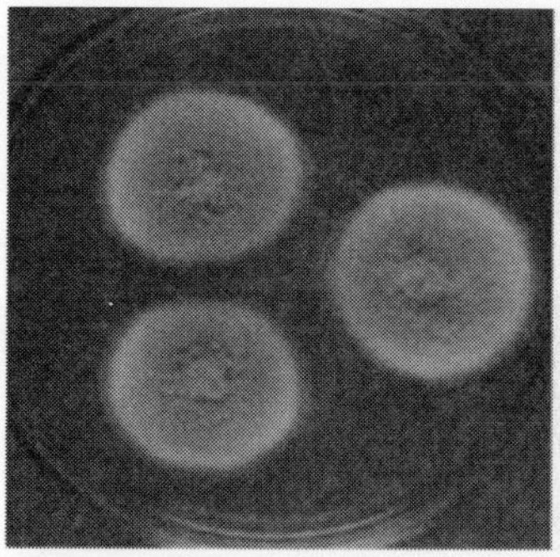 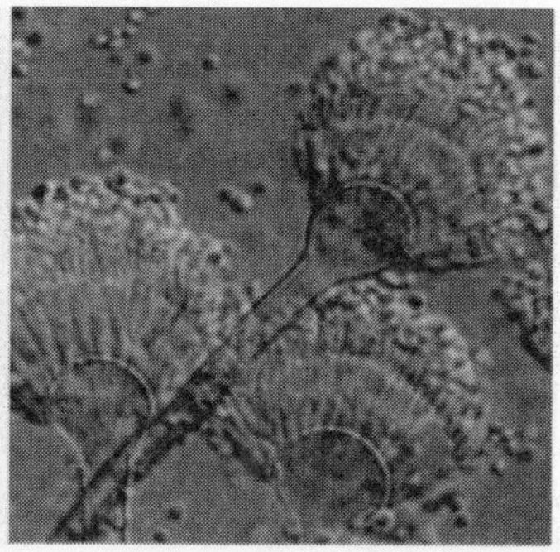 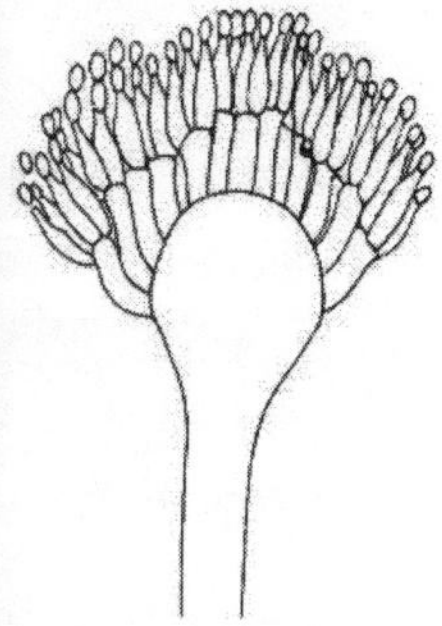

**Colour:** Cinnamon -brown

**Vesicles:** Hemiphaerical

**Metulae:** Crowded, parallel

**Phialides:** Biseriate, closely packed

**Conidia:** Globose to subglobose, smooth

**Sclerotia:** Absent

**Mycotoxins:** Terric acid, patulin, terrtrim B, emodin

**Diagnostic features:** Colonies cinnamon-buff to wood-brown, two series phialides, globose to sub globose, smooth conidia

**xxiv)** ***A. thermonutatus*** **(Paden) Petersan.**

**Teleomorph:** *Neosartorya pseudofischeri* Peterson

**Sub genus:** Fumigatin

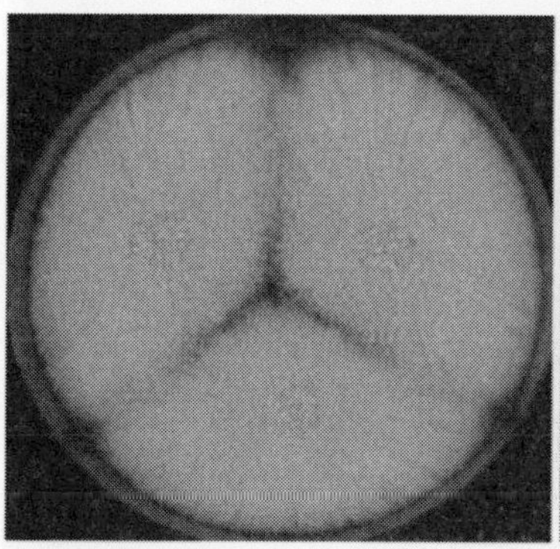 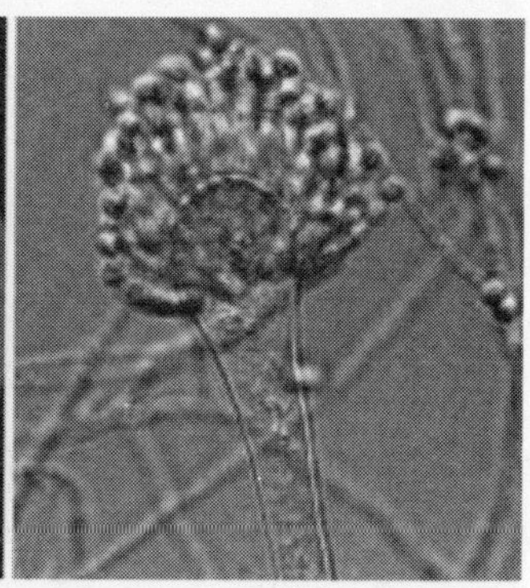

**Section:** Fumigati

**Teleomorph:** *Neosartorya pseudofischeri* Peterson

**Colony:** Velutinous

**Colour:** White to pale-creamish

**Reverse:** Clear or faintly yellowish

**Conidial head:** Columnar

**Vesicle:** Subglobose

**Phialides:** Uniseriate

**Conidia:** Globose to subglobose, smooth.

**Ascocarp:** Cleistothecia white, ascospores, subglobose, with two equatorial crests with raised flaps resembling triangular projections.

**Mycotoxins:** Gliotoxin, pyripyropins, cytochalasins live compound.

**Diagnostic features:** Distinctly ornamented ascospores.

### xxv) *A. ustus* (Bain) Thom & Church

**Synonymys =** *Sterigmatocetus ustus* Bainier (1881)

= *A. humus* Abbot (1926)

= *A. minutus* Abbot (1927)

**Subgenus:** Nidulantes.

**Section:** Usti

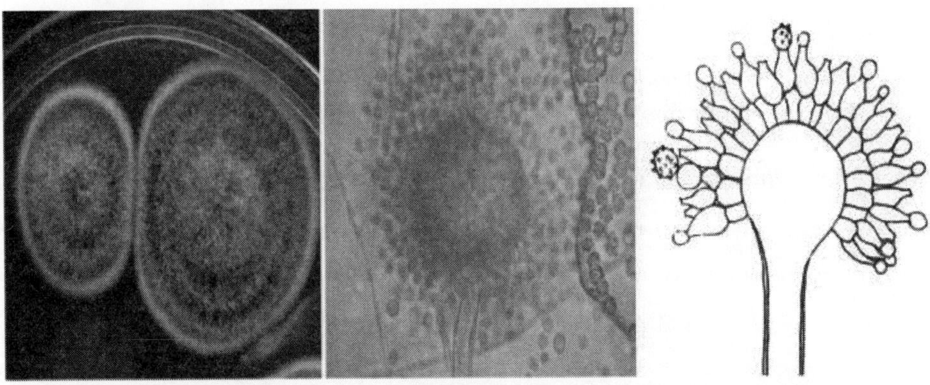

**Teleomorph:** Absent

**Colony:** Slow growth

**Colour of the colony:** Greyish-brown to dark-brown.

**Conidiation:** Moderate

**Reverse colour**: Yellow, olive-brown to dark-brown.

**Colony texture:** Floccose, plane, sulcate or umbronate.

**Conidial head:** Radiate to hemispherical

**Conidia:** Globose, roughened, greenish to dark yellow-brown.

**Hulle cells:** Irregularly ovoid or elongate, scattered

**Mycotoxins:** Ustic acid, austocystin, versicolorins, austalida, sterigmatocystin and ridulol.

**Diagnostic features:** No growth at 37°C, good growth on creatine

**xxvi) *A. versicolor* (Vuill) Tirab.**

**Synonymys** = *Sterigmatocystis versicolour* Vuill (1903)

=*A.* amoenu *Robwerg (1931)*

**Subgenus:** Nidulartes.

**Section:** Versicolores

**Teleomorph:** Absent.

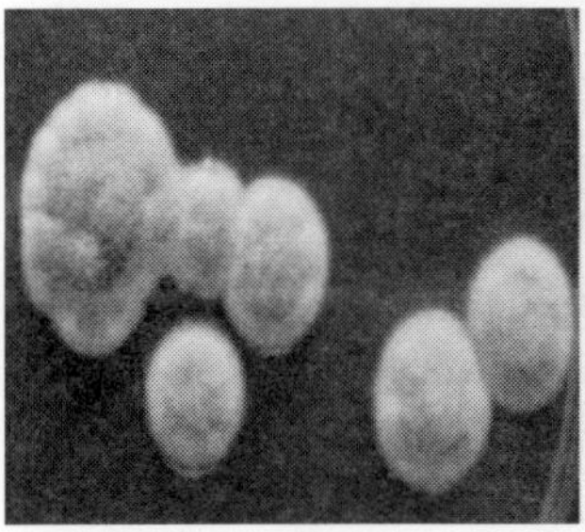

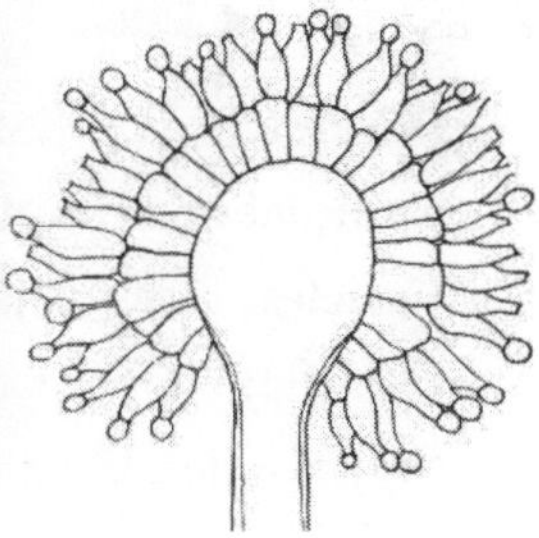

**Colony:** Sulcate, centrally raised, sporulating well.

**Colour:** Pale grayish-green near tea green.

**Reverse:** Vinaceous or brown or scarlet.

**Conidiophores:** Smooth occasionally lightly-tuberose, hyaline to yellow with brownish shades.

**Vesicle:** Pyriform to spathulate, biseriate.

**Metulae:** Covering half to entire vesicle

**Phialides:** Fragmentory heads resembling penicillate frunctfilations.

**Conidia:** Spherical to sub spherical, occasionally, elliptical, finely roughened wall.

**Hulle cells:** Globose in NARL5219-not in others.

**Mycotoxins:** Sterigmatocystin, versicolorin A.

**Diagnostic features:** Pale grayish-green to tea green, pyriform to spathulate, uniseriate with biseriate phialides, spherical to sub spherical elliptical finely rounded conidia.

**xxvii) *A. viridinutans* Ducker & Thrower**

**Synonymys** = *A. fumigatus* var. *sclerotiorum*

**Subgenus:** Fumigati

**Section:** Fumigati

**Teleomorph:** Absent

**Colony:** Velutinous, center raised

**Colour:** Niagara green

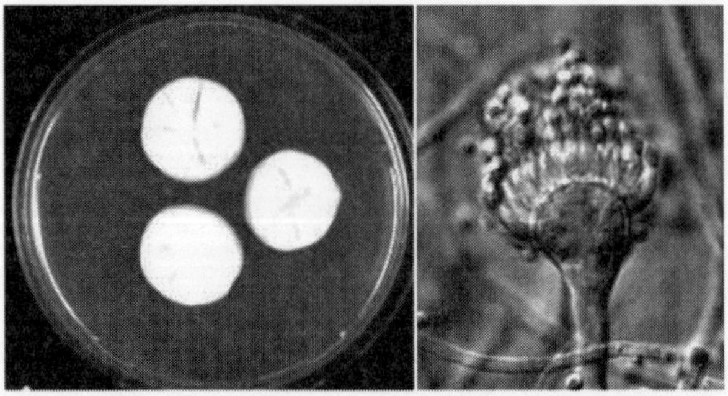

**Reverse:** Colourless, yellowish-green to light-brownish, olive.

**Conidial head:** Columnar

**Vesicles:** Flask shaped to subglobose

**Conidia:** Globose, delicately roughed

**Mycotoxins:** Viriditoxin, phomaligin, Visidin A, Variatin.

**Diagnostic features:** Nodding conidial heads, Niagara green colony colour.

**xxviii)** ***A. wentii*** **Wehmer**

**Synonymys** = *Aspergilus orachaeoflavus* Blochwitze (1933)
= *A.wentii* var. *minimus* Nakaz *et al.* (1934)

**Subgenus:** Circumadati

**Section:** Wenti

**Teleomorph:** Absent

**Colony:** Dense floccose masses of aerial mycelium

**Colour:** White or tinged with pink or rose

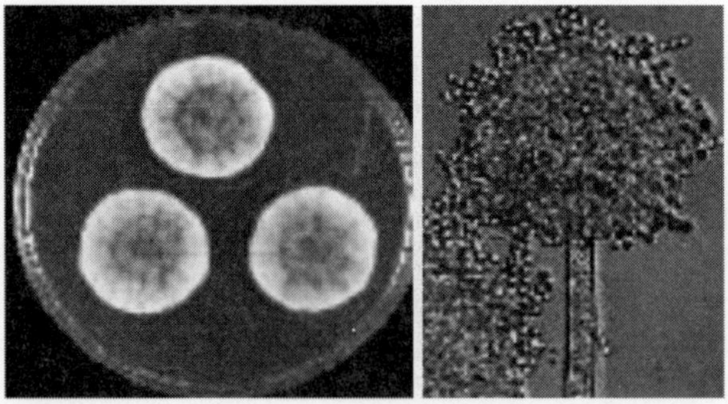

**Conidial heads:** Large yellow to coffee-coloured

**Vesicles:** Globose, fertile over the entire surface

**Phialides:** Biseriate, metulae

**Conidia:** Globose or slightly elliptical usually rough but occasionally almost smooth, Yellow or yellowish-brown

**Mycotoxins:** Emodin

**Diagnostic features:** Colonies floccose, white or tinged with pink or rose color, globose vesicle with entire surface fertile bearing biseriate phialides conidia globose, smooth yellow to yellowish-brown colour.

### 6. *Beauveria* Vuvill

*Beauveria* species are white entomopathogenous produce unicellular conidia that are typically. Colonies. Colonies hydrophobic and very small are typically white or off-white on artificial culture media. The conidia are formed holoblastically from basally inflated conidiogenous cells. After conidium production, the conidiogenous cell elongates before producing another conidium on a small denticles, which results in the formation of a distinctive, slender, zig-zag rachis. Three species of Beauvaria are recognized, two of which are well known parasites of insects. B. bassiana is the most common species and is best known as the causal agent of the disastrous muscardine in silkworms. Beauveria species are occasionally isolated in the clinical laboratory as saprophytic contaminants.

***B. bassiana* (Bals. Criv.) Vuvill**

**Synonymys** = *Botrytis bassiana* Bals. –Criv
= *Stradylidium bassianum* Bals. –Criv (Mont)
= *Spicarai bassiana* (Bals. –Criv) Vuvill
= *Sporotrichum globuliferum* Speq
= *Beaveria globulifera* (Speq) Picafrd
= *Botrytis acrediorum* (Brongn & Delacr
= *Spicaria acrediorum* (Brongn & Delacr) Vuvill
= *Botrytis delacroiseic* Sacc
= *Isaria vexans* Peitt
= *Beaveria vexans* Peitt (Petch)
= *Botrytis effuse* Beau
= *B. necans* Mars
= *B. stephanoderris* Balty
= *Beaveria laxa* Pctch
= *B. doryphora* Poisson

**Colony:** White spreading fluffy to powdery parasitic on insect, mycelial strands cylindrical margin, a septate.

**Conidiophore:** Single, irregularly grouped or in verticellate clusters in some species inflated at the base, tapering to a slender spore bearing portion which appears zigzag after spores are produced more or less comparable to cymose with scorpiod, helicoid dichotomy, sense, globose.

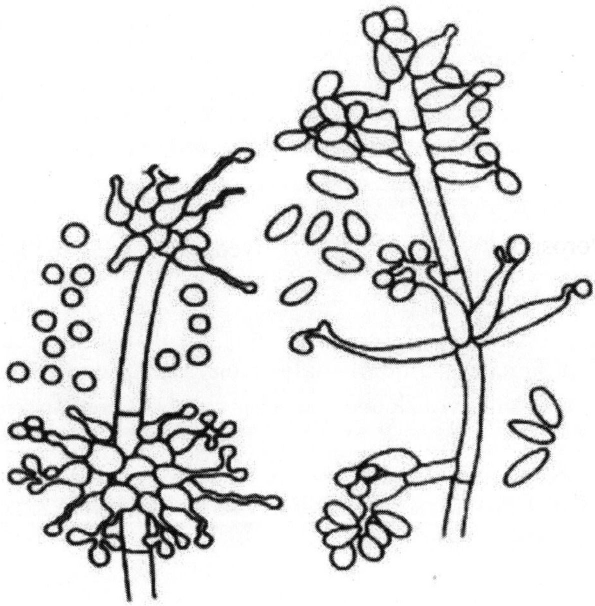

**Conidia:** Hyaline, rounded to ovoid, 1-celled, borne singly on small sterigmata in a several bar fungal spore bearing portion of conidiophore becoming zigzag after bearing several spores.

**Mycotoxins:** Oosporin, Isoosporin, Chaetomidin.

**Diagnostic features**: Insecticolous, conidiophores in verticellate clusters.

### 7. *Bipolaris* Shoemaker

The genus Bipolaris belongs to Hypomycetes. Its sexual stage is Cochliobolus typified by C. heterostrophus is now linked with the type species of Bipolaris maydis. The sexual state of Bipolaris is not common in nature, but it is occasionally produced under laboratory conditions. Even though Cochliobolus is the oldest name, Bipolaris is more frequently used by plant pathologists in disease reports and widely applied in taxonomic literature, due to its predominance.

Species in Bipolaris were initially described in the genus Helmisporium Link, which was typified by Helmisporium velutinum altered the spelling of the name to Helminthosporium Link accepted the altered spelling as an appropriate orthographic variant and Helminthosporium has since been widely used. The graminicolous species described in Helminthosporium were to be different from the type species H. velutinum showing the morphological distinctiveness of graminicolous Helminthosporium species divided into two subgenera Cylindro-Helminthosporium and Eu-Helminthosporium. Species with straight cylindrical conidia that germinate with one or more germ tubes from any cell were placed in the former subgenus Cylindro-Helminthosporium *and present* Drechslera, whereas species with fusiform and curved conidia germinating only from end cells were placed in the latter and present name *Bipolaris*.

**B. *maydis* Shoemaker**

**Teleomorph:** *Cochliobolus heterostrophus*

**Colony:** Spreading, olive-brown

**Conidiophores:** Mononematous, macronematous, erect, pigmented, geniculate, acroporogenous porospores

**Conidia:** Porospore, brown, smooth walved, distoseptate acropetalous, bipolar germination, 2-9 sepatate

**Mycotoxins:** Sterigmatocystin (*B.* sorokiniana), reveulin

**Diagnostic features**: Conidia porospores, brown, 2-3 distoseptate and germinate through two polar cells. Teleomorph *Cochliobolus*. However, existence of this genus is doubtful as many expert feel that all graminicolous species of *Helmenthosporium* be transferred to genus *Drechslera.*

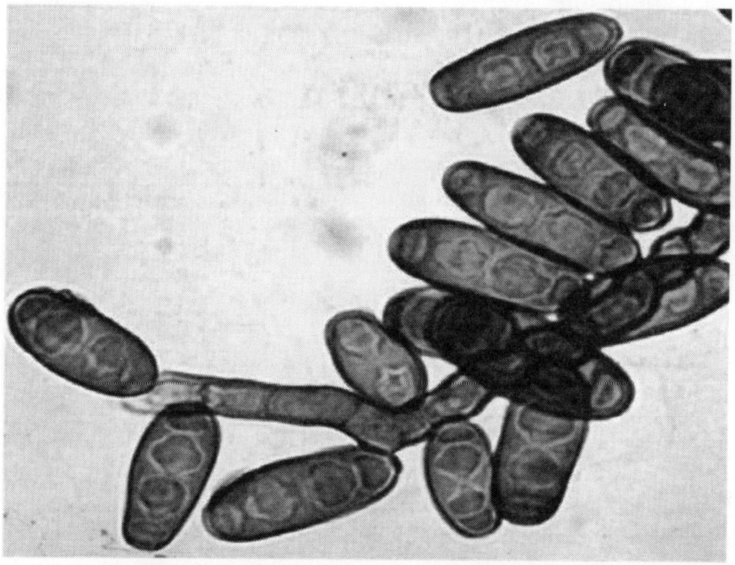

The genus Helminthosporium is divided into *Drechslera, Bipolaris, Exserohilum* based on the following characters (Table 3.12).

**Table 3.12.** Distinguishing features of three genera

| Anamorph | Main characters | Teleomorph | Main characters |
|---|---|---|---|
| *Drechslera* | Conidia cylindrical, germinating from any cell | *Pyrenophora* | Ascomata immersed and erumpent not setose, ascospores very large, dictyoseptate with a gelatinacious sheath |
| *Bipolaris* | Conidia fusiform-ellipsoidal, central cells not much darker and boarder than the | *Cochliobolus* | Ascomata superficial erumpent, setose, ascospores very large, form many |

| | | | |
|---|---|---|---|
| | distal ones, hilum not prtruberent, germination bipolar. | | celled, coiled in the ascus |
| *Curvularia* | Conidia with 2-3 broader and central cells, often curved, with or without a prominent hilum, cylindrical, germination bipolar. | *Cochliobolus* | |
| *Exserohilum* | Conidia fusiform-cylindrical to obclavate, with procumbent hilum germination bipolar. | *Setospora* | Ascomata superficial, setose, ascospores fusiform pharagmosporous with a gelatinous sheath and similar distal. |

## 8. *Byssochlamys* Westling

*Byssochlamys* is a genus of the Trichocomaceae family and Deuteromycotina and produces mycotoxin patulin, byssochlamic acid and mycophenolic acid. Colonies on CYA 40-50 mm diam low and quite sparse, white to slightly-grey; reverse pale to mid-brown at 37°C, 50-70 mm diam low to floccose, moderately dense, white to cream, reverse pale to brown, sparsely produced, with short stipes bearing conidiogenous cells sometimes on irregular metulae, and with conidiogenous cells also sometimes borne solitarily from hyphae. Conidiogenous cells are cylindrical then gradually tapering. Conidium ellipsoidal to pyriform, smooth walled. Sarcomata comprising single asci borne from, but not enveloped by wefts of contorted white hyphae, asci 8-11 µm, ascospores 4-6 µm maturing in10-14 days at 25°C and in 7-10 days at 30°C, but rarely found at 37°C. Anamorphs are of two kinds: 1. "Aleurioconidia", borne singly, common at 30 and 37°C, spherical to pyriform, 2. "penicilli"

### *B. nivea* Westling

**Anamorph:** *Paecilomyces ehrlichii*

**Colony:** Spreading, powdery, penicillate

**Conidiophores**: Penicillate, bearing group of sterigmata, bearing 1- celled conidia in chains

**Conidia:** Conidia 1- celled produced in basipetal chains, hyaline, one celled, globose to elleptical.

**Ascocarp:** Ascocarp lacking peridium, axiglobose and ascospores are ovate to elliptical without enduring peridium inconspicuous very scanty loose weft of hyphae surrounds the clusters of asci.

**Asci:** naked in clusters, Asci globose, 8- spored

**Ascospores:** Ovoid to elliptical, hyaline, smooth, one celled.

**Chlamydospores :** Terminal, thick-walled, one-celled

**Mycotoxins:** Patulin (*B. nivea, B. gymnoascus*), Byssochlamic acid, Malformins.

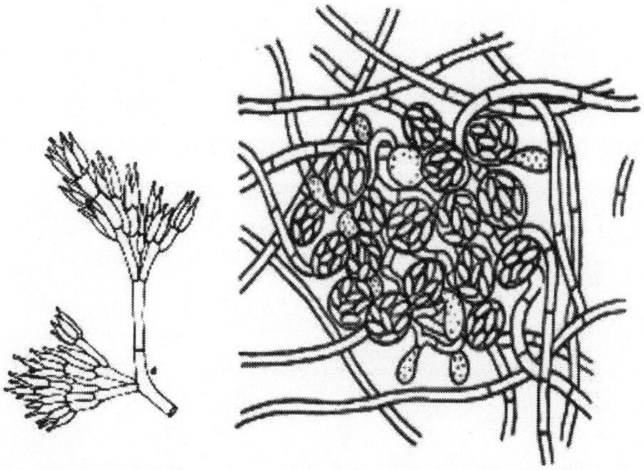

**Diagnostic features:** Asci globose and arise in clusters, ascocarp without distinct peridium

### 9. *Ceratocystis* Ellis and Halst

Ascomata single to aggregated, superficial or immersed in the substratum, ostiolate or non-ostiolate, globose, subglobose or obpyrium, hyaline to dark-brown or black: ostiolar necks when present, usually very long and cylindrical, often with a ring of hyphae around ostiole. Centrum containing pseudoparanchyma cells. Ascus globose to subglobose, produced at different levels in the centrum, 8-spored, evanescent at maturity. Ascospores hyaline, 1-celled, with or without a gelatinous sheath, variable in shape, curved, lunate, reniform, or hat-shaped.

Anamorphs: *Chalara, Hyalorhinocladiella, Leptographium, Phialographium.*

#### *C. fimbriata* Ellis and Halst

**Colony**: White turning to black with the production of ascomata

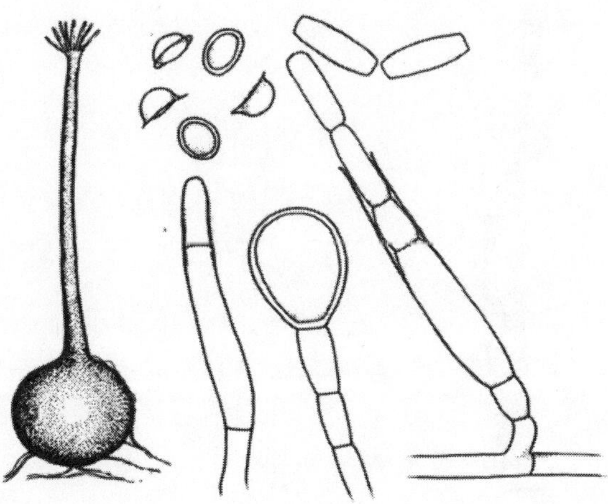

a. Mature ascoma with ostiolar neck, b. Scattered globose asci, c. Ascospores, d. Conidia and conidiophore

**Ascomata:** Single to aggregated, superficial or immersed in substratum, ostiolate or non-ostiolate, globose to sub-globoase, hyaline to dark brown, ostiolar neck is long and cylindrical

**Ascus**: Globose to sub-globose, scalled 8-spored and evanescent

**Ascospores**: Hyaline, 1-celled with or without gelatinous sheath, variable in shape, curved, lunate, remiform or hat shaped

**Anamorph:** *Chalara*

**Mycotoxins**: Naphthoquinones, isocoumarins

**Diagnostic features:** Globose to sub-globose perithecium with long neck ascus with single celled ascospore, remiform or hat shaped

## 10. *Chalara* (Corda) Robenhorst

Chalara is characterized by simple conidiophores with terminal integrated phialides. The phialides are mostly sharply differentiated into basal center of variable shape and on elongated cylindrical collarette in which conidia are produced in long basipetal chains conidia mostly cylindrical with truncated or somewhat rounded ends. Hyaline or slightly pigmented, 1-celled or septate.

### *C. fimbriata*

**Teleomorph:** *Ceratocystis fimbriata*

**Colony:** White, spreading

**Conidiophores:** Short, micronematous, manonematous cylindrical cell with an uppermost cell giving rise to an elongate conidiogenous cell.

**Conidiogenous cells:** Fragmenting to form arthroconidia sometimes phialides

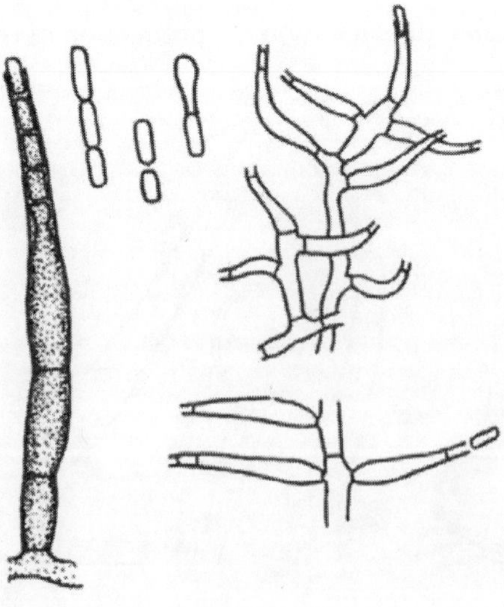

**Conidia:** Terminal or intercalary, determinate, catenate, cylindrical rarely with a hyaline longitudinal slit, sometimes slightly inflated in the lower half, long and tubular in the upper half or semi endogenous, cylindrical, truncate at both ends, hyaline or pigmented, variable in width and length, produced in long fragile chains

**Mycotoxins:** 8 hydroxy-6-methoxy-3-methyl isocoumari

**Diagnostic features:** Short micromat at cylindrical cell producing arthroconidia or phialoconidia, conidia catenate and cylindrical.

**Ascocarp:** Perithecia, globose with very long apical ostiolate, glaberous wall, brown or black no paraphyses

**Ascus**: Large number, globose, evarencent

**Ascospores**: One celled, hyaline, hat or bean shaped, rarely ellipsoidal without germ pores or slits

**Mycotoxin:** 4-Ipomeanol, Isofusidienol A, B, C, and D.

### 11. *Chaetomium* Kunze ex Fr

Members of this genus are characterized by the ascomata mostly superficial attached to the substratum by rhizoids, subglobose to vasiform with lateral and terminal, septate hairs which may be variously branched or contorted with a mostly coarse ornamentation. Ascomata wall consisting of several pseudoparenchymatous layers. Asci are thin walled without apical structures with a slender stalk either cylindrical with uniseriate ascospores or clavate with uniseriate spores, deliquescing simultaneously at an early age. Ascospores one- celled, pigmented, smooth walled of varying shape, mostly provided with one germpore which is apical and binuecleate. Most species are homothallic *Acremonium* anamorph occurs in several species particularly the heterothallic ones, with solitary aculeate phialides scattered over the aerial hyphae producing pyriform to dacryoid conidia in chains or heads. Chlamydospores or aleuroconidia of *Botryotrichum* or *Humicola* type occur in some species.

*Chaetomium* is one of the familiar genera of Ascomycetes associated with agricultural commodities due to their cellulolytic activity. Species of this genus are characterized by dark-coloured perithecia with short neck which are covered with irregularly coiled or tightly spiraled hairs or with tuff, simple or branched hairs and evanescent asci which are of taxonomic importance. On the other hand, Arxe Von (1986) gave due importance to ascospores morphology and suggested several synonymys. However, recent *Chaetomium* taxonomists have selected ascospores morphology as a major taxonomic character.

Carter (1981) divided the genus into four subgenera *Chaetomium, Murora, Bomerella* and *Achaetomium*. The Subgenus *Chaetomium* consists of six sections (*Chaetomium, Bostrychodes, Torulosa, Brasiliensia, Crispata* and *Indica*), while *Murora* has five sections (Murora, Aura, Atrobruno, Spiralia and Cuniculosa).

Out of 160-180 species of the genus, 56 species have been described in recent years. Ames (1963), Seth (1972), Haksworth and Wells (1973) and Millner (1977) have reviewed the taxonomy of this genus. While most authors use the branching and coiling of the terminal hairs as the principle criteria for a sub-division of the genus. Sorgel (1960) emphasized the realibilty of ascospores features and suggested a sub-division into 13 species groups which has been followed with some modifications by Dreyfuss (1984). The shape of the asci is generally a clear cut criterion for a sub-division of the genus, but in two species both clavate and cylindrical asci have been observed in the same ascoma. The wall structure is of potential value in the taxonomy of the genus. The occurance of morphologically similar homothallic or heterothallic isolates in some cases have been taken as criterion for species separation. The interaction between self-fertile and self-sterile isolates has, as yet only been studied for a few species. The species of *Chaetomium* are active in cellulose degradation in soft-rot in wood and fruit-rot of apples, dung and straw.

Oosporin (*Chaetomium aureum* and *C. trilaterale*), chaetoglobosin C (*C. globosum, C. cochliodes*) chaetoglobosin B (A-D, F and J (*C. globosum*)), chaetocin (*C.minutum)* and chaetomin (*C. cochliodes* and *C. globosum*) are some of mycotoxins reported to be produced by several species of this genus.

On the whole, these species are important components in the decomposition of herbaceous and lignified plant matter reaching the soil. Some species occur commonly on birds feathers.

The toxicity of *C. globosum* to roots was first demonstrated by Christensen *et al.* (1996). Subsequently two toxins cochliodinol and chaetomin were isolated from cultures of *Chaetomium* (Bewer *et al.,* 1968; Brewer *et al.,* 1972). Umeda *et al.* (1975) have isolated group of new cytochalasins and chaetoglobosins from rice cultures of *C. globosum.* Oosporin was isolated from cultures of *C. aurum* and *C. trilaterale* by Cole *et al.* (1974). Bsewer and Taylor (1978) reported chaetoglobins production by 3 strains of *C. globosum* and chaetomin production by *C. cochliodes, C. fimicola, C. globosum* and *C. umbonatum.* Mollicellins A-H are reported to be produced by *C. mollicellum* (Stark *et al.,* 1978).

At present, much more work is still required to indicate how frequently toxigenic *Chaetomium* infested material are used as foods and feeds and more information on different factors contributing to human and animal mycotoxicoses. In *Chaetomium,* a very well defined correlation is given between particular species and the mycotoxins they produce, so that secondary metabolism could readily find a place as an adjunct to taxonomy. Further, identification of *Chaetomium* to species level will provide a good basis for recognization of the toxins produced. Mycotoxins elaborated by species of *Chaetomium* are listed in Table 3.13.

Good sporulation of species is obtained in most substrates of high C/N ratio where the carbon source is provided in an insoluble form such as starch or cellulose and little soluble carbohydrate is added. Suitable media for the growth of *Chaetomium* are CMA (sometimes with 00.3 and yeast extract), OA, PCA or 1-2% MFA with cellulose.

Filter paper strips added to PCA, proved to be particularly valuable. Very rich media encourage the production of aerial mycelium rendering perithecial hair less amenable to examination under dissecting microscope. Asci are studied when it is young and the perithecial hair can well be studied in 2 weeks old. Spores mounted in lactic acid must be heated to counteract shrinkage. Of several species, key to only mycotoxigenic species is provided (*C. tetrasporum, C. elatum, C. funicola, C. fimeti, C. globosum, C. subglobosum* and *C. cochliodes*).

## Synaptic key

Terminal hairs, dichotomously branched ........................– Group I *C. elatum, C. funicola*

Terminal hairs, straight or flexed ......................................................– Group II *C. fimeti*

Terminal hairs, long undulate, sinuous to loosely coiled. ................................................*C. globosum, C. subglobosum, C. olivaceum*

Terminal hairs, regularly spirally coiled. ............................*C. cochliodes, C. tetrasporum*

**Table 3.13.** Mycotoxins produced by different species of *Chaetomium*

| Species | Chaeto-chromin | Chaetog-lobosins A,B,C,D,F, G & J | Chaeto-min | Cochli-odin | Sterigma-tocystin | Chae-tocin |
|---|---|---|---|---|---|---|
| *C. abuense* | - | - | - | + | - | - |
| *C. coarctatum* | - | ± | - | - | - | - |
| *C. cochliodes* | - | + | + | + | - | - |
| *C. elatum* | - | - | + | + | - | - |
| *C. fimeti* (Chaetomidium) | - | - | - | + | - | - |
| *C. funicola* | - | - | + | - | - | - |
| *C. globosum* | - | + | + | + | - | - |
| *C. globosum* var. *rectum* | - | + | - | - | - | - |
| *C. megalocarpum* | - | + | - | - | - | - |
| *C. ochraceus* | - | - | - | - | - | - |
| *C. olivaceum* | - | - | - | - | - | - |
| *C. subuffine* | - | + | - | - | - | - |
| *C. subglobosum* | - | - | + | - | - | - |
| *C. terissimum* | + | - | + | - | - | - |
| *C. tetrasporum* | + | - | - | - | - | - |
| *C. umbonatum* | + | - | + | - | - | - |
| *C. thielavioidarum* | + | - | + | - | + | - |
| *Achaetomella virescens* | + | - | - | - | + | - |
| *A. udgawae* | + | - | - | - | + | - |
| *C. caprinum* | + | - | - | - | - | - |
| *C. minutum* | - | - | - | - | + | + |

Chaetocin-*C. minutum*; Oosporein- *C. aureum* and *C. trilaterale*

## Key to species

1. Terminal hairs dichotomously branched ................................................................ **2**
2. Branches of terminal hairs essentially straight, dichotomously branched at right angles ascospores large ............................................................ ***C. elatum***

2a. Primary terminal hairs branched at acute angles with short internodes and abrupt tips, coarsely roughened .......................... ***C. funicola***

1a. Terminal hairs straight or flexed ........................................................................................... **3**

1b. Terminal hairs flexed and with simple tips, short awl -shaped, unbranched ascospores avate ......................................................... ***C. fimeti***

3. Terminal long undulate, sinous loosely coiled ascospores globose to subglobose......................................................................... ***C. megalocarpum***

3a. Terminal hairs regularly spirally coiled ........................................................... **4**

4. Perithecia globose or subglobose to ovate ................................................................ **5**

4a. Perithecia subglobose, asci4-spored ............................................***C. tetrsporium***

5. Terminal hairs in reverse loops

5a. Straight terminal hairs absent, undulate to loosely coiled hairs present ........................................................................................ - ***C. cochliodes***

**i) *C. cochliodes* Pallise**

**Synonymys=** *C. flexmosum* Palliser

**Colony:** Broadly spreading, dark olive-green to dark-olive-grey

**Perithecium:** Dark-olive-brown to black, sub-globose, ostiolate

**I.** **Terminal hairs:** Dark-olive-brown at first, straight below and sinuous above at maturity two types.

a. Straight below spirally coiled above

b. Slender varied from irregularly twisted or undulated to loosely coiled rather light-colored, minutely roughened

**II.** **Lateral hairs:** Dark olive-brown, straight or flexed, gradually interwoven with the terminal hairs, roughned, septate.

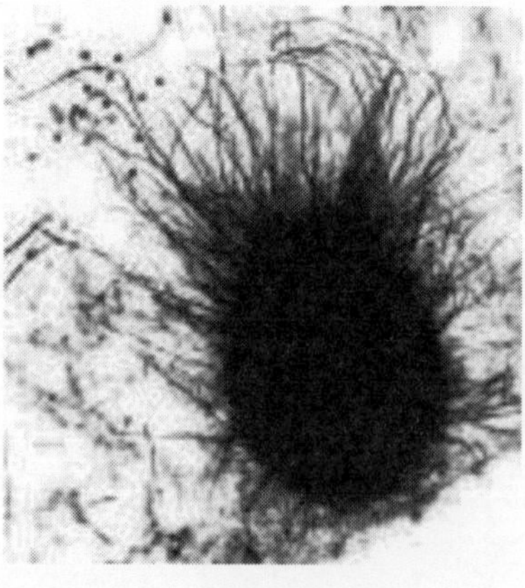

**Ascus:** Club- shaped, 8 spored

**Ascospores**: Single celled, olive-brown, lemon shaped, sub- apiculate, at both ends. **Mycotoxins:** Chaetoglobosins, chaetosin, chaetomin and cochliodin

**Diagnostic features:** Colonies dark- olive brown - olive- brown, dark-olibe-grey, dark olive brown terminal hairs two types (1) Straight below and spirally coiled above and slender varied from irregular twisted or (2) undulated to loosely coiled and minutely roughed.

**ii) *C. elatum Kunze* ex. Fries**

**Synonymys** = *C. comatum* (Tode) Fr.
= *Psilona cylindrical* (Pers.)
= *C. pannosum* Waller.
= *Sphaeria chaetomium* Lib.
= *Chlorodium ovoideum* Corda
= *C. lageniforme* Corda
= *C. affine* Corda
= *C. cylindricum* (Pers.) Rabenh
= *C. graminis* Rabenh
= *C. graminicolum* Fuckel
= *C. fieberic* Fuckel
= *C. libertii* Roum. et Pat.
= *C. velutinum* Ell. et Ev.
= *C. atrum* var. *distinctum* Roum
= *C. glabrescens* Ell. et Ev.

**Colony:** Spreading broadly, dark-greenish-grey to greenish-black, consisting of a matted basal felt with aerial hyphae, white to pale-yellow.

**Perithecium:** Black, ovate, ostiolate

**i) Terminal hairs:** Dark-olive-brown, unbranched hairs obscure, branched hairs straight below, several times dichotomously branched above with branches at right angles and often reflexed, coarsely roughened trough out, stiff with pointed tips, later formed hairs similar but with branches at acute angles, more narrow and sinuous than those of the former

**ii) Lateral hairs:** Dark olive-brown, straight or flexed, unbranched with rounded tips roughened, obscurely septate

**Ascus:** Club shaped, 8-spored, very evanescent

**Ascospores:** Dark-olive-brown, lemon shaped, apiculate at both ends

**Mycotoxins:** Chaetomin and cochliodin

**Diagnostic features:** Colonies dark –greenish- grey to greenish black perithecia terminal hairs dark-olive-brown, branched at right angles and reflexed, ascus club shaped ascospores dark- olive- brown lemon shaped, apiculate at both ends.

**iii) *C. funicola* Cooke Grevilllea**

**Synonymys** = *C. stesumllis* et Ev

= *C. bartholomai* Sacc. et. Syd.

**Colony:** Spreading broadly, greenish-black on cellulose agar

**Perithecia**: Dark olive-brown, subglobose, ostiolate, terminal hairs dark-olive-brown to black forming a dense head.

(A) Straight, unbranched, similar to the lateral hairs but coarsely roughened, narrowly projecting above the mass of ascospores and branched hairs.

(B) Dichotomously branched at acute angles with the short internodes, straight or slightly curved, coarsely roughened, septate, with abrupt tips, later formed hairs profusely branched, narrow, alternately constricted and inflated slightly roughened.

Lateral hairs dark -olive brown to black below fading above straight, unbranched tapered to collapsed tips, smooth to slightly roughened.

**Ascus**: Club shaped, 8-spored, very evanescent

**Ascospores**: Olive-green to dark- blue –green, ovate to ellipsoid, more or less apiculate at the ends.

**Mycotoxins:** Chaetomin, cochliodin

**Diagnostic features:** Colonies are spherical to ovoid, olivaceous to dark grey, with well-developed brown rhizoids. Peridium thin-walled, composed of brown angular cells 8-14 μm diam, with dimorphic hairs formed from the upper part. Hairs either seta-like, at the base, thick-walled, long and tapering and smooth to warted or short, narrow, strongly verrucose and repeatedly dichotomously branched. The aerial mycelium grey or white and sometimes with a yellow reverse.

**iv) *C. globosum* Kunze ex Fr Fries**

**Synonymys** = *C. fieberi* corda

= C. *lanosum* Peck

= *C. orientalae* Cooke

= *C. kunzeanum* Zopf

=*C. cymatotrichum* Cooke

=*C. oospore* Beavev

=*C. elasticae* Koord

=*C. subterraneum* Saito etrah

=*C. japanicum* Saitoetokazaki

=*C. destrum* Bat.et pontual

**Colony:** Spreading broadly, variously colored, olive-grey, greenish, olive-green, dark, olive-green and olive- black. Reverse usually uncolored.

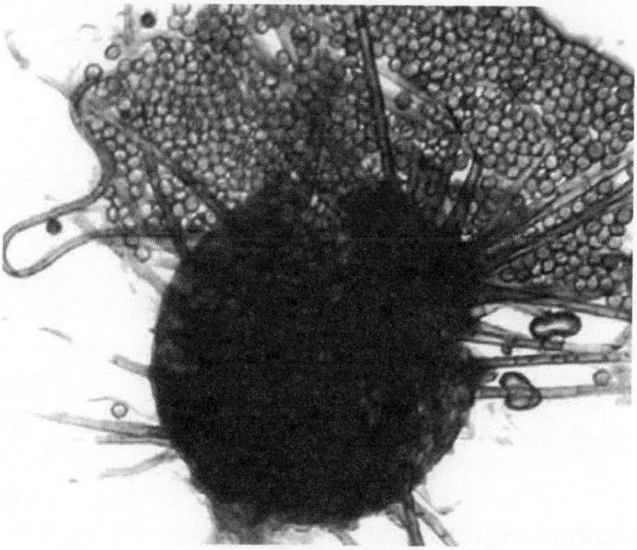

**Perithecium:** Dark-olive-black, sub-globose to ovate, often with pointed bases, ostiolate.

**Terminal hairs:** Numerous, forming densely interwoven mass, olive to brown, flexuous from the base, usually becoming sinuous or very loosely coiled towards the ends in some strains long undulate throughout, slender, slightly tapering to pale rounded tips, minutely roughened, septate.

**Lateral hairs**: Light olive-brown, straight or slightly flexed to undulate finally intermingled with the terminal hairs

**Ascus:** Club shaped, 8-spored

**Ascospores:** Olive-brown, lemon shaped, sub-apiculate at both ends.

**Mycotoxins:** Chaetoglobosins, chaetosin, chaetomin and cochliodin.

**Diagnostic features:** Colonies varingly colored, olive- grey, olive-green, olive-black, olive- brown, flexuous becoming sinulate towards the ends minutely rounded terminal hairs, asci club shaped, ascospores olive – brown, lemon shaped, sub-apiculate at both ends.

**v) *C. olivaceum* Cooke et Ellis**

**Colony:** Spreading broadly, olive-grey to dark-olive- green

**Perithecium:** Dark-olive–brown, globose to subglobose to ovate, often bluntly pointed at the base

**Terminal hairs**: Numerous and interwoven, olive-brown, straight or flexed below, becoming long undulate above with somewhat blunt or pointed tips, minutely roughened throughout, septate.

**Lateral hairs**: Similar but only slightly flexed

**Ascus**: Club shaped, 8-spored, evancent

**Ascospores:** Dark olive-brown, broadly ovate to subglobose, very finely apiculate at both ends.

**Mycotoxins:** Chetomin

**Diagnostic features:** colonies are olive-grey to dark-olive- green, terminal hairs are numerous and interwoven, olive- brown, straight or flexed below becoming long undulate, club shaped ascus dark olive- brown, broadly ovate and finely epiculate at both the ends.

**vi) *C. trilaterale***

**Colony:** Broadly spreading, olive-yellow to olive-brown, reverse brownish

**Perithecium:** Globose to ovate with bluntly pointed base

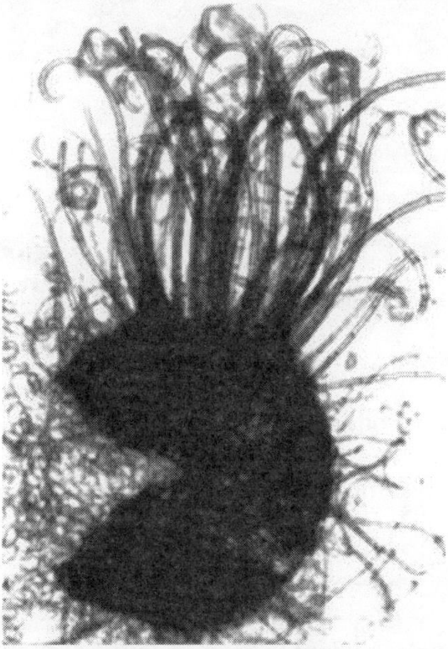

**Terminal hairs:** Dark coloured below fading above, minutely roughened, septate, arcuate from the base and usually increased 1-3 spiral *convulate* tips.

**Lateral hairs:** Narrow, regularly and distinctly septate, golden-yellow at base, minutely roughened near the tip, smooth with loosely one or two spiral convulate

**Ascus:** club shaped, 8-spored, evanescent

**Ascospores:** Dark, hemi-ovoid or evenly ovoid, flattened one side

**Mycotoxins:** Oosporein

**Diagnostic features:**Olive-brown, to olive yellow colonies reverse brownish terminal hairs coloured below, minutely roughened, usually with 1-3

spiral convulate tips ascus club shaped with 8 –dark, ovoid flattened ascospores.

### 12. *Cladosporium* Link ex Fries

**Colony:** Effuse, olivaceous, buff, dark-brown.

**Conidiophores:** Macronematous, or semi-macronematous, straight or flexuous, mostly unbranched, olivaceous-brown or brown, smooth or verrucose

**Conidiogenous cells:** Cicatrized, Scar prominent, polyblastic, sympodial.

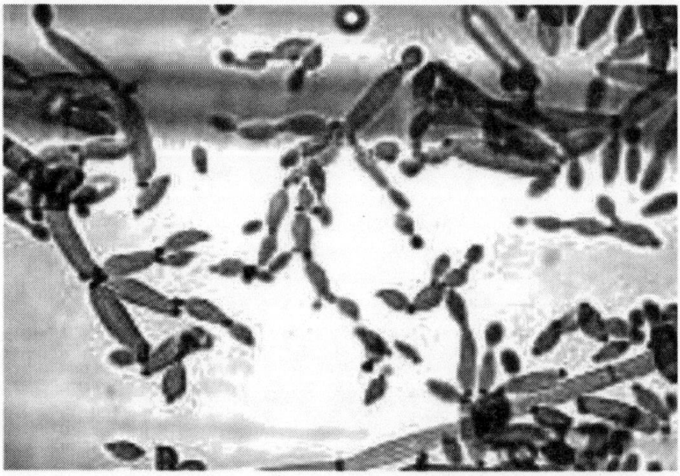

**Conidia:** Catenate, produced by apical budding often in a branched chains, forming tree-heads, 0-3 septate, cylindrical, doliform, ellipsoidal, fusiform, oval with distinctly protuberant scar at each end or just at the base.

**Key to species**

1. Conidia with rounded ends, without prominent scars ............*Amorphotheca resinae (1)*

1a. Conidia apiculate prominent scars ............................................................... 2

2. Conidia usually not extending 4.5 μm in width, smooth or slightly roughened; conidiophores not elongating sympodially ........................................ 3

2a. Conidia usually exceeding 5.0 μm width, distinctly verrucose, conidiophores commonly smooth-walled elongating sympodially ............... 4

3. Most conidia sub-globose, 3-4.5 μm diam, finely roughened. ......*C. sphaerosperum (2)*

3a. Most conidia elongate, 3.7X 2-7 μm, smooth –walled or almost so ........................................................................... *C. cladosporides (3)*

4. One–celled conidia mostly 5.5-13 X3.8-6.0 μm; somewhat larger 2-3 celled ones alsopre-sent. ......................................................... *C. herbarum (4)*

4a. One–celled conidia mostly7-15 X 9-9 μm; the commonly occurring 2-3 celled conidia considerably larger. .............................*C.macrocarpum (5)*

**i) *C. cladosporioides* Fres de varies**

**Colony:** Reaching 3.0-4.0 cm in dia. in 10 d. old on MEA at 20°C, olivaceous – green to olivaceous –brown velvety, reverse olivaceous –black.

**Conidiophores**: Short, branching, acropleurogen ausly apex bearing neumarous conidial chins arising arising below septa but without–swellings and sympodial elongatous.

**Conidia:** Ellipsoidal to leumen shaped, mostly smooth walled, rarely minutely verruculose, ovlivaceous –brown, 1-celled. Ramot conidia 0-1 septate, smooth or occasionally verruculose.

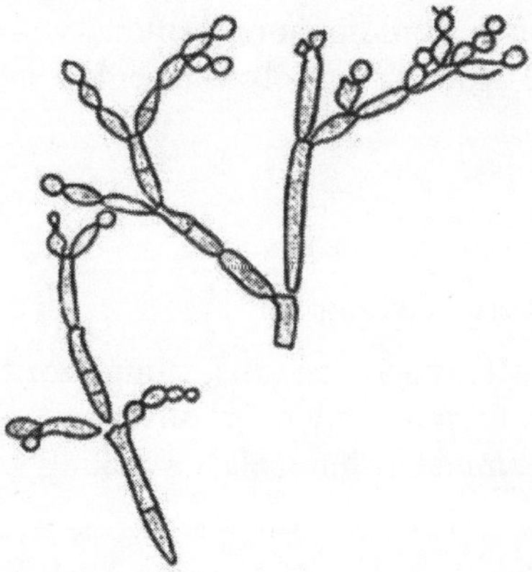

**Mycotoxins**: Cladosporic acid, cladosporol B-E, N-methylcarbamate2, N-methylcarbamate 2, cladosporin, isocladosporin 3 and calphostin C4.

**Diagnostic features:** Acropleurogenous, bearing numerous conidial chains.

**ii) *C. herbarum* (Pers.) Link ex S.F. Gray**

**Colony:** Effuse, olive,-green to olivaceous-brown, velvety. Reverse on malt agar greenish black

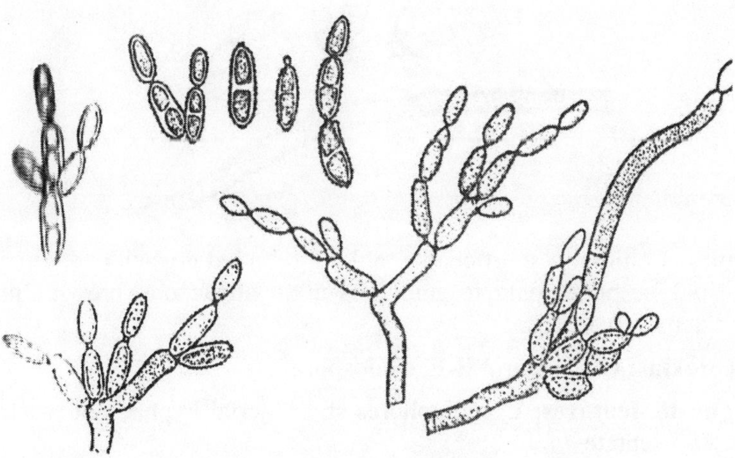

**Stroma**: Often well developed, specially on herbaceous stems

**Condiophore:** Mostlymacronematous, straight or flexous, sometimes geniculate, often nodose, pale to mid-olivaceous -brown or brown, smooth, terminal or intercalary vesicular swellings present

**Conidia:** Fairly long, often branched chains, ellipsoidal or ablong rounded at the ends, pale to mid-brown or olivaceous brown, rather thick walled and distin-guishly verruculose with low warts, 0-1 septate additional septa are formed occasionally, scars at one and or both ends small but clearly protuberant

**Mycotoxins:** Citreoviridin A, Herbarin A and B and Herbaric acid

**Diagnostic features:** Conidiophores often with terminal or intercalary swellings, short conidia densely verruculose

**iii) *C. macrocarpum* Pereuss**

**Colony**: Effuse, olive-green, velvety

**Stroma:** Sometimes well developed

**Conidiophores**: Mostly macronemtous, straight or flexus, often geniculate and rounded, pale to mid-brown or olivaceous brown, smooth or in part verruculose, terminal or intercalary swelling are present.

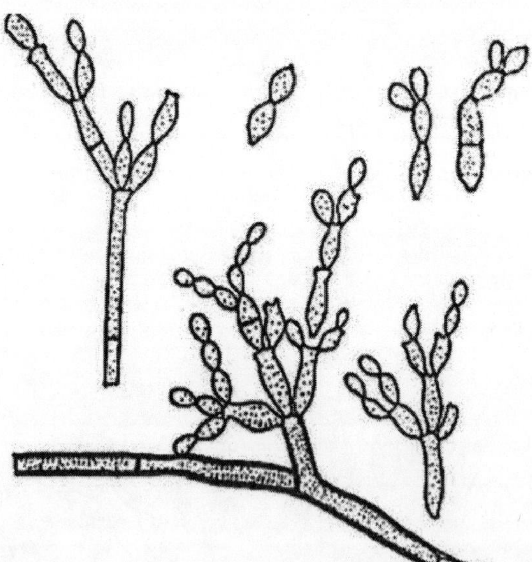

**Condia:** Conidia in short chains, oblong rounded at each ends to ellipsoidal, 0-3 sesptate, pale to mid–brown or olivaceous brown, thick walled, densely verucose

**Mycotoxins:** Cladosporol B-E, cladosporin

**Diagnostic features:** Conidiophores short, aerial hyphae not twisted, conidia, 0-3 septate

**iv) *C. sphaerospernum* Periz.**

**Colony:** Olive-green to olivaceus–brown, pulverulence, often funalised reverse on malt agar greenish-black

**Conidiophores**: macronematous, micronematous, mononematous, generally short pale to mid or sometimes dark –olivaceous-brown, smooth or verruculose

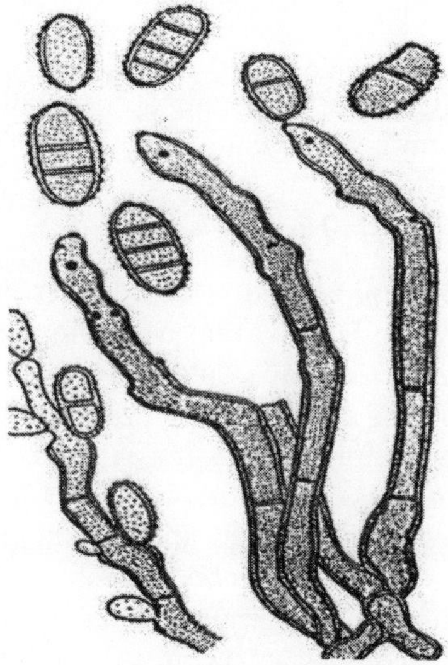

**Conidia:** Mostly globbose to sub-globbose, mid to dark–olivaceous brown, verrucose warts more clearly visible in air bubbles or water much less distinct in lactic acid or lacto-phenol

**Mycotoxins:** Cladosporol B-E, Quinolactacin A

**Diagnostic features:** Conidiophores short, conidia sub-globbose pale to mid or sometimes dark –olivaceous-brown, smooth or veruculose

## 13. *Clavariopsis* De wild

Colonies white and diffuse, hyphae hyaline branched. Conidiophores long single branched slender, hyaline simple. Conidia hyaline, mainly broadly, clavate to pyriform, 2-celled tree branches from the upper cells widely divergent at angles of above 120°.

***C. aquatica***

**Colony**: Spreading, hyaline, moist.

**Conidiophore**: Hyaline, slender, septate branched in upper part

**Conidia**: Apical, single, branched, main axis broadly clavate or pyriform, 2-celled, the three branches from the upper cell, widely divergent at angles of about 120°.

**Mycotoxins**: Clavarin

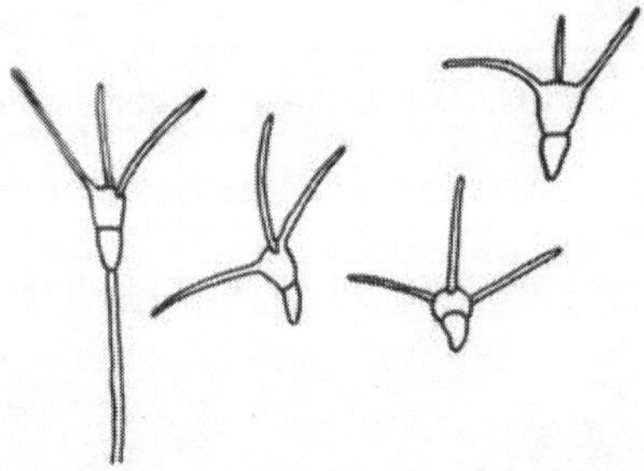

**Diagnostic features**: Conidiophores slender, septate, branches bearing single clavate or pyriform 2-celled conidia with three branches.

## 14. *Claviceps*

The members are plant pathogenic ascomycetes. Under normal field conditions the sclerotium is the structure which is important from the point of view of mycotoxins. This compact fungal plectenchyma may contain up to approximately 1% (wt/wt) of ergoline alkaloids. Two types of sporulation may be exhibited by the ergot fungus. The asexual sphacellial sporulation occurs shortly after infection of the host ovary and contributes a large number of conidia to the honeydew stage. The various morphological characteristics of the conidia and conidiophores form the basis of the taxonomy of the anamorphic genus *Sphacelia*. Teleomorphic stage of ergot fungus is exclusive to and synonymous with the genus *Claviceps* and detailed variations in the morphology of the hypocreaceous stromata, in which the perithecia are embedded, contribute to the acknowledged existence of several distinct species. Opinions differ on the validity of assigning species status to a number of the forms which have been described in the literature. However, in mycotoxicological terms, the principal distinctive species of ergot fungus can be easily recognized by the morphology of sclerotia body and there is generally no need to annotate the sexual stage in order to make an unequivocal identification. The characteristics of the principal ergot fungi are summarized in Fig. 3.1 and Table 3.14.

**Anamorph:** *Sphacelia*

**Mycotoxins:** Paspali, paspalicine, paspaline.

**Diagnostic features**: It is characterised by a special sclerotia shape and elongate and produces a circular, one celled embedded conidia. Sclerotia produces perithecium in which hyaline single celled 8-ascospores are present in a cylindrical ascus. The sclerotia are rich in fatty acids and triglycerides and also contain characteristic ergotamine alkaloids.

Whereas identification of the species solely by examination of whole ergot sclerotia is not difficult, it is quite possible that small fragments of sclerotia may need to be identified in milled feed stuffs. Conidia of the sphacelial stage can usually be found on the surface of the sclerotium. Thus, microscopic examination of an aqueous suspension of suspect material can be expected to reveal characteristic spores (Table 3.14). Ergot sclerotia are rich in triglyceride oils but only *C. purpurea* and *C. sulcata* oils have a sufficiently distinctive fatty acid composition which serves as diagnostic feature (Table 3.14). *C. purpurea* oil is the sole microbiological source of ricinoleic acid and *C. sulcata* is exceptionally rich in 9, 10-dihydroxysteric acid, which is absent from all other species. The sclerotia of the principal species of ergot fungi contain distinctive combination of ergot alkaloids. Its qualitative spectrum is usually a sufficiently stable characteristic for the alkaloid composition to same as useful component for a diagnostic key.

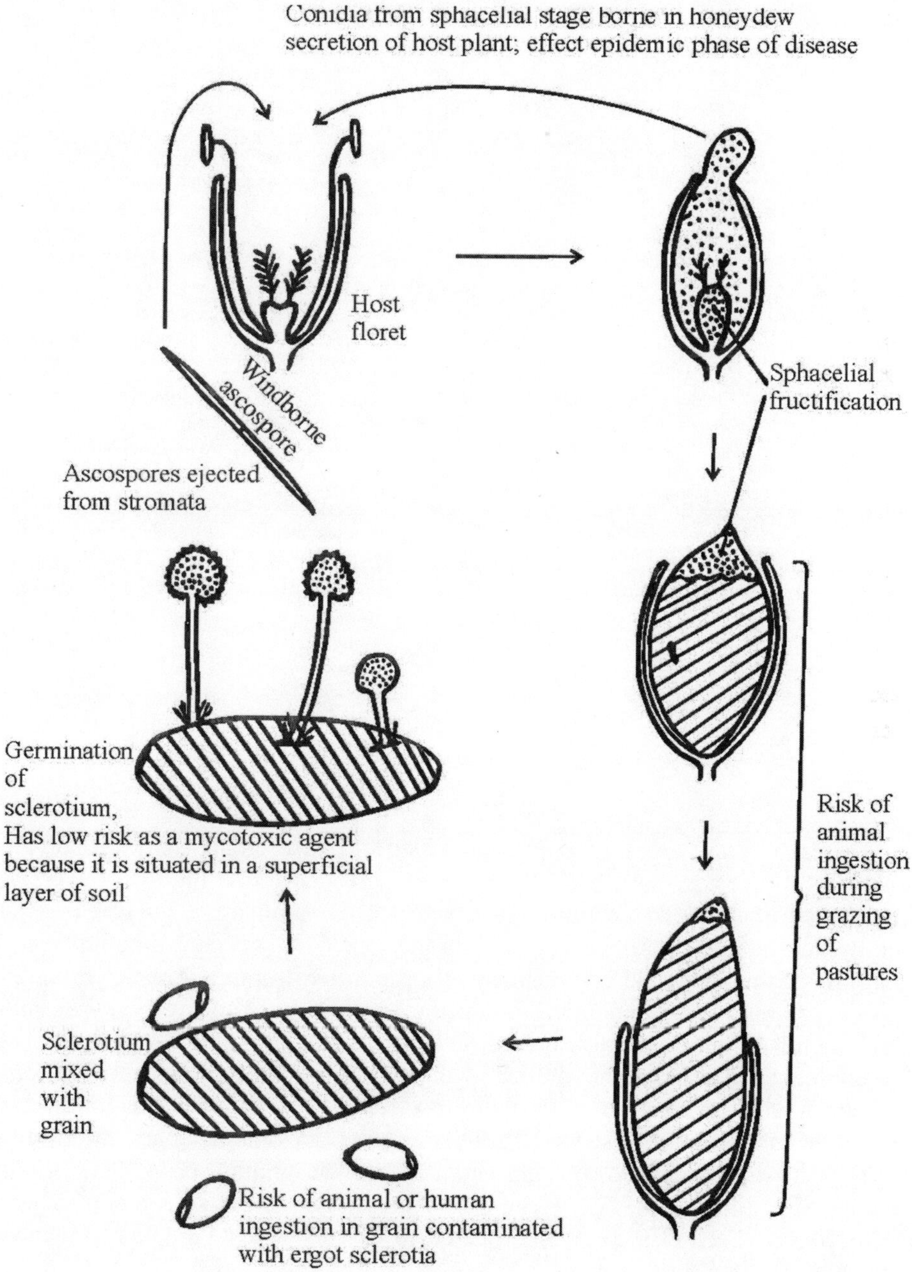


**Fig. 3.1.** Generalised scheme of the life cycle of ergot fungi, indicating the stages of mycotoxicosis risk

**Table 3.14:** Diagnostic characteristics by which the principal Mycotoxin-Producing Ergot fungi may be identified

| Ergot fungus | Diagram -matic morpho- logy of sclerotia | Approxi- mate dimen- sions of sclerotia (mm) | Color of sclerotia | Types of adhe- ring conidia | Specific diagno- stic fatty acid of trigly- ceride oil | Alkaloids commonly present | Types of poisoning | Rating as a myco- toxic agent (- to +++) |
|---|---|---|---|---|---|---|---|---|
| *C. purpurea* | --- | 2-20 X 1-6 | Purple-black | One type | Ricinoleic acid | Ergotamine group, ergotoxine (and isomers) Ergochrome EC (Secalonic acid A) | Anorexia, ame-ness, peripheral gangrene, reproductive failure, agalactia, intestinal haemorrhage | +++ |
| *C. paspall* | -- | 2-4 diameter | Buff | One type | None | Lysergic acid α-hydroxyethylamie, lysergic acid amide | Ataxia, reporoductive failure agalactia | ++ |
| *C. fusiformis* | -- | 2 X 4 | Purple – red | --- | None | Agroclavine, elymoclavine, chanoclavine | Acute central stimulation, reproductive failure, agalactia | ++ |
| *C. gigantea* | --- | 5 X 20 | White – buff | ---- | None | Festuclavine, pyroclavine, chanoclavine | None reported | ± |
| *Sphacelia sorghi* | --- | 2 X 2-4 | Buff-pink | ---- | None | Dihydroergosine, festuclavine, pyroclavine, dihydroelymo-clavine | None | ± |
| *C. sulcata* | ---- | 3 X 5-15 | Light gray | ---- | 9, 10 – dihydroxy -stearic acid | None | | - |

The detection of ergot in compounded feed stuffs, when the characteristic gross morphology of the components has become obscured, or in the ingests or excreta of animals, may possibly be achieved by microscopic examination. Such tedious and generally unsatisfactory procedures may, in the case of *C. purpurea*, be replaced by an analysis of the fatty acid composition of a saponified sample. The sclerotia of *C. purpurea* are unique among microorganisms in containing a large component of triglyceride oil in which ricinoleic acid is the principal (30-40%) component. Thus, the presence of ricinoleate in a feed stuff known to be free from castor oil is diagnostic for the presence of *C. purpurea* sclerotia. The sample is saponified, the free fatty acids thereby released are methylated with diazomethane and the resulting methyl esters are analyzed by gas liquid chromatography (GLC). Qualitative detection, of as little as 0.3% ergot in 1 to 2 g of feed stuff has been confirmed by this procedure where efficiency is expected to extend to even smaller amounts of ergot contamination.

## 15. *Colletotrichum* Corda

**Colony:** Spreading, white, black, pinhead acervuli appear on the colony with age

**Hyphae:** Hyaline, branched

**Conidiomata:** Acervulus, saucer shaped surrounded by stiff black, unbranched setae.

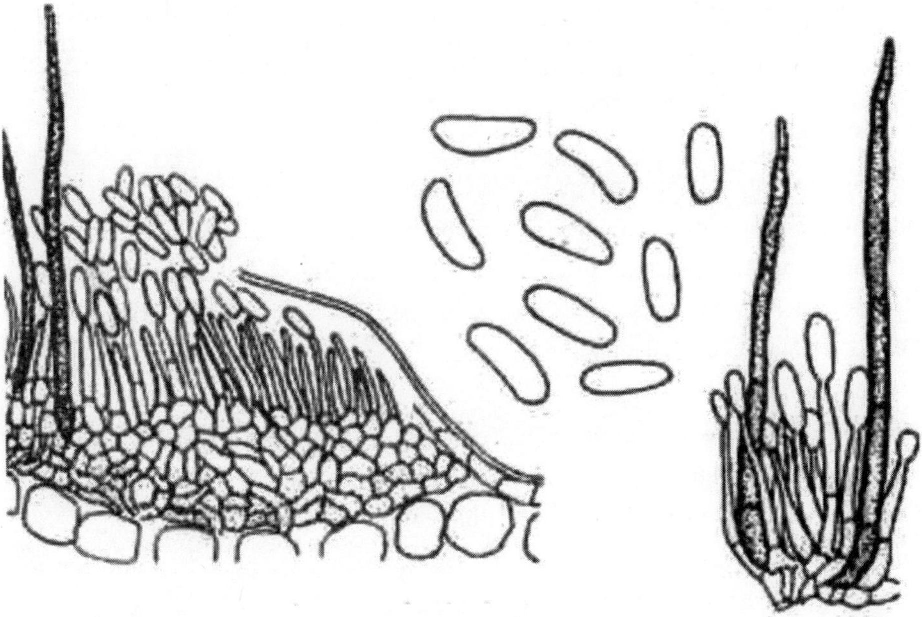

**Conidiophores:** Closely packed, simple, short hyaline

**Phialides:** Palisade layer

**Conidia:** Aseptate, fusoid, somewhat curved or sickle shaped or cylindrical, smooth, 1-celled, hyaline, aggregated in cream orange or red or brownish slimy masses appearing as glistering masses

**Teleomorph**: *Glomurella*

**Mycotoxins:** Colletodiol (Polyketide)

**Diagnostic features**: Conidiomata, acervulus with dark-brown setae surrounded the acervulus, conidia single celled, sickle shaped, hyaline.

## 16. *Curvularia* Boedijn

**Synonyms**= *Malustela* Batista & Lima.

**Teleomorph:** *Cochlibolus*

**Colony**: Effuse, brown, grey or black, hairy, cottony velvety.

**Mycelium**: Immersed in natural substrata

**Stroma**: Often large, erect, black, cylindrical sometimes branched

**Conidiophores**: Macaronematous, mononematous straight or flesuous, often geniculate, sometimes nodose, brown usually smooth.

**Conidiogenous cell**: Polytrtic, integrated, terminal sometimes later becoming intercalary, sympodial cylindrical or occasionally swollen, cicatrized.

**Conidia:** Solitary, acropleurogenous, simple, often curved, clavate, ellipsoidal broadly fusiform, obovoid or pyriform with or more septa, pale or dark-brown, end cells usually the end one paler than the others sometimes with darker bands at the septa, smooth or verrucose, hilum protrubent. In many species occasionally triradiate stauroconidia are formed besides normal conidia

**Key to species**

Conidia with a distinctly protuberant .......... 1

Conidia with hilum scarcely or not at all protuberant .......... 9

1. Conidia sigmoid. .......... *C. deightonii (1)*

1a.Conidia not sigmoid .......... 2

2.Conidia predominant l 3-septate .......... 3

2a.Conidia predominantly 4-septate .......... 7

2b. Conidia 3-5-septate, very variable.......... *C. stapeliae (2)*

3.Conidia usually straight or only slightly curved .......... 4

3a.Conidia usually distinctly curved at the third cell from the base .......... 5

4.Conidia clavate .......... *C. borreriae (3)*

4a.Conidia cylindrical or ellipsoidal .......... *C. harveyi (4)*

5. Conidia 45-66μ long .......... *C. andropogonis (5)*

5a. Conidia less than 40μ long .......... 6

6. Parasitic on *Ophioglossum vulgatum* .......... *C. crepinii (6)*

6a. On *Trifolium* and various grasses .......... *C. trifolii (7)*

6b. On Gladiolus .......... *C. trifolii* f. sp. *gladioli (8)*

7. Conidia over 40μ long.......... 8

7a. Conidia 27-38μ long.......... *C. protuberata (9)*

8. Conidia clavate or sometimes ellipsoidal .......... *C. cymbopogonis (10)*

8a. Conidia obclavate to ellipsoidal ..........*C. comoriensi (11)*

9. Conidia becoming rough-walled .......... 10

9a. Conidia remaining smooth-walled .......... 12

10. Conidia 4-septate .......... *C. verruciformis (12)*

10a. Conidia 3-septate.......... 11

11. Conidia straight.......... *C. tuberculata (13)*

11a. Conidia usually curved, asymmetrical.......... *C. verruculosa (14)*

12. Conidia predominantly 3-septate.......................................................................... 13

12a. Conidia predominantly 4-septate...................................................................... 22

12b. Conidia 3-4-septate, thick-walled.............................................. *C. prasadii (15)*

13. Middle septum usually truly median.................................................................. 14

13a. Middle septum not median ................................................................................ 16

14. Conidia symmetrical ........................................................... *C. eragrostidis (16)*

14a. Conidia asymmetrical........................................................................................ 15

15. Conidia 19-26 x 10-14μ ........................................................ *C. brachyspora (17)*

15a. Conidia 27-40 x 13-20μ ......................................................... *C. intermedia (18)*

16. All conidial cells usually pale or very pale brown ............................................ 17

16a. Some conidial cells always mid or dark brown ................................................ 18

17. Conidia usually straight or only slightly curved............................. *C. pallescens (19)*

17a. Conidia strongly geniculate or uncinate ........................................ *C. leonensis (20)*

18. Conidia straight or almost straight .................................................................... 19

18a. Conidia curved ................................................................................................. 20

19. Conidia symmetrical, obclavate to ellipsoidal.................................... *C. oryzae (21)*

19a. Conidia symmetrical, clavate ........................................................ *C. clavata (22)*

19b. Conidia often asymmetrical.......................................................... *C. ovoidea (23)*

20. Conidia 29-42 x 13-20 μ ............................................................ *C. penniseti (24)*

20a. Conidia 18-32 x 8-16 μ .................................................................................... 21

21. Stromata very rarely formed in culture, colonies on p.d.a. not markedly zonate -.................................................................................................... *C. lunata (25)*

21a. Stromata regularly and abundantly formed in culture, colonies on PDA p.d.a. usually zonate ........................................................ *C. lunata* var. *aeria (26)*

22. Conidia predominantly uncinate ..................................................... *C. uncinata (27)*

22a. onidia not uncinate ........................................................................................... 23

23. Conidia averaging 24 μ long ............................................................ *C. senegalensis (28)*

23a. Conidia averaging over 45 μ long, over 2 ................................................ *C. robusta (29)*

23b. Conidia averaging over 30 μ long, under 17 μ thick.......................................... 24

24. Conidia straight or slightly curved, only tapering a little at each end........................................................................................ *C. inaequalis (30)*

24a. Conidia curved or straight, tapering gradually towards ...................................... 25

25. Conidia often distinctly geniculate, stromata not formed in culture ........................................................................................ *C. geniculata (31)*

25a. Conidia often curved but seldom geniculate; stromata regularly and abundantly formed in culture ......................................................................... 26

26. Stromata simple, conidia averaging 32 x 10 μ ........................................ *C. affinis (32)*

26a. Stromata branched, conidia averaging 30 x 12-2 μ ................................. *C. fallax (33)*

**i) *C. lunata* (Wakker) Boedin**

**Synonyms** = *Curvularia aeria* (Bat. Lime & Vascone) Tsuda
= *Mallestela aeria* Bal-lima & Vasume.

**Teleomorph**: *Cochliobolus lunatus*

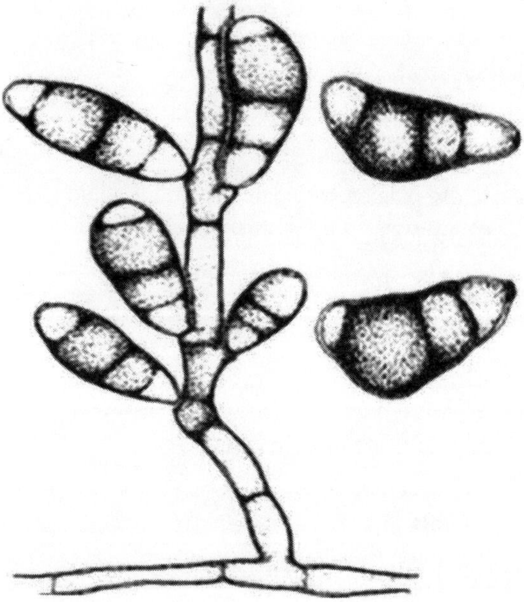

**Colony**: Dar-grey, usually-zonate

**Stromata:** Regularly and abundantly formed in culture

**Mycelium:** Branched, septate

**Conidiophores:** Macronematous, mononematous, straight geneictulate, smooth, brown, eptate.

**Conidia**: Solitary, acropleurogenous, 4 celled, middle cells darker, curved pseudoseptate.

**Mycotoxins:** Brefeldin A, Radicinnin

**Diagnostic features:** Acrogenous, porospore, 4 celled, 2 middle cells darker, curved, with protruberant hilum

## 17. *Cylindrocarpon* Wollenweber

Colonies fast growing, hyaline, or bright coloured, velvety, felty or woody aerial mycelium and differed or sporodochial sporulation. Conidiophores consisting of single phialides or repeatedly verticellately or almost penicillately branched. Phialides cylindrical to subulate with a single apical opening and a short collarets producing, hyaline, smooth walled conidia which adhere in singly masses. Macro-conidia, always present, cylindrical with rounded ends, straight or curved with one to several transverse septa, 1-celled microconidia distinct from the macroconidia are formed in many species, hyaline to brown, chlamydospores are present in one species.

**Colony:** Colonies on PDA are non aerial or centrally aerial, yellowish brown, purplish-tinted, zonate, marginally resupinate, withage, dark purplish brown

**Mycelium:** Hyaline

**Stroma:** Sclerenchymatous

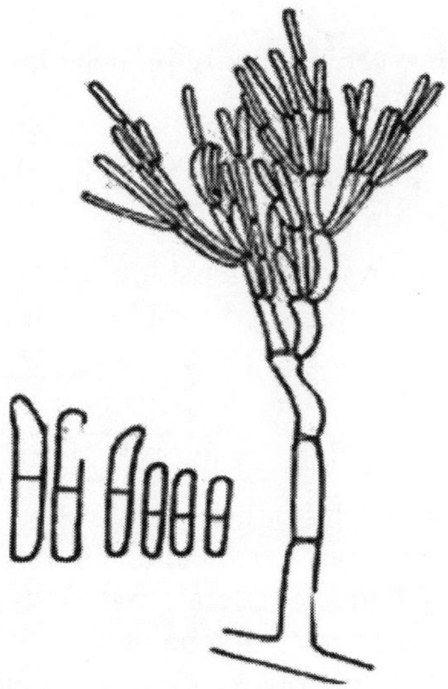

**Conidiophore:** Simple, irregularly branched, free fluccose, caesetose or sporodochial, terminating phialides.

**Microconidia:** Usually, 1-celled, ovate, fusiform, claviform, straight or curved

**Macroconidia:** -0-many septate, cylindrical, straight or curved, sometimes obtuse.

**Mycotoxins:** Roridin H, isororidin E and H 7B, 8B epoxyisoridin

**Diagnostic features:** Conidiophores irregularly branched terminating in phialides microconidia and macroconidia are produced. Macroconidia cylindrical, straight or curved, o-many sepatate, hyaline.

### 18. *Dendrodochium* Bon

Colonies spreading, white producing bulbous sporodochia, sporodochia caution like bursting out of the substratum. Conidiophores hyaline, verticellately branched, phialides in terminal whorls, bearing conidia terminally, conidia hyaline, 1-celled, ovoid to oblong, dry in mass

**i) *D. rubellum v. microsporum***

**Sporodochia**: Cushion like, burting light out of the substratum

**Conidiophores**: Hyaline, verticellatetly branched, phialides in terminal whorls bearing conidia.

**Conidia:** Conidia hyaline, 1-celled, ovoid to oblong, dry in mass

**Mycotoxins:** Dendrodochin

**Diagnostic features**: Sporodochin cushion like, conidiophore verticallately branched, comidia single celled ovoid to oblong

### 19. *Diaporthe* Nitschke

Ascomata an ostiolate perithicium formed in clusters in a stroma composed of both host and fungal tissues (pseudostroma), with the outer margine of the stroma delimited by a broad blackned zone. Perithecium in groups or sometimes separate, subglobose to obpyriform, oblique to lateral, with ostiolar necks converging and erumpent through outer tissues of host. Ascomatal wall two-layered, the outermost cells isodiametric, thick-walled and black, the inner cells hyaline thin walled and flattened. Ascus unitunicate, clavate to ellipsoid, with non-amyloid, refractive apical ring, 8-spored, wall persistent, but with deliquescent base so that mature asci lie free in perithecium at maturity. Ascospores hyaline 2-celled, septum median, constricted or not, cylindric to fusoid or cylindrical, straight or slightly curved and inaequilateral, with short, narrow appendages in some species.

**i) *D. phaseolorum***

**Colony:** White turning to dark–born with the productive of conidiomata and ascomata.

**Ascomata:** Ostiolate, formed in clusters in a stroma composed of both host and fungal tissue, sub globose to pyriform, oblique to lateral, ostiolar necks converging and erumpent through outer tissues of host, two layered wall, outer wall with iso-diametric, thick and black, inner layer cells hyaline.

**Ascus:** Unitunicate, clavate to ellipsoid non amyloid refractive apical ring, 8-spored, persistent but with deliquescement base.

**Ascospores**: Hyaline, 2- called constricted or not, cylindrical to fusoid, straight or slightly curved and in a equilateral with short narrow appandages

**Anamorph:** *Phomopsis*

**Mycotoxins:** 3-hidroxypropionic acid

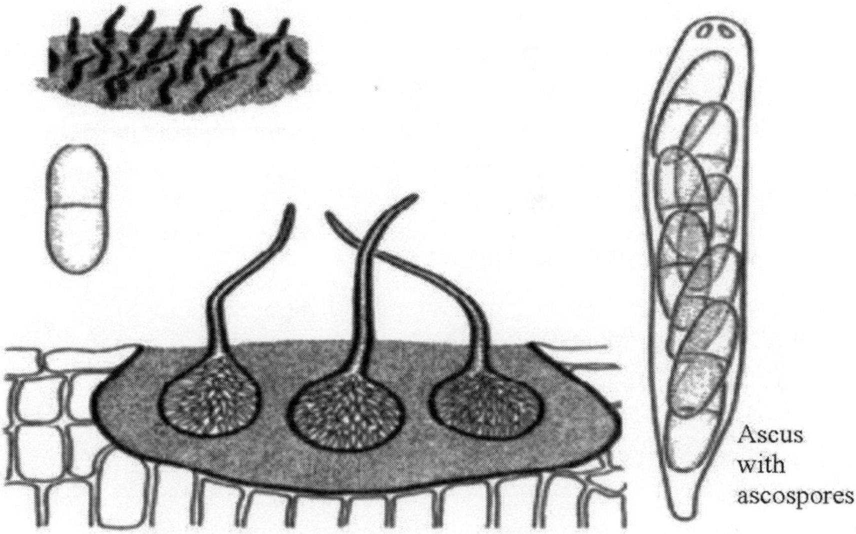


Section through stroma with perithecia

**Diagnostic features:** Ascospores formed in clusters in stroma of host and foreign tissue ostiolar neck in rumpent ascus well persistant, ascospores 2-celled hyaline

### 20. *Diplodia* Sacc.

The fungus causes a seedling blight, stalk-rot and ear-rot of maize wherever the crop is grown intensively. The ear rot usually progresses upward from the base of ear and is visible as a conspicuous, coarse and white to greyish-brown mycelial growth over the husk and kernels. The husks are often glued to the kernels by the white mycelium and infected kernels have a lusterless appearance and a dull-gray to light-brown color. Black pycnidia are produced late in the season on the invaded husks and kernels as well as on the rotted stalks. Pycnidia are subglobose or flask-shaped, 150 to 350 µm in diameter, completely immersed with a distinct, slightly protruding ostiole. Conidiophores are phialidic and formed from the cells of the inner pycnidial wall and are elongate, straight or curved, hyaline, simple, aseptate, 10 to 20 x 2 to 3 µm. Conidia are formed from the apex of the conidiophores and are cylindrical or ellipsoidal, narrowing at both ends to a rounded apex and truncate base, straight or curved, brown, smooth-walled, 0 to 2 but usually 1-septate, slightly or not constricted, 15 to 45 x 3 to 8 µm.

The fungus grows easily on media such as corn meal, potato dextrose, or malt extract agars, producing a thick, coarse, white, woolly aerial mycelium. The reverse of the colony becomes yellowish-brown with age, and some strains produce a greenish pigment. Dark-brown to black-globose, pycnidia containing conidia are gradually produced on the surface and become immersed in the agar. Considerable variation occurs in conidial shape and size but conidia are typically light-brown, 1-septate, and somewhat smaller than conidia formed on the host. Certain strains in culture produce slender, thread like, hyline, scolecospores, 21 to 45 µm long, in separate pycnidia or sometimes in the same pycnidium as brown conidia.

*D. maydis* (Berk.) Sacc. (=*D. zeae* (Schw.) Lev.) occurs in nature as an important pathogen of maize (*Zea mays* L.). Maize ears infected by *D. maydis* cause field outbreaks of a mycotoxicosis also known as diplodiosis in cattle and sheep.

In addition to *D. maydis,* two other species of *Diplodia* are known to cause ear rot of maize, i.e., *D. macrospora* and *D. frumenti* they can be identified with the help of following key.

**Key to species of mycotoxigenic species**

1. Conidia cylindrical or ellipsoidal, 0- to 3-septate, light-brown, smooth-walled, mycelium white. .......... 2

1a. Conidia ellipsoidal, 0- to 1-septate, dark-brown, longitudinally striate, 15 to 31 x 10 to 15 μm, mycelium becoming brown.......... *D. frumenti*

2. Conidia cylindrical or ellipsoidal, 0- to 2-septate, 15 to 45 x 6 to 8 μm .......... *D. maydis*

2a. Conidia cylindrical, 0- to 3-septate, 43 to 98 x 6 to 14 μm produce diplodiol .......... *D. macrospora*

**i) *D. macrospora***

**Colony:** Thick reverse, cottony, aerial mycelium. Reverse of the colony becomes yellow brown

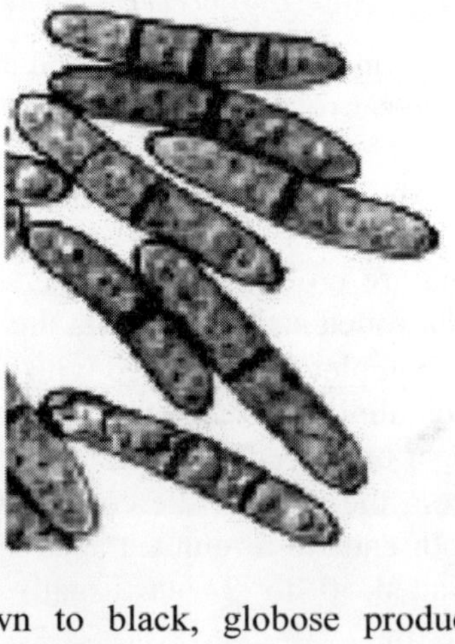

**Pycnidia:** Dark-brown to black, globose produced on surface later gets immersed

**Conidia:** 1-3 septate, cylindrical to elongate

**Mycotoxins:** Chaetoglobosin K.

**Diagnostic features:** Produce cylindrical conidia with 1-3 septa.

**ii) *D. maydis* Sacc.**

**Synonyms =** *Stenocarpella maydis*

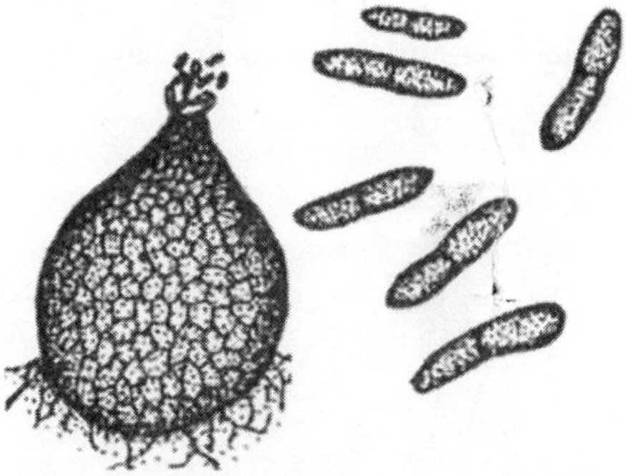

**Colony:** A thick course, white wooley aerial mycelium. Reverse of the colony becomes yellowish-brown with age. Some strains produce, greenish pigment.

**Pycnidia:** Dark-brown to black, globose, produced on surface but becomes immersed in the agar

**Conidia:** Light brown-,1-septate, some strains produce slender,thread like hyaline sclerospores.

**Mycotoxins:** Chetoglobosin K, Diplodia toxin

**Diagnostic features:** Pycnidia subglobose or flask shaped. Conidia cylindrical or ellipsoidal 0-3 septate, light –brown, smooth walled. Some strains produce slender thread like hyaline sclerospores.

### 21. *Dichotomomyces* Saito ex Scott

**Synonymy:** =*Talaromyces cejpii* Milko

=*Dichotomomyces albus* Saito

=*Royella albida* Dwiveli

**Colony:** White to cream coloured, spreading vegetative hyphae often bears rhizomorphs.

**Conidiophores:** Homothallic, subglobose to pyriform, smooth walled Conidiophore apices dichotomously branched.

**Conidia**: Produced from these branched racquet hyphae are frequently produced.

**Cleistothecium**: Variable in size, spherical, white to cream coloured.

**Ascospores:** Lenticular, with two closely appressed very thin equatorial crests and convex walls smooth.

***D. cejpii* (Milko) Scott**

**Colony**: Colonies on malt and oat meal agar grow moderately well forming white floccose to lanose mycelial felt.

**Ascomata:** Cleistothecium arises from a closely wound hyphal coil which is produced as a side branch for repentant hyphae. The peridium composed of single layer of thin walled, hyaline, polygonal cells, numerous asci are borne at all levels within the ascocarp and are produced by a typical crozier formation.

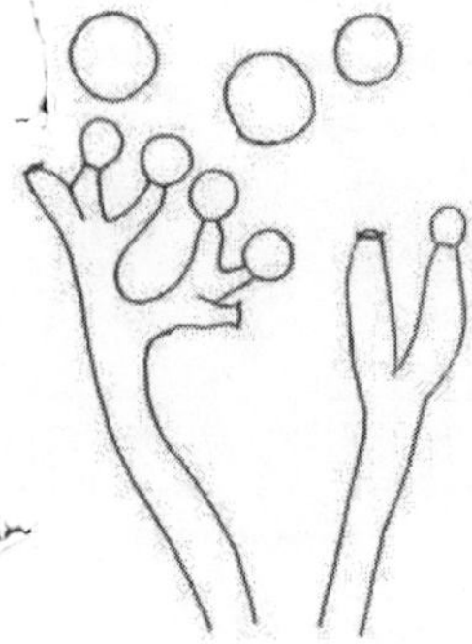

**Asus**: The asci are globose to subglobose, 8-spored, evanescent

**Ascospores**: Ascospore hyaline, lenticular with two closely appressed, very thin equatorial ridges.

**Conidiophores:** Conidiophores macronematous arise as long smooth walled septate, branches from aerial hyphae. Conidiophore apices are dichotomously branched, one or rarely more than one septa between these short branchlets.

**Conidia**: Conidia and the main axis of the conidiophore are usually having septa. Aleurospores are produced singular in short chains of 2-5.

**Mycotoxins:** Gliotoxin

**Diagnostic features**: Hyaline, thick walled, smooth sub-globose to pyriform.

### 22. *Drechslera*

**Synonymys** = *D. sorokiniana* (Socc.) Subranum and Jain

= *Helminthosporium sativum.* Stri. Bull. 116:180

**Telomprph:** *Cochliobolus sativus* (1 to α kuribayashi) Drechster ex Dastun

**Colony**: Spreading dark-olive-brown, black

**Conidiophores**: Mononematus, macronematous, geniculate, pale to mid-dark brown, septate, genicuilate, acrogenous.

**Conidia:** Curved, fusiform broadly ellipsoidal, dark-olivaceous-brown, smooth 3-12 pseudoseptate.

**Mycotoxins**: Sterigmatocystin.

**Diagnostic features:** Conidiophore acrogenous, geniculate at the apex, adistinct scar is present at the point of conidial origin, conidia broadly ellipsoid 3-12 pseudoseptate protrubent hirm.

**i) *D. dematioidea* (Bubak & Wroblewski)**

**Synonyms:** *Helminthosporium dematioideum* Bubak & Wroblewski.

**Colony:** Olive-brown to black

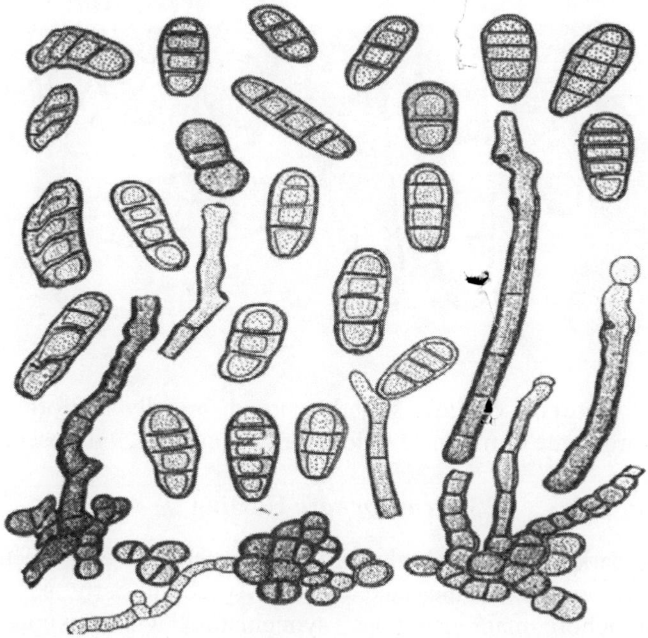

**Conidiophores:** Macronematous, mononemetous, arising single or in small groups, erect or flexous, geniculate with several conidial scars, brown.

**Conidia**: Straight, cylindrical to clavate, rounded at the ends, golden-brown to dark-brown, smooth, thick walled 2-7 pseudosepta, distinct hilum.

**Mycotoxins:** Stregmatocin, cytochalasins A, B&F

**Diagnostic features**: Conidia straight, cylindrical to elevate rounded at the edges, golden–brown to dark-brown, smooth, thick walled, 2-7 pseudoseptate.

**ii) *D. victoriae* Meehan & Murphy**

**Synonymys =** *Helminthosporium victoriae* Meehan & Murphy

**Teleomorph:** *Cochliobolus victoriae* Nelson

**Colony:** Spreading, olive –green to greenish black

**Conidiophores:** Mononematous, macronematous, solitary or in groups, straight or flexuous, septate, geniculate, pale to mid-brown.

**Conidia:** Slightly curved, broadly fusiform or obclavate, fusiform, pale to mid-pale-golden brown, smooth 4-11 pseudosepta.

**Mycotoxins:** Sterigmatocystin, cytochalasins

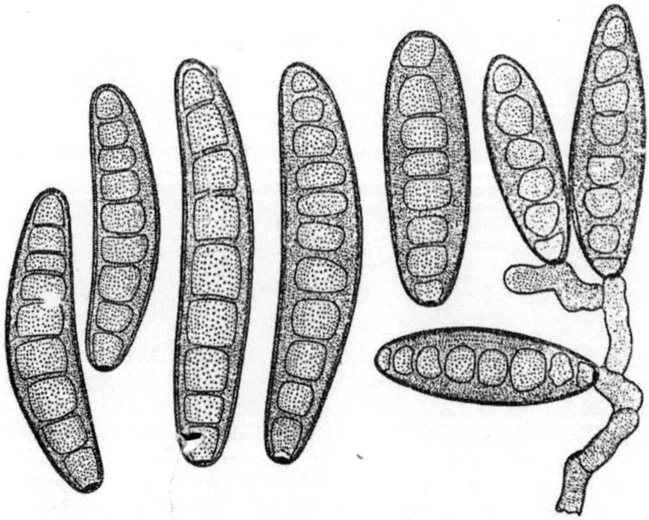

**Diagnostic features:** Conidia slightly curved, broadly fusiform or obclavate fusiform, pale or mid pale golden-brown, smooth, 4-11 pseudoseptate.

### 23. *Dematophora* Harting

Colonies effuse, dark-chocolate to blackish-brown, with several erect dark-brown synnemata clearly visible to the naked eye. Mycelium party superficial party immersed, conidiophore macronematous, synnematous with flexuous intertwined repeatedly branched often dichotomously towards apex, pale to mid-brown or olivaceous brown bearing conidiogenous cells. Conidiogenous cell polyblastic, integrated and terminal on branched or discrete, sympodial and geniculate, denticulate, with short thin walled separating cells which break across the middle leaving a minute frill or collar at each geniculation which corresponds to a frill at the base of each conidium. Conidia solitary, dry, acropleurogenous, simple, ellipsoid or obvoid, colourless to pale olivaceous-brown, single celled, smooth.

**(i) *D. nectarix* Harting**

**Telomorph:** *Rosellinia necatrix* (Haring) Berl.ex.prill.

**Colony:** Effuse, dark-chocklet blackish brown with neumarous erect dark brown synnemata

**Stroma:** None

**Septae and hyphopodia:** Absent

**Conidiophores:** Macronematous, synnematous components in threads flaxeous, interwined, repeatedly branched often dichotomously branched towards the apex, playing out to form head, pale to mid brown or olivaceous-brown.

**Conidiogenous cells:** Polyblastic, integrated and terminal on branches or diseriate, sympodial, geniculate which break across the middle leaving minute fill or collar at each geniculation which crossponds to a fill at the base of each conidium.

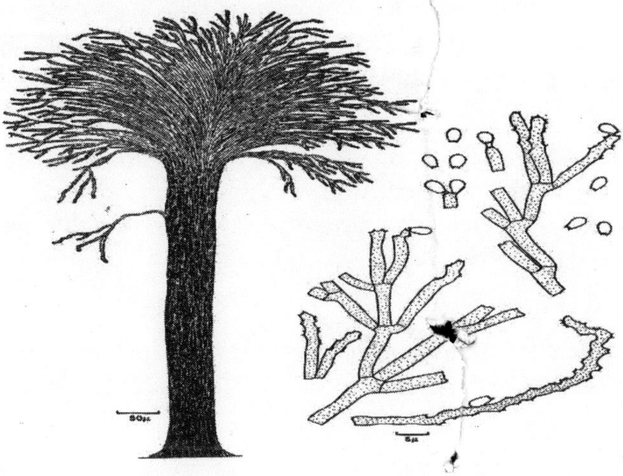

**Conidia**: Solitary, dry, acropleurogenous, simple ellipsoidal or obvoid, colour less to pale –olivaceous, brown, (0-septate), smooth

**Mycotoxins:** Cytochalasins-E

**Diagnostic features:** Conidiophores synnematous, conidiogenous cells poly-blastic, integrated, sympodial, geniculate, denticulate with short thin walled separating cells which break across the middle leaving minute fill or collar at each geniculation which crossponds to a fill at the base of each conidium acropleu-rogenous, ellipsoidal or obvoid (0-septate).

### 24. *Epichloe* (Fr.) Tul. & C. Tul.

**Ascomata:** Formed in a stroma composed of fungal tissues; stroma effuse, superficial on host, often encircling stem, initially white and byssoid, becoming fleshy and brightly colored.

**Perithecium:** Ostiolate, immersed in stroma, but often with ostiolar neck free, obpyriform. wall thin, composed of elongated, thin-walled cells.

**Ascus:** Lining inside of perithecial wall Unitunicate, long cylindrical, with a thickened, non-amyloid apex, 8-ascospores.

**Ascospores**: Hyaline, filiform, septate multi, not fragmenting 1-celle, dark-brown, ellipsoidal.

**Anamorph:** *Acremonium*

**(i) *Epichloe* sp.**

**Stroma:** Bright yellow to orange, tuberculate dry covered with perithecia

**Perithecium:** Pyriform short neck with many asci

**Ascus**: Asci with apical thickening and pore, containing 8-ascospores

**Ascospores:** Filamentous, hyaline, ejected in chains, developing upto 20-septa after germination.

**Anamorph:** *Neotyphodium* sp or *Spacelia*

**Colony:** White, cottony, smooth or slightly rough, aerial mycelium smooth, septate. Reverse pale–brown and lighter towards the margins. Sporulate in abundance.

**Conidiogenous cells:** Discrete, asrising solitary from the aerial mycelium , septate at the base more or less cylindrical over the lower one-third hyaline, smooth tapering towards the tip septate at the base

**Conidia:** Hyaline, (0-septate), lunate to reniform.

**Mycotoxins:** Lolitrem -B

**Diagnostic features:** Stromatic prithecium with short neck, ascial pore present, filamentous hyaline ascospore, *Neotyphodium* sp as its anamorph

**(ii)** ***E. typhina***

**Colony:** White spreading stromata

**Acomata:** It develops in crustose stroma, ascocarp wall fleshy, brightly coloured

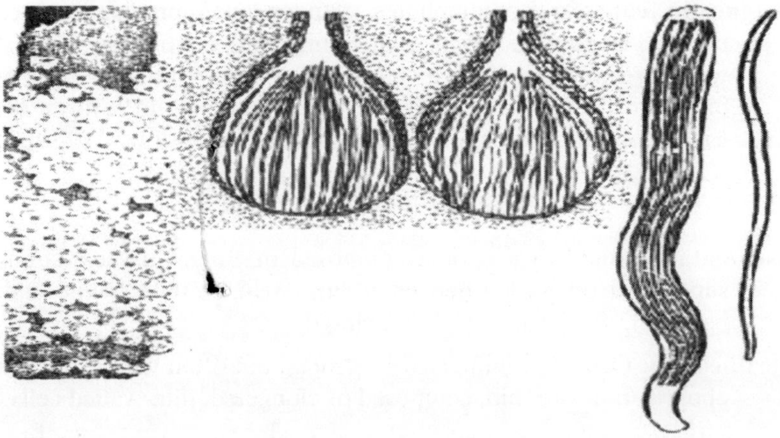

**Ascus:** Apiculate, apical body performed by a narrow canal, clavate, unitunicate

**Ascospores:** Ascospores single celled, filiferous liberated from apical canal of ascus.

**Anamorph:** *Spaecelia typhina*

**Mycotoxins:** Lolitrem -B

**Diagnostic features**: Ascomata develop on crustose stroma, wall fleshy, asci with narroe canal, ascospores filiferous or fun form apiculate.

## 25. *Epicoccum* Link ex Schlecht

Colonies are fast growing, lanose to felty, yellow, orange, red or orange usually sterile and produce no chlamydospores. Sporodochia pulvinate, blastoconidia are formed singly or densely, compacted, slightly pigmented, conidiogenous isodiametric conidia globose to pyriform with funnel shaped based and broad attachment seeding with or without protrubent basal cell wall golden brown to dark – brown verrucose, obsening the septa which divide the conidia in variant directions. Conidia are actively discharged often simultaneously from the whole sporodochia.

### *i) E. purpurascens* Link Ex Schlecht.

**Synonyms=** *E. purpurascens* Ehrenb

=*Torulopdea tobaica* Svilv

=*E. nigrum*

**Colony:** Colonies on the MEA, OA etc. fast growing, lanose to faint yellow, orange, red or brown sometimes also greenish, reverse similar.

**Sporodochium:** Pulvinate, black, not convoluted

**Hyphae:** Mostly immersed

**Stroma:** Pulvinate

**Setae and hyphodia:** Absent

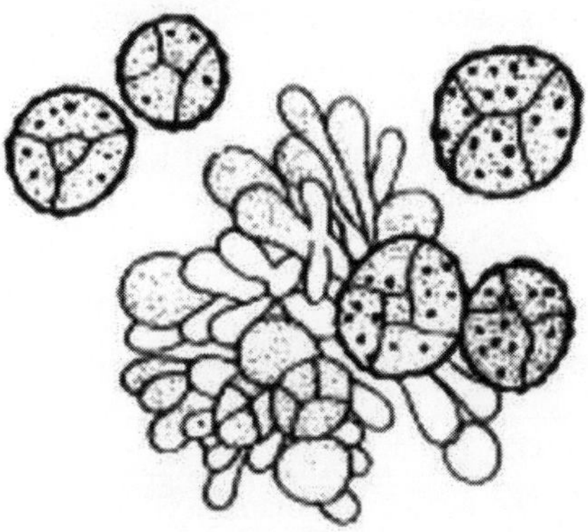

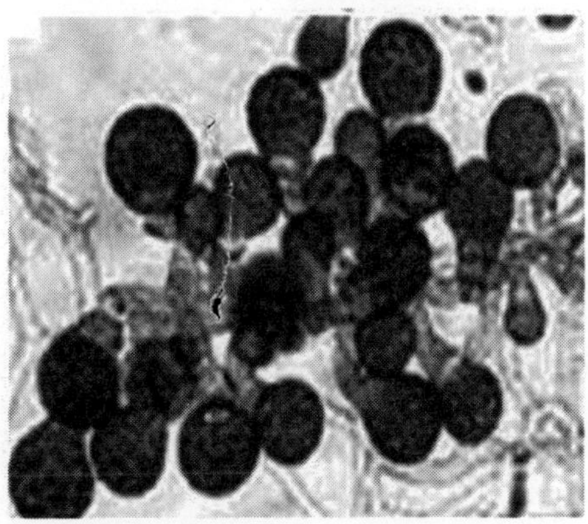

**Conidiophores:** Macronematous, mononematous, solitary or in dense clousters, unbranched or occasionally branched, short, straight or flexuous, cloloureless, tip pale-brown, smooth or verrucose.

**Conidiogenous cells:** Monoblastic, integrated, terminal, determinate, cylindrical

**Conidia:** Solitary, dry, clustered in masses, acrogenous, subspherical or pyriform, dark-golden-brown, muriform but with septa obscured in mature conidia by the rough opaque wall often with a pale-protruberant funnel shaped basal stalk cell having broad attachment scar

**Mycotoxins:** Flavipin, epicorazines A and B, rhodoxantin

**Diagnostic features:** Conidiophores in clusters, unbranched, conidiogenous cell monoblastic, integrated, conidia solitary, acrogenous, muriform funnel shaped basal stalk with scar.

### 26. *Fusarium* Link ex. Fr.

*Fusarium* is one of the most important genera of plant pathogenic fungi on earth with a record of devastating infections causing wilts, blights, root-rots and cankers in legumes, coffee, pine trees, wheat, corn, carnations and grasses. Importance of *Fusarium* species with reference to mycotoxins, infections may sometimes occur in developing seeds especially in cereals and also in maturing fruits and vegetables having potential to produce mycotoxins in foods.

At one time, the taxonomy of *Fusarium* species may have been a subject of purely academic interest. However, ever since it was proved that species of this genus incited grave disorders in man (Joffee and Plati, 1977) and animals (Ueno *et al.*, 1972, Joffe, 1978) by the production of toxins, determination of the exact identity of these species and strain has become a matter of acute importance. The taxonomy of the sporotrichella section is a matter of some dispute. The various species in this section are bunched by Snyder and Toussom (1965) under *F. tricinctum* and species are not of phytopathological importance. Therefore, the Snyder-Hansen system is not acceptable.

The fusaria were reported to be the cause of Alimentary Toxic Aleukea (ATA), human mycotoxicosis epedimic in the USSR which killed about one lakh people between 1942 and 1948 (Joffe, 1978). It also occured in Russia during 1913 and 1932 and also earlier. According to Morassas (1981), ATA also occurred in other countries including England during 16$^{th}$ to 18$^{th}$ Centuries. Recent researches have shown that *Fusarium* species are capable of producing a bewildering array of mycotoxins. Formost among these are at least about 140 trichlothecenes which are known to be produced by fusaria and suspected to be responsible for acute and chronic diseases of man and animals (Table 3.15).

**Table 3.15.** Different species of *Fusarium* producing mycotoxins

| Name of the species | Name of the mycotoxin |
|---|---|
| *Fusarium acuminatum* | Diacetoxyscirpenol, HT-2 toxin, Neosolaniol, T-2 toxin |
| *F.anguioides* | Diacetoxyscirpenol |
| *F.avenaceum* | Acetyl T-2 toxin, Acetyldeoxynivalenol, Acetylneosolaniol Diacetoxyscirpenol, Acetoxyscirpenediol, Fumonisin B1, HT-2 toxin, Moniliformin, Monoacetoxyscirpenol, Neosolaniol, Nivalenol |
| *F.bacteridioides* | Diacetoxyscirpenol |
| *F. crookwellense* | Zearalenone |
| *F. culmorum,* | Acetoxyscirpenediol, Acetyl T-2 toxin,Acetyldeoxynivalenol, Acetylneosolaniol, Deoxynivalenol, Diacetoxyscirpenol, Fumonisin B1, HT-2 toxin, Moniliformin Monoacetoxyscirpenol, Neosolaniol, Nivalenol, Zearalenone |

| | |
|---|---|
| *F.diversisporum* | Diacetoxyscirpenol |
| *F.episphaeria* | Diacetoxyscirpenol, Fusarenon-X, Zearalenone |
| *F. equiseti* | Acetoxyscirpenediol, Acetyldeoxynivalenol, Acetylneosolaniol Acetyl T-2 toxin, Diacetoxyscirpenol, Fusarenon-X, Moniliformin, Monoacetoxyscirpenol, Neosolaniol, Nivalenol, Zearalenone |
| *F. graminearum* | Deoxynivalenol, Nivalenol, Zearalenone, Diacetoxyscirpenol |
| *F.lateritium* | Diacetoxyscirpenol |
| *F. moniliforme* | Acetoxyscirpenediol, Acetyl T-2 toxin, Acetyldeoxynivalenol, Acetylneosolaniol, Deoxynivalenol diacetate, Deoxynivalenol, Diacetoxyscirpenol, Fumonisin B1, Fusaric acid, HT-2 toxin, Moniliformin, Monoacetoxyscirpenol, Neosolaniol, Nivalenol, T-2 toxin, Zearalenone |
| *F. nivale* | Acetoxyscirpenediol, Acetyldeoxynivalenol, Acetylneosolaniol Acetyl T-2 toxin, Deoxynivalenol diacetate, Fumonisin B1, HT-2 toxin, Moniliformin, Monoacetoxyscirpenol, Deoxynivalenol, Nivalenol, Diacetoxyscirpenol, T-2 toxin, Fusarenon-X, Zearalenone |
| *F. oxysporum* | Acetoxyscirpenediol, Acetyl T-2 toxin, Acetyldeoxynivalenol, Acetylneosolaniol, Diacetoxyscirpenol, Fusarenon-X, Moniliformin, Monoacetoxyscirpenol, Nivalenol, T-2 toxin, Zearalenone |
| *F.poae* | Diacetoxyscirpenol, HT-2 toxin, Neosolaniol, T-2 toxin |
| *F.rigidiusculum* | Diacetoxyscirpenol, Neosolaniol, T-2 toxin |
| *F. roseum* | Acetoxyscirpenediol, Acetyl T-2 toxin, Acetyldeoxynivalenol, Acetylneosolaniol, Diacetoxyscirpenol, Fusarenon-X, HT-2 toxin, Moniliformin, Monoacetoxyscirpenol, Neosolaniol, Nivalenol, T-2 toxin, Zearalenone |
| *F.sambucinum* | Diacetoxyscirpenol |
| *F.scirpi* | Diacetoxyscirpenol, HT-2 toxin,T-2 toxin, Zearalenone |
| *F. semitectum* | Deoxynivalenol, Fusarenon-X, Neosolaniol, Zearalenone |
| *F. Solani* | Diacetoxyscirpenol, HT-2 toxin, Neosolaniol, T-2 toxin |
| *F. sporotrichioides* | Diacetoxyscirpenol, HT-2 toxin,T-2 toxin |
| *F. sulphurium* | Diacetoxyscirpenol, HT-2 toxin,T-2 toxin |
| *F. tricinctum* | Deoxynivalenol, HT-2 toxin, Neosolaniol, T-2 toxin, Zearalenone |

The taxonomy of *Fusarium* is complex because many of species, even when grown from monosporic cultures, exhibit extreme variability in their morphological characteristics. Most of the species of *Fusarium* have teleomorphs in one of three genera (*Calonectria, Nectria* and *Gibberella).* In these Samson *et al.* (1984) treated eight species, while, Pitt and Hocking (1985) recorded 13 species as important in food microbiology. This led early taxonomists, to distinguish between hundreds of so-called species and varieties foremost among them were Wollenweber (1935) and Wollenweber and Reinking (1935). Later classifications by Subramanian (1952),

Bilai (1950), Gordon (1960), Seemuller (1968), Gerlach (1970), Booth (1971), Joffe *et al.* (1973), Joffe (1974) and Joffe and Palti (1975), were largely based on Wollenweber and Reinking's earlier system.

The taxonomy of *Fusarium* has undergone a phenomenal change in species concept due to multitude problems during last 60 years. Wollenweber and Reinking (1935) recognized over 100 species, while Snyder and Hansen (1945) and Tousoun and Nelson (1968) have accepted only nine species each with many formae species. On the other hand, Booth (1971), treated 44, Nelson *et al.* (1983) accepted 54 species and Gerlach and Nirenberg (1982) have recognized about 75 species. Identification of *Fusarium* species is ideally carried from the growth on carnation leaf agar, an effective medium for macroconidium production (Nelson *et al.*, 1985). The level of agreement between *Fusarium* taxonomist, however, gradually increasing and probably in future a general accepted system containing about 50 species will be evolved.

Determined attack on the taxonomy of these fungi by International collaborative group led by Nelson has resolved most of the conflicts and developed most acceptable classification with 30 valid species (Nelson *et al.*, 1985). An attempt is made to present precisely the characteristics of those *Fusarium* species that are known to produce mycotoxins.

The Snyder and Hansen classification of *Fusarium* chiefly regard to the distinction between toxigenic *Fusarium* sections and species, the most important being the *Sporotrichiella* section which is quite inadequate and leads to wrong identification. The species of *Sporotrichiella* have attracted world wide attention because of their capacity to produce toxins dangerous to man and animals. The most prominent of these are *F. poae* and *F. sporotrichoides*, which grow in the Soviet Union on over wintered cereals and cause the fatal disease, Alimentary Toxic Aleukia in man (Joff, 1977), *F. tricinctum*, grows in the United States on mouldy corn, wheat and fescue hay and produces toxic symptoms in many animals (Ueno *et al.*, 1972). Metabolites of *Fusarium* have received a great deal of attention from research scientists because of their mycotoxic potential (Smalley and Strong, 1974).

A direct consequence of confusion in taxonomy is the confusion in species mycotoxin associations. *Fusarium* isolates producing a particular toxin have been given different names as a result of the different taxonomic systems used or simply as a result of misidentification. Marassas *et al.* (1984) after intensive study of about 200 toxigenic *Fusarium* isolates provided accurate information for species identification and toxins produced by them. They listed 24 species of which they considered *F. sporotrichoides*, *F. equiseti, F. graminearum* and *F. moniliforme* to be most important from human health point of view.

Of the sixteen sections of *Fusarium* accepted by Gerlach and Nirenbreg (1982), 12 have one or more toxigenic species. But three sections Sporotrichella, Liseola and Roseae are rich in toxigenic species.

Species belonging to section Sporotrichella produce trichothecenes and butanelides. Species of section Liseola produce zearalenone, while species placed in Rosea produce both trichothecenes and zearalenone.

The taxonomy of species of the Sporotrichiella section is a matter of some dispute. The various species in this section are bunched together by Snyder and Toussoun (1965) under *F. tricinctum*. This section includes species which are not of particular phytopathological importance and therefore, the Snyder-Hansen system is not acceptable.

Most significant change in the species composition is of *F. moniliforme* along with two other species which formed a Liseola section of Wollenweber is reduced to one species *F. moniliforme* and one sub-species *F. moniliforme* var. *subglutinans* by Booth (1971). Another interesting nomenclatural change is that of *F. roseum* which was used by Snyder and Hansen (1945) and commonly used in phytopathological literature has been applied to five species and one variety in three sections of *Fusarium* (Gerlach and Nirenberg, 1982). Nirenberg (1976) replaced *F. moniliforme* with the older name *F. verticilloides* and expanded the old concept of *F. moniliforme* to include in addition to *F. verticilloides, F. proliferatum* and *F. lactus*. She divided the old *F. moniliforme* var *subglutinans* among *F. sacchari, F. sacchari* var. *subglutinans*, *F. neoceras* and *F. anthophilum*.

Another problem with regard to *Fusarium nivale* is reported to have teleomorph in *Micronectriella nivalis* and *Calenectria nivalis*. The teleomorph of this is also reported to be *Monographella* and anamorph superficially resembles *Fusarium.* Now it is placed in *Gerlachia* (Games and Muller, 1980) and subsequently in *Microdochia* (Samuels and Hallett, 1983) which produces trichothecenes.

Gams and Muller (1980) feels that *Fusarium* species be typified by material producing only microconidia, while other argue that macroconidia should be taken as a character for classifying species.

Another group of *Fusarium* of considerable toxicological interest is the *Discolor* section, comprising such species as *F. graminearum (Gibberella zeae), F. sambucinum (G. pulicaris)* and *F. culmorum*. Snyder and Hansen (1965) devoted much attention to these species, which cause well-known cereal diseases. Largely on the basis that they all attack cereals, species as disparate as these *Discolor* species were grouped together with *F. equiseti (G. intricans)* of the *Gibbosum* section, and *F. avenaceum (G. avenancea)* of the *Roseum* section. All these species belonged to different sections (Discolor, Gibbosum, Roseum), but were included by Snyder and Hansen (1945) in their Roseum section. This grouping is completely unacceptable in the view that it disregards important maphological and physiological features is completely unacceptable. On the other hand, Snyder and Hansen's treatment of other sections of interest to toxicologists, such as the Arachnites *(F. nivale)*, Liseola *(F. moniliforme)*, and Martiella *(F. solani)* sections, deviates only in less important details from the generally accepted schemes based to some degree on Wollenweber and Reinking (1935). A toxicologist or other scientist who wishes to identify a species of *Fusarium* for toxicological purposes may have difficulty utilizing Snyder and Hansen's (1965) system, since it is over simplified and tailored for use in plant pathology.

According to Nelson *et al.* (1985) there are 12 sections or groups in *Fusarium*. Only four sections contain most common toxic species. They are (1) Sporotrichella (*F. sporotrichoides, F. poae*), (2) Gibbosum (*F. equiseti*) (3) Discolor (*F. graminearum, F. culmorum*) and (4) Liseola (*F. moniliforme, F. proliferatum* and *F.*

*subglutinans*). Samson *et al.* (1995) have provided excellent keys for identification of species of *Fusarium*.

The toxigenicity of *F. graminearum, F. sambucinum, F. culmorum, F. equiseti* and *F. avenaceum* has been studied fairly extensively, especially with regard to the estrogenic and emetic effects observed in swine and some other animals in the United States and toxicoses in humans, animals, and plants. *F. nivale,* which is highly lethal to animals, induce pathological changes, degenerations and necroses in various organs and produces cytotoxic and strongly irritant effects on rabbit skin. *F. moniliforme* is the most prevalent fungus on corn kernels and causes acute toxicity in different animals. *F. moniliforme* produces a cytotoxic compound with antitumor activity. Neurotoxic syndrome (Wilson and Maronpot, 1971) and necrosis in the brain (leukoencephamalacia) in horses in South Africa (Kellerman *et al.*, 1972). *F. solani,* a soil fungus causes numerous destructive plant diseases. *F. solani* also cause intoxication of horses in Japan, characterized by disturbances of the central nervous system, and hemorrhages and crust formation on rabbit skin (Ueno *et al.*, 1972). It was responsible in some cases for corneal disease (keratitis) in a man in Florida (Jones *et al.*, 1969). Characteristic features of different sections are given below:

1. Section *Arachnites* Wr. (*F. nivale, F. larvarum*) Macroconidia on aerial mycelium, rarely in sporodochia or pionnotes, curved, apedicellate. Chlamydospores and sclerotia absent.
2. Section *Sporotrichiella* Wr. Joffe *(F. poae, F. sporotrichioides, F. sporotrichioides* var. *tricinctum, F. sporotrichioides* var. *chlamydosporum)*: The basis for classification in this section is (a) shape of microconidia, whether lemon or pear shaped, globose, ellipsoid or elongate, dispersed in aerial mycelium or formed in false heads; (b) the relative frequency of micro – and macroconidia. Macroconidia sparse, small, oblong, narrowly fusoid to falcate, pedicellate, formed in aerial mycelium or in sporodochia. Chlamydospores intercalary, terminal, in chains or knots, occasionally with plectenchymatous sclerotia. Perithecial states absent.
3. Section *Roseum* Wr. (*F. avenaceum, F. arthrosporioides*): Cultures yellow, ochre, carmine, purple or red. Microconidia absent or sparse, chlamydospores absent, macroconidia in sporodochia, pionnotes, or on aerial mycelium, sublunate, slender almost filiform, falcate with thin walls, narrowing at both ends, pedicellate. Sclerotia white, yellow, purple to brown. Stroma yellow-red.
4. Section *Liseola* Wr. (*F. moniliforme, F. moniliforme* var. *subglutinans, F. moniliforme* var. *anthophilum):* Cultures white-cream-brown, orange, violet. Microconidia on aerial mycelium, usually in long chains or small false heads, oval, fusiform, oblong, rarely pyriform. Macroconidia thin walled in sporodochia, pionnotes, or on aerial mycelium, sublunate, spindle shaped to cylindrical, straight or curved with narrow apex and base cells, typically 3-septate.Chlamydospores absent.
5. Section *Gibbosum* Wr. em. Joffe (*F. equiseti* and *F. equiseti* var. *acuminatum, F. equiseti* var. *compactum, F. equiseti* var. *caudatum, F.*

*equiseti* var. *longipes*): Culture white-pale pink, pale ocher, olive, carmine-red. Microconidia absent or sparse in aerial mycelium. Macroconidia in pionnotes and sporodochia, falcate, narrowing at both ends with elongated apical cell and well-developed pedicellate foot cell, dorsiventral, parabolic or hyperbolic, typically 5- rarely 3-septate. Chlamydospores intercalary, abundant, smooth or rough walled, single or in chains and knots, yellow-brown.

6. Section *Discolor* Wr. (*F. heterosporum, F. graminearum, F. sambucinum, F. sambucinum* var. *coeruleum, F. sambucinum* var. *trichothecioides, F. cumorum, F. tumidum) :* Cultures white-rose, peach, grayish, rose, red to brown. Microconidia absent; macroconidia thick-walled, in aerial mycelium, sporodochia and pionnotes, either broad, falcate with short apical cell, and well developed foot cell, spindle or sickle-shaped with elongated gradually narrowing apical cell, well-marked basal cell, typically 5-septate. Chlamydospores intercalary, sometimes terminal, often in knots and chains. Sclerotia purple-blue, brown to dark. Stroma yellow.

7. Section *Martiella* Wr. em. Joffe Palti (*F. solani, F. solani* var. *coeruleum, F. solani* var. *ventricosum, F. javanicum*): Cultures white, cream, orange-blue to brown. Microconidia abundant, oval or oblong, hyaline. Macroconidia in aerial mycelium, sporodochia or pionnotes, fusoid, cylindrical, curved or elongate, with thick walls and short, rounded apical cells and foot cells. Chlamydospores globose, oval, smooth or rough-walled, terminal and intercalary, single or in pairs, short chains or knots.

## Key to mycotoxigenic species of *Fusarium*

1. Microconidia very heterogeneous, singly or in false heads or in short chains. ............ 2

1a. Microconidia abundant, broadly oval, elongated, singly or in false heads. ................ 7

2. Microconidia elongated in chains usually with a broad and rounded apex ................. 8

2a. Microconidia absent or rare ........................................................................ 11

3. Microconidia pyriform cylindric, clavate, lemon shaped or oval................................ 4

3a. Microconidia globose, spherical with basal papilla, rarely pear-shaped .................... 3

4. Microconidia pyriform to ellipsoid, globose or elongate ............................................ 7

4a. Microconidia narrowly clavate with rounded apex, pyriform to fusiform, not spherical or globose ........................................................................ 6

5. Microconidia abundant. Macroconidia very sparse, small, curved, falcate, usually without foot cell, typically 3-septate, 19 to 38 x 3.5 to 6.8 µm. Sporodochia and pionnotes absent ...................................................... ***F. poae* (1)**

5a. Microconidia often as numerous as macroconidia. Macroconidia formed in sporodochia and in aerial mycelium, curved, falcate, elongated, and narrowly fusoid. Typically 3-to 5-septate, 22 to 45 x 3.6 to5.2 µm ......................................................................... ***F. sporotrichoides* (2)**

6. Microconidia more abundant than macroconidia. The latter formed in sporodochia, rarely in aerial mycelium, curved, spindly ellipsoid, falcate, 3-to 5-septate, 23 to 50 x 3.3 to 4.4 µm. Typically 3-septate ................................................ ***F. tricinectum*(3)**

6a. Microconidia more or less numerous than macroconidia, the latter formed in aerial mycelium, curved, fusoid with narrowly pointed apex and marked foot cell. Sporodochia absent or very rare, 3- to 5-septate, 28 to 42 x 3.2 to 4.5 µm. Typically 3-septate .................. ***F. chlamydosporum* (4)**

6b. Macroconidia elliptical, sausage-shaped, elongated, slightly curved, with round apical and round basal cells or marked foot cell, 3 to 5-septate, 27 to 52 x 4.4 to 6.8 µm. Typically 3-septate .................................... ***F. solani* (5)**

7. Culture peach to violet. Macroconidia spindly, elongated, cylindrical, straight, or slightly falcate with narrow curved apical cell and marked foot cell, 3-to 5-septate, 29 to 67 x 2.6 to 4.0 µm. Typically 3-septate Microconidia monophialidic ................................................ ***F. moniliforme* (6)**

7a. Microconidia are produced from polyphialidic conidiophores ....... ***F. proliferatum* (7)**

8. Culture light-cream, dark-yellow to brown. Microconidia rare, short, oval, spindle-shaped or clavate, 0 to 2-septate. Macroconidia falcate to parabolic with more or less gradually or sharply narrowing elongated, curved or non-curved apical cell and well-marked foot cell, 3 to 7-septate, 27 to 62 x 3.5 to 5.5 µm. Typically 5-septate ......................***F. equiseti* (8)**

8a. Cultures from below sulphureous conidia 3-4 septate, 23 – 32 x 3.5 – 4.0 µm ................................................ ***F. sulphureum* (9)**

9. Culture sometimes whitish, mostly strongly colored, rose, red, dark red, or crimson, Microconidia rare, appear initially in young aerial mycelium, short elliptical to spindle-shaped. Macroconidia elongated up to 90 µm, long narrow, filiform or slightly hyperbolic, curved, with elongated narrowing apical cell and well-marked foot cell, 3 to 7-septate, 31 to 92 x 2.5 to 4.1 µm. Typically 5-septate........................................ ***F. avenaceum* (10)**

9a. Microconidia absent. Macroconidia usually in aerial mycelium, small, often sparse, 1 to 3-septate, curved, sometimes aseptate .................................... 10

10. Microconidia with an ovate or hooked apical cell macroconidia curved, beaked or elongated apical cells ................................................ ***F. lateritium* (11)**

10a. Culture light-colored, white, light-rose, or yellow to place, slow-growing. Macroconidia short, elliptical, curved, sickle-shaped with pointed apex and wedge-shaped or round basal cell, 0 to 3-septate, 8 to 32 x 2.2 to 4.8µm ................................................ ***F. nivale* (12)**

11. Microconidia absent. Macroconidia typically 5-septate, mostly abundant, slightly curved, strongly dorsiventral, with a short and wide pointed apex and well-marked foot cell ................................................ 12

11a. Macroconidia 3-to 7-septate, broad, 25 to 68 x 4.8 to 82 µm.......... ***F. culmorum* (13)**

12. Macroconidia less broad, cylindrical, curved, strongly dorsiventral, falcate, elliptic or sickle-shaped with short beaked apical cell or slightly curved and gradually narrowing, elongated apical cell, and well-marked foot cell................................................ 13

12a. Macroconidia cylindrical, curved, strongly dorsiventral, fusoid with short beaked apical cell, 3-to 7-septate, 25 to 53 x 3.8 to 5.9 µm. .... ***F. sambucinum*** **(14)**

13. Macroconidia elongated up to 78 µm, spindle-or sickle-shaped, slightly dorsiventral, curved with elongated apical cell, 3 to 7-septate, 28 to 78 x 3.2 to5.0 µm ........................................................................ ***F. graminearum*** **(15)**

13a. Apical cell of macroconidia elongated narrowing eventy to a point with chlamydospores formed in knots or chains. ............................ ***F. acuminatum*** **(16)**

13b. Microconidia present or absent macroconidiate usually present macroconidia with wedge shaped foot cell, cultures from below beige brown becoming dark-brown ...................................................... ***F. semitectum*** **(17)**

**i) *F. acuminatum* Ellis & Everhart**

**Synonyms** =*E. Scirpi* Lamb. & Fautr. var acuminatum

= *F. roseum* Link emend. Snyd. & Hansen

**Teleomorph:** *Gibberella acuminate* Borth

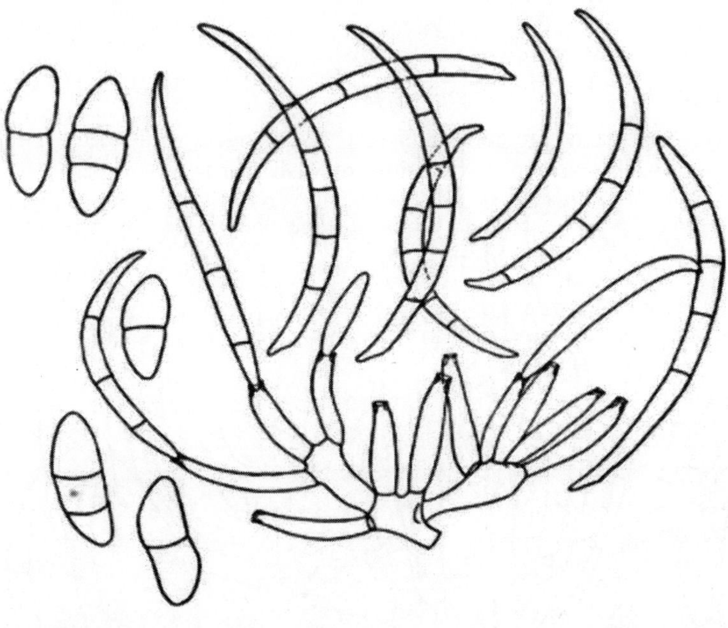

**Colony:** Strongly caramine-red but may be saffron to bay, floccose, white developing a brownish tinge in the center and later reddish – brown *Sporodochial* stromata develop.

**Conidiophores:** Simple phialides develop from aerial mycelium and produce microconidia occasionally. But these are soon replaced by loosely branched conidiophores, the terminal cells of which produce 1-4 phialides. These phialides show a marked tendency to proliferate and proliferation may be repeated 3-4 times, so as to form a string of successive phialides.

**Macroconidia:** Broadly falcate, strongly dorsiventral, 3-7 septate with an incurved elongation of the apical cell.

**Chlamydospores**: Intercalary, smooth walled globose to oral.

***Gibberella acuminate***: Perithecia, oral velobpyri form with smooth wall and undergo a pinched lateral or paniform collapse on drying. The inner penithecial wall consists of 2-3 layers hyaline cells

**Asci:** Clavate, shortly stipitate, 4-8 ascospores

**Ascospores:** Smooth, hyaline, ellipsnidal to oral but becoming slightly constricted at the central septin

**Mycotoxins:** Diacetrcyscirpenol, HT-2 toxin, Neosoloniol, T-2 toxin.

**Diagnostic features:** Colonies strongly caramine-red but may be saffron sporodochia stromatic, phialides simple develop from mycelium macroconidia falcate, 3-7 septate with an incurved elongation of epicalcell, chlamydospores intercalary.

**ii) *F. avenaceum* (corda ex Fr.) Sacc.**

**Synonyms** = *Fusiporium avenaceum* Fr.

**Teleomorph:** *Gibberella avenacea* Cook.

**Colony:** Superficial mycelium is rose red fringed with white and becoming yellowish – brown in the centre reddish – brown reverse.

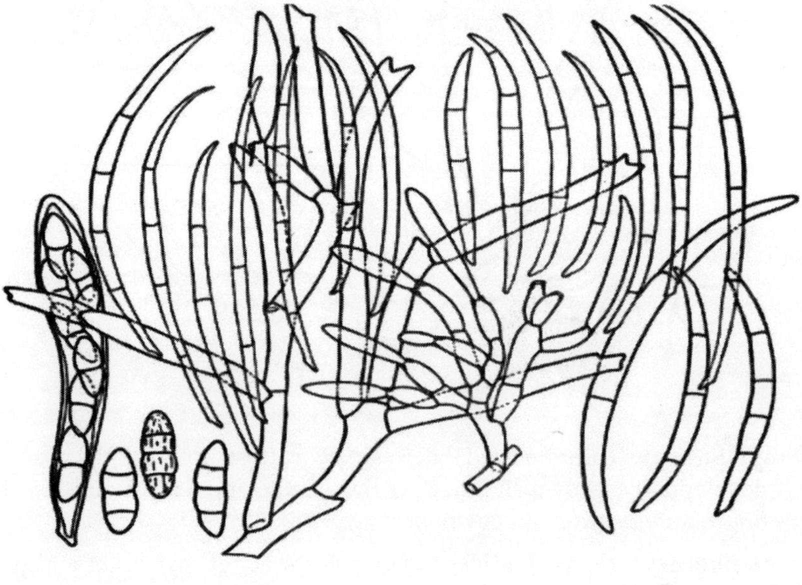

**Microconidia:** None

**Conidiophores:** Develop throughout the aerial mycelium, initially as simple lateral phialides or as polyblastic conidiogenous cell. Sporodochia develop after 7 days piomatial like with masses of short simple phialides covering the surface.

**Conidia :** Normally fusoid, curved, 4-7 septate with an elongated apical cell and well marked foot cell.

**Chlamydospores:** Absent.

**Mycotoxins:** Acetyl T-2 toxin, acely (deoxynivallenol, acetylsolaniol, diacetoxy scirpenol, HT-2 toxin, moniliformin, monoacetoxy scopenol, neosolaniol, nivalenol, fumanisin B).

**Diagnostic features:** Superficial mycelium rose-red fringed with white and becoming yellowish-brown, reddish – brown reverse. No microconidia and chlamydospores, after 7 days and simple phialides covers, conidia fusicoid 4-7 septate with elongated apical cells well branched foot cell.

**iii) *F. culmorum* (W.G. Smith) Sacc.**

**Synonyms** = *Fusisporium culmorum* W.G. Smith

**Teleomorph**: Absent.

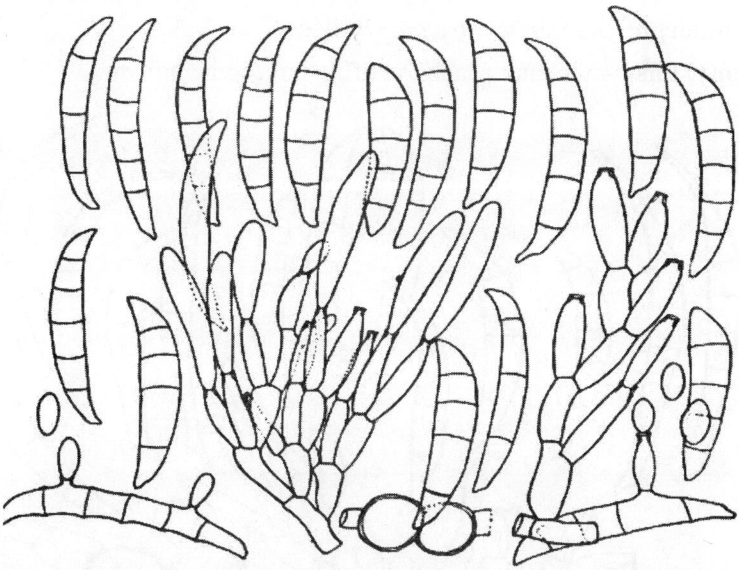

**Colony**: Floccose, becoming yellow after 2-3d. A red, pigmentation develops which diffuses into the medium which finally becomes reddish – brown. This pigmentation changes from golden yellow on acid media through brilliant carmine becoming red to purple on alkaline media (there pigness – are called rubrofusarin and aurofusarin).

**Microconidia:** Absent.

**Conidiophores:** Loosely branched conidiophores bear phialides. But later phialides are confined to Sporodochia. Sporodochia often form as pionnotes.

**Macroconidia:** 3-5 septate when mature slightly curved and strongly dorsiventral with a pointed apex and well marked foot cell.

**Chlamydospores:** Oval to globose, generally inter-calary but occasionally terminal, smooth to rough walled. They form singly, in chains or clumps.

**Mycotoxins:** Acetoxyscirpenol, Acetyl, T-2 toxin, Acetyldeoxynivalenol, Acetylneosolaniol, Deoxynivalenol, Diacetoxyslirpenol, HT-2 toxin, moniliformin, monoacetoxy scirpenol, neosolaniol nivalenol, zearalinone, Fumonisin B.

**Diagnostic features:** Floccose colonies becoming yellow after 2-3 days a red, pigmentation develops which diffuses into the medium finally becomes reddish – brown. Microconidia absent and chlamydospores oval to globose, generally inter calary occasionally terminal, smooth walled. They from singly in chains or clumps.

**iv) *F. equiseti* (corda) Sacc.**

**Synonyms** = *Selenosporium equiseti* Corda

=*Fusarium scirpi* Lamb & Fautr.

**Teleomorph:** *Gibberella intricans* Wollenw.

**Colony:** First white later peach to buff later become brown

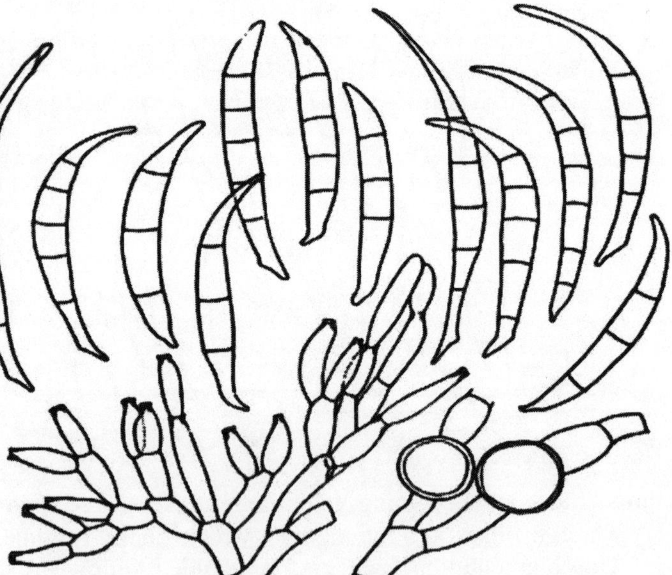

**Conidiophores:** Penicillately branched. Each branch beans 2-4 phialides or 2-4 branches may form at the apex each of which may produce several phialides or further series of methylae on which the phialides are normally borne, phialides are obclavate, not stromatic.

**Sporodochia:** Generally absent in some isolates pionnote or effuse sporodochia are formed.

**Conidia**: Falcate, pedicellate foot cell and an attenuated apical cell which is bent inwards, 4-7 septa.

**Chlamydospores**: Intercalary, solitary, in chains or in knots, globose.

**Ascoarp:** Penithecia solitary to gregarious, ovoid with nupulose outer wall, Asci clavati 4-8 mano or obliquely distichous ascospores. The ascospores are 3 or rarely 1-2 septate, hyaline fuloid.

**Mycotoxins:** Sarpenol, nivalenol, solaniol, Fusarione – X, Moniliformin, neosolaniol, T-2 toxin, Zearalenone, nivalenol.

**Diagnostic features:** Colonies white later turn to peach, later become brown. Sporodochia either absence or pionnote or effuse sporodochia are formed. Conidia falcate, pedicellate foot cell and an atteunated apical cell, 4-7 septa.

**v) *F. graminearum* Schwabe**

**Synonyms** = *Sphaeria zeae* Schweinitz

= *Fusarium roseum* Link emendsyryder & Harisen

**Teleomorph**: *Gibberella zeae* (Schwein.) petch

**Colony:** Greyish-rose or livid red to crime on often becoming vinaceous with a brown tinge but they may sector to grayish yellow or white. Pigmentation depends on medium and on pH.

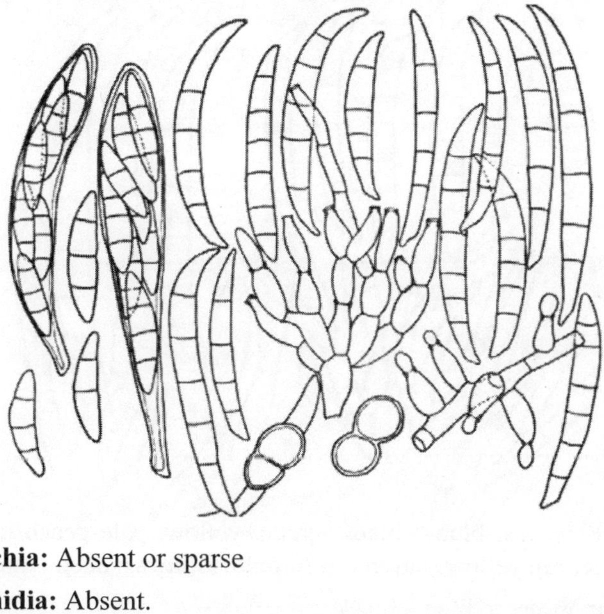

**Sporodochia:** Absent or sparse

**Microconidia:** Absent.

**Conidiophores**: Multi branched which terminate in short doliform phialides, after first production of conidia, the conidiophores may proliferate further by growth through the apex of some of the phialides to produce a further series of new phialides and aggregate in to *Sporodochia.*

**Phialides:** Doliform, sometimes proliferating percurrently but not symipoplially.

**Macroconidia:** Falcate, with or without an elongated apical cell to sickle – shaped or markedly dorsiventral and have well marked foot cell which in certain strains in very pedicellate (Fort shaped), 3-7 septate.

**Chlamydospores**: Intercalary, single, in chains or occasionally in clumps.

**Mycotoxins:** Deoxynivalenol, Nivalenol, Zearalenone, Diacetoxyscirpenol.

**Diagnostic features:** Greyish-rose, livid red to crimson often becoming vinaceous, sporodochia and microconidia absent. Multi branched conidia which terminate in short doliform phialides, macroconidia falcate, marked foot cells 3-7 septate.

### vi) *F. lateritium* Nees

**Synonyms** = *Sphaeria baccata* wallroth

=*Gibberella pulicaris* (Fr.) Sacc.

=*G. baccata* (Wallr.) Sacc. var. *mosicole* (De Not) Wollenw.

=*G. baccata* (wallr.) Sacc. var. *evonymmi* (Fickel) Wollenw.

=*G. lateritium* (Nees) Synder & Hansen.

**Teleomorph:** *Gibberella lateritium* (Nees)

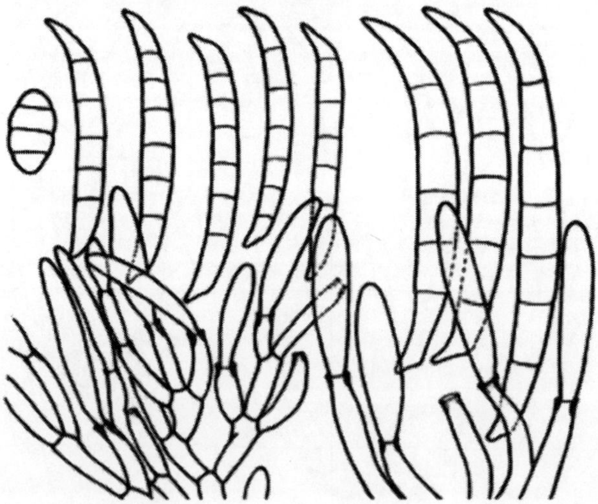

**Colony:** Floccose, blue – black, green, yellow, pale peach near white, felted mycelium yellow, vinaceous to reddish-brown deep orange pigment.

**Stromata:** Violet to blue – black.

**Conidiophores:** Initially simple lateral branch terminating in a 2-3 metulae each of which terminate in 1-3 simple cylindrical to doliform phialides. Sometimes sporodochia develops.

**Microconidia**: Absent.

**Macroconidia:** Short, straight or curved 3-5 septate with a short beak

Sometimes long straight or falcate conidia with 5-8 septate.

**Chlamydospores:** Sparsely developed, mono or bicellular chlamydospores may form in conidia

**Mycotoxins:** Diacetoxyscirpenol.

**Diagnostic features:** Floccose, blue–black, vinaceous to reddish-brown deep orange pigment. Stroma voilet to blue–black, simple lateral branch, terminating in a 2-3 metulae each of which terminate in 1-3 simple doliform phialides, microconidia absent. Macroconidia short straight, 3-5 septate, with a short beak

**vii) *F. moniliforme* Schelon Rep.**

**Synonymys** = *Lisea fujiKuroi* Sawada

= *Gibberella moniliforme* (Sheld.) Wineland

= *F. verticilloides*

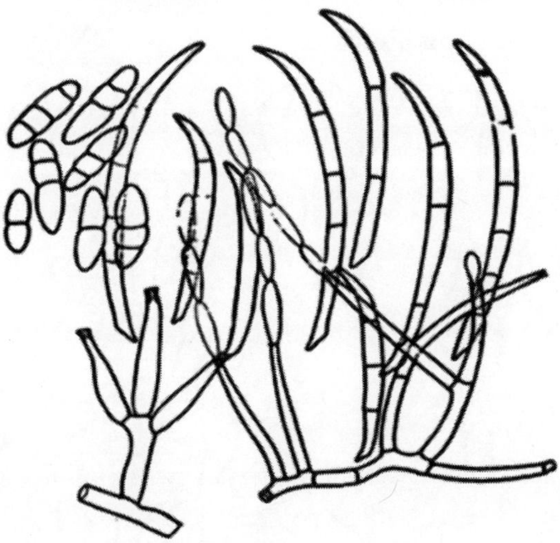

**Teleomorph***: Gibberella fujikuroi*

**Colony:** Rapidly growth dark-voilet from below occasionally paler, lilac, vinaceous or even-cream.

**Microconidiophores:** Simple lateral, subulate phialides, formed on the aerial hyphae, rarely in short lateral branches which produce microconidous in short chains ellipsoid to clavate with a flattened base.

**Macroconidiophores:** Monophialidic

**Macroconidia:** Sickle shaped to almost straight 3-5-7 septate

**Chlamydospores**: Absent

**Mycotoxins:** Acetoxyscirpenediol, Acetyl T-2 toxin, Acetyldeoxynivalenol, Acetylneosolaniol, Deoxynivalenol diacetate, Deoxynivalenol, Diacetoxyscirpenol, Fumonisin B1, Fusaric acid, HT-2 toxin, Moniliformin, Monoacetoxyscirpenol, Neosolaniol, Nivalenol, T-2 toxin, Zearalenone.

**Diagnostic features:** Dark-violet colonies, simple lateral, subulate phialides produce microconidia in chains, ellipsoid to clavate. Macroconidia sickle shaped almost straight 3-7 septate

**viii) *F. nivale* (Fr.) Ces.**

**Synonyms** = *Larosa nivalis* Fr.

**Teleomorph:** *Micronectriella nivalis* (Schaffna) Borth
*Calonectria nivalis* Schaffn.
*Griphosphaeria nivalis* (Schaffn.) Mullar & Arx
*Calonectria graninicola* Wr.

**Colony**: White to pale peach to apricot with little discoloration of the agar.

**Sporodochia:** Small sporodochia.

**Conidiophores:** Branched conidiophores that arise as lateral branches.

**Phialides:** Subulate to pyriform

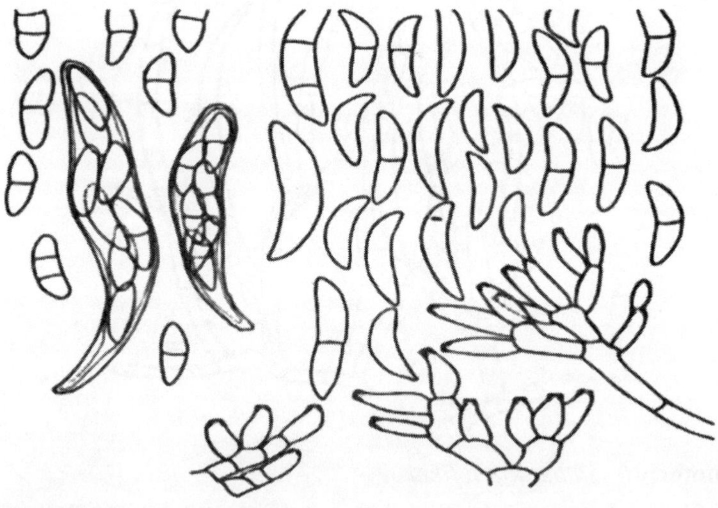

**Microconidia:** Absent.

**Macroconidia:** Curved, broadly falcate with pointed apex and flattened, wedge shaped base 1-3 septate.

**Chlamydospores:** Absent.

**Mycotoxins:** Acetoxy scirpenol, Acetyldeoxynivanenol, Acetylneosolaniol, Acetyl $T_2$-toxin, Deoxynivalenol diacetate, Fumonisin $B_1$, $H^2T_2$ toxin, moniliformin, Monoacetoxyscirpenol, nivalenol, diacetoxyscirpenol, Fusarenon-x and zearalenone.

**Diagnostic features:** White to pale peach to apricot, sporodochia small branched, branched conidiophores that arise as lateral branches, phialides pyriform to subulate, Macroconidia and chlamydospores absent, macroconidia falcate with pointed apex wedge shaped base 1-3 septate.

**ix) *F. oxysporum* Schlecht**

**Synonymys** = *F. angustum* Sherb.
= *F. bostrycoides* Wollenb.
= *F. bulbegenum* Cooke & Massea
= *F. conglutinans* Wollenw.
= *F. orthroceros* Appel & Wollenw
= *F. vasinfectu* Atkinson
= *F. lini* Bolley
= *F. dianthi* Prill.

**Teleomorph:** *Gibberella* spp

**Colony:** White or peach usually with purple linge sparse to abundant sparasely branched conidiophores.

**Microconidiophores:** Simple arising laterally on the hyphae or short sparsely branched.

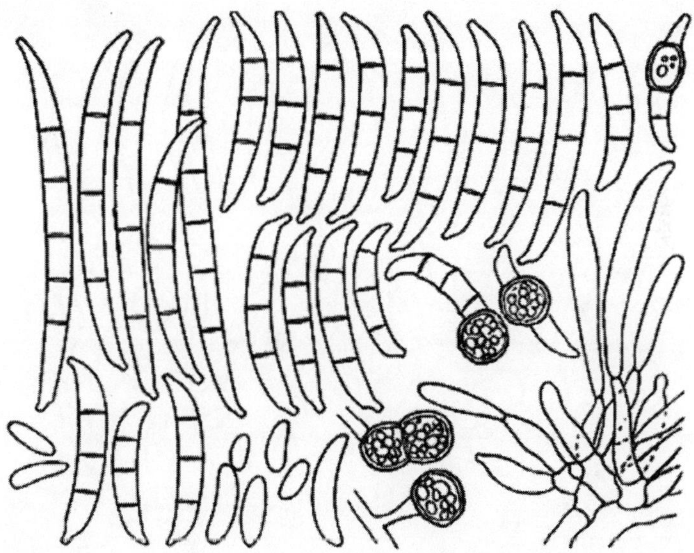

**Microconidia***:* Abundant oval-ellipsoid, cylindrical, straight or curved.

**Macroconidiophores***:* Elaborately branched or on Turbercularia like sporodochia

**Macroconidia***:* Sparse in some strains, 3-5 septric fusoid-subulate and pointed at both ends fusoid to falcate macrocosidic are formal with somewhat hooked apex and a pedicellate base.

**Chlamydospores:** Absent

**Mycotoxins:** Scirpenol, solaniol, T-2 toxin, Deoxynivalenol, fusarinone-X, maniliformin, zeasalenone.

**Diagnostic features:** Colony white or peach usually with microconidiophores simple sparsely branched, microconidia single celled oval cylindrical straight or curved, macroconidiophore elaborately branched or on turbercularia like sporodochia, macroconidia 3-5 septate pointed at both ends. No chlamydospores

**x) *F. poae* (Peak) Wolleniweber.**

**Synonyms** = *Sporitrichum poal* peak

= *F. tricinctum* (Corda) Sacc.

= *F. poae* (Pk.) Wr. F. Pallens

**Teleomorph:** *Gibberella*

**Colony:** Hairy to felted and assumes a powdery appearance with the formation of microconidia. Conidiophores inistiall they are lateral pores or from small globose to dolii form lateral phialides with an apical colour subsequently formation of lateral branch results compact conidiophores which under micro sense resembles a bunch of grapes.

**Microcondia:** Ampulliform to globose.

**Conidiophores:** They develops from a sporodochia individual slender but branched at the apex and bears 1-4 phialides. Phialides cylindrical to subulate at the end of each branch.

**Macroconidia:** Curved falcate and slightly wider above the media septum, 3-reptate.

**Chlamydospores:** None

**Mycotoxins**: Diacetoxy scirpenol, HT-Z toxin, Neosolaniol, T-2 toxin.

**Diagnostic features:** Hairy to felted and assumes a powdery appearance with the formation of microconidia, macroconidia. Ampuliform to globose developed sporodochia, branched bearing 1-4 phialides.

**xi) *F. sambucinum* Fuckel**

**Synonyms**= *Sphaeria pulicoris* Fries

= *Selenosporium sareochorum*

= *Fusarium sarcoehorum*

**Teleomorph:** *Gibberella pulicaris*

**Colony:** Rose colour becoming bay to blood red with age Bluish pigments may develop on neutral or slightly alkaline medium. Aerial mycelium floccose, white and tinged with rose.

**Sporodochia:** Stromatic, erumpent, rose to ochraceous pale-salmon

**Microconidia:** Absent

**Conidiophores:** Initially scattered in the aerial mycelium but later become complexed and form *sporodochia.* The conidiophores begin as a lateral branch which after becoming 2-4 cells long divides usually several times to form the metulae each of which forms 2-4 phialides at the apex.

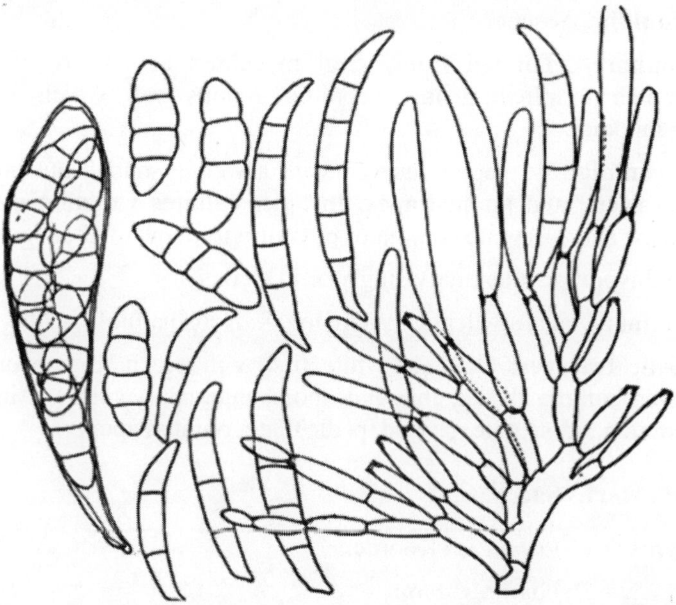

**Phialides:** Cylindrical, obclavate to dolii form with a broad apical pore surrounded by a collar conidiophores formed in *Sporodochia* are much reduced and present the typical palisade of phialides over the surface of the sporodochium.

**Macro conidia:** Curved, fusoid, lanceolate, strongly dorsiventral with appointed apex and well naked foot cell, 3-5 septate.

**Chlamydospores:** Form sparsely as single globose cells or in chains or clumps are either intercalary or terminal on short lateral branches or in cells of the macroconidia.

*G. pulicaris* is heterothallic

**Mycotoxins:** Deacetoxy scirpenol

**Diagnostic features:** Colony rose colour becoming blood red coloured, sporodochia stromatic, micro conidia absent, chlamydospores single globose cells or in chains or clumps, intercalary or terminal, macro conidia curved, fusoid, lanceolate, 3-5 septate

**xii) *F. semetectum* Berk. & Rav**

**Synonyms** = *Pseudofusarium semitectum* (Berk. & Ravenel)

= *Fusisporium incarnatum* Roberge ex Desm.

= *Fusisporium pallidoroseum* Cooke, Grevillea

**Colony:** White first with peach tinge and peach coloured from below.

**Mycelium:** Floccose peach gradually changing to a vellaneous finally becoming buff-brown.

**Sporodochia**: Absent

**Microconidia:** Absent

**Conidiophores:** Formed from aerial mycelium as loosely branched, each branch terminating in a conidia genous cell which is polyblastic sympodial.

**Macro conidia:** 3-5 septate curved with a wedge shaped but not pedicellate basal cell and pointed apex. In older cultures variation in conidial sze shape and 1-septate conidia of pysiform to obovate after spores globose.

**Chlamydospores:** Intercalary singly or in chains.

**Mycotoxins:** Deoxynivalenol, Fusarenon-X, Neosolamiol, T-2 toxin.

**Diagnostic features:** Colonies white first with peach tinge sporodochia and microconidia absent, chlamydospores intercalary singly or in chains, acro conidia 3-5 septate, curved, pedicellate, pointed apex.

**xiii) *F. solani* (Mart.) Sacc.**

**Synonyms** = *F. jaranicum* Koorden

= *Hypomyces solani*

**Teleomorph**: *Nectria haematococca.*

**Colony:** Striate, sparse to dense and floccose grayish – white. Agar medium typically develop blue or blueish – brown.

**Conidiophores**: (1) **Micro conidiophores** are later conidiophores which initially may be merely elongated lateral phialides – later formed conidiophores are elongated and sparsely branched.

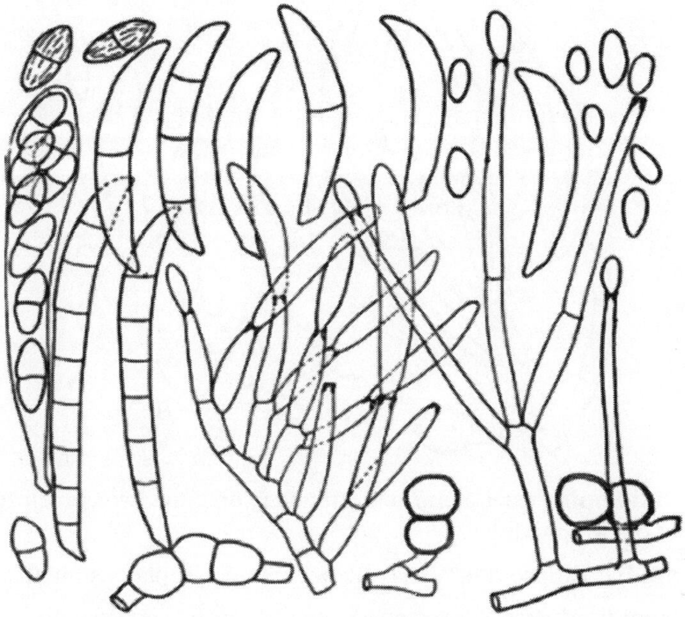

**Phialides**: Each branch of lateral conidiophores terminates in a single cylindrical to barely subulate phialide.

**Microconidia:** Broad and more oval in a shape with somewhat thicker walls, 1-septate.

**(2) Macro conidiophores:** They are short multi branched which soon form effuse sporodochia. Macroconidia inequally fusoid with many of conidia having dest diameter in the penultimate cell, many have rounded foot cell apical cell is pointed and somewhat beaked.

**(3) Chlamydospores:** Develop after 7-14 days on a weak medium globose to oval, smooth to rough walled.

**Mycotoxins:** Diacetoxyscirpenol, HT-2 toxin, neosoloniol, T-2 toxin.

**Diagnostic features:** Colonies floccose grayish to white, microconidiophores are merely elongated lateral phialides cylindrical to subulate, microconidia oval, 2-celld, macroconidiophores multibranched soon form effuse sporodochia macroconidia fusoid with round foot cell apical cell is pointed

**xiv) *F. sporotrichoides* Sterb.**

**Synonyms** = *F. tricinctum* (Corda) Sacc.

= *F. sporotrichiella V. sporotrichioides* (Corda) Bilary

**Colony:** Profuse mycelia growth, aerial mycelium floccose livid red below but white on the surface later becoming tinged with brown, reverse deep livid- red agar remains colourless.

**Micro conidiophores:** Lateral and branches once or twice each terminates in one or two phialides.

**Microconidia**: Ellipsoid to obclavate becoming 1-septate, smooth walled.

**Macro conidiophores**: Profusely branched or in sporodochia. Each branch terminates in 2-4 phialides

**Macroconidia:** Falcate to curved, fusoid, with an apedicellate basal cell, 3-septate.

**Chlamydospores:** Intercalary, solitary or in chains hyphae after 2-3 weeks.

**Mycotoxins:** Diacetoxy scirpenol, HT-2 toxin T-2 toxin.

**Diagnostic features:** Colonies with profuse mycelium, floccose, microconidiophores are lateral and branches once or twice each terminated by flask shaped, phialides. Microconidia ellipsoid to obclavate, 1-septate, macroconidiophores profusely branched or in sporodochia, each branch terminates in 2 or more phialides. Macroconidia falcate to curved, fusoid, 3-septate, chlamydospores intercalary, solitary or in chains hyphae after 2-3 weeks.

**xv) *F. sulphureum* Schlechten**

**Synonyms** = *Phaeria cyanogena* Desm.

= *Gibbera* sanibinetii Mont

= *Gibberella sanibinatii* (Mont) Sacc

**Teleomorph:** *Gibberella pulicaris* (Fr.) Sacc.

**Colony:** Cream or light brown

**Conidiophores:** Aerial mycelium is sparse to floccose, felted appearing powdery. Conidial formation from conidiophore in the aerial mycelium. Simple palmate conidiophore from a lateral branch, more complex.

**Macroconidia**: Thin walled, curved, fusiform with a curved pointed apicall cell and marked a pedicellate foot cell.

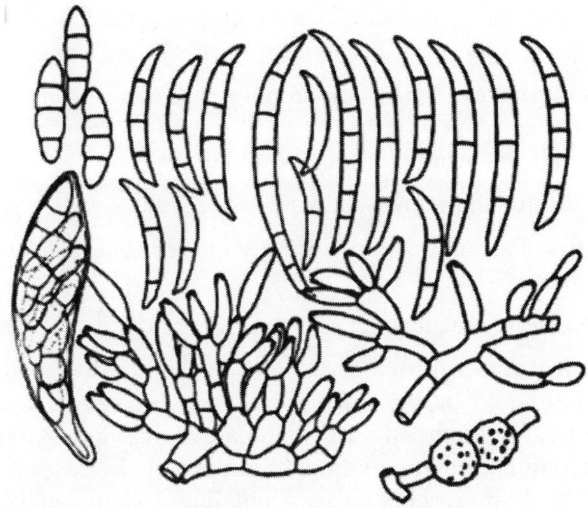

**Conidiophores:** Sparse, globose with smooth walls are formed single or in short chain in the mycelium.

**Mycotoxins:** Diacetoxyscirpenol, HT-2 toxin,T-2 toxin

**Diagnostic features:** Aerial mycelium, conidiophore farm from a lateral branch and Chlamydospores globose.

**xvi) *F.tricinctum* (Corda) Sacc.**

**Synonyms** = *Sclenosporium tricinatum* Corda

= *F. citriforme* Jamalainen.

**Teleomorph**: *Gibbarilla tricinatum*

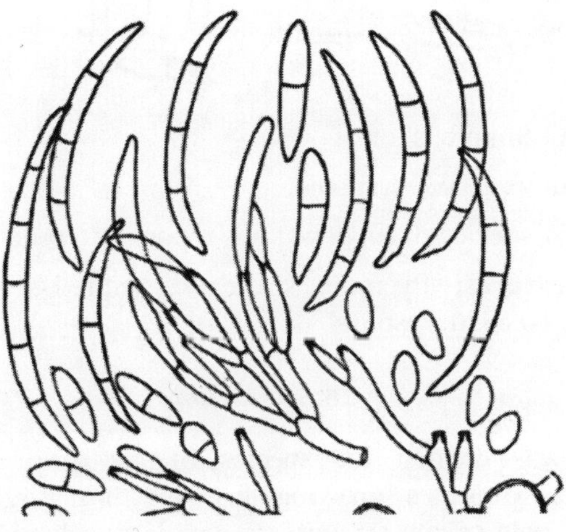

**Colony:** Aerial, sparse to floccose, white becoming carmine-red to purple.

**Microconidiophores:** Microconidiophores initially simple bearing one or two simple cylindrical to subulate phialides. Later becomes profusely branched.

**Microconidia**: Oval to pyriform with a minute aperulum. Some later becomes septate.

**Macroconidiophores**: Macroconidiophores develop from sporodochia which are convex pencil form pale to orange often become effuse each conidiophores terminating in 2-3 phialides.

**Phialides:** Shorter and more obclavate

**Macroconidia**: Falcate or more strongly curved with a well marked foot cell, 3-5 septate.

**Mycotoxins:** Deoxynivanol, HT-2 toxin, Neosoloniol, T-2 toxin.

**Diagnostic features:** Aerial mycelium sparse to floccose, white becoming carmine-red to purple, microconidiophores simple bearing one or two phialides, cylindrical to subulate phialides, oval to pyriform microconidia, macroconidiophores develop from sporodochia, phialides short and more obclavate, phialides short and more obclavate macroconidia fulcate or strongly curved 3-5 septate.

## 27. *Fusicoccum* Corda

**Colony:** White, spreading

**Pycnidia:** Spherical or flatted, erumpent, stroma dark, one to several pycnidia per stroma, opening separately with a common pore.

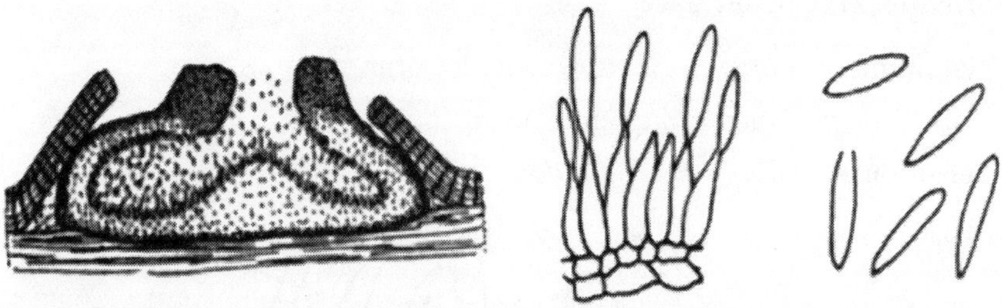

**Conidiophore:** Simple, short.

**Conidia:** Hyaline, 1-celled, fusoid.

**Mycotoxins:** A species of *Fusicoccin* is reported to produce.

**Diagnostic features:** Pycnidia stromatic, with 1-several locules, conidia hyaline, 1-celled, fusoid.

## 28. *Gerlachia*

Genus *Gerlachia* was created for species of *Fusarium* producing annellidic, conidiogenous cells. Colonies growing fast reaching slightly more than 1cm diameter, white to pale-peach or apricot, aerial mycelium or felted sporulating sparsely in the aerial mycelium or in pale-orange sporodochia. Conidiogenous cells short, dolii form to obpyriform showing an apical annulated zone. Conidia broadly falcate with almost equal ends, the base only shortly flattened 1-3 septate, chlamydospores absent.

**i) *G. nivale***

**Synonymys=** *Fusarium nivale* (Fr.) Cesrabench Herb. Myco. Ed.1 or 1439

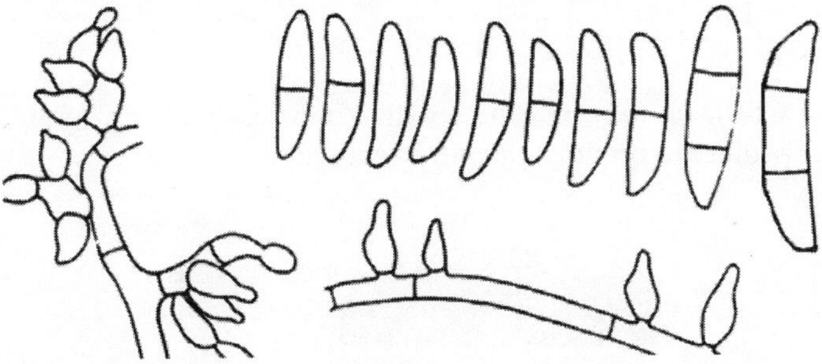

**Colony:** White to pale- peach to apricot with little discoloration of agar

**Conidiophores**: Borne on aerial mycelia but after 7-10 days they formed abundantly or small sporodochium marker by the pale orange many of conidia. Phialides subulate to obpyiform

**Conidia:** Curved, broadly falcate with a pointed apex and flattened, wedge shaped base, 1-3 septate, chlamydospores are not formed or rarely formed.

**Mycotoxins**: Acetoxy scirpenol, Acetyldeoxynivanenol, Acetylneosolaniol, Acetyl $T_2$-toxin, Deoxynivalenol diacetate, Fumonisin $B_1$, $H^2T_2$ toxin, moniliformin, Monoacetoxyscirpenol, nivalenol, diacetoxyscirpenol, Fusarenon-x and zearalenone.

**Diagnostic features**: Conidiophores are borne directly on aerial hyphae or grouped to form sporodochial conidial with pointed apex, wedge shaped base, 1-3 septate.

## 29. *Gibberella* Sacc.

**Ascomata:** An ostiolate perithecium, borne free or associated with psedoparenchymatous or effuse tissue.

**Perithecium:** Superficial to partially immersed, scattered to caespitose, fleshy, ovoid to globose, with an ostiolar papilla, blue to violet. Ascomal wall pseudoparenchymatous centrums with apical paraphyses.

**Ascus:** Unitunicate, with an undifferentiated apex, non-amyloid, ellipsoid to broadly clavate, 8-spored.

**Ascospores:** Pharagmosporous, but often variable in septation, ellipsoid, fusoid or cymbiform, straight or curved, smooth, hyaline to slightly yellowish

**Anamorph:** *Fusarium*

**i) *G. fujikuroi* (Sawada)**

**Anamorph:** *Fusarium moniliforme*

**Colony:** Hyaline, spreading, turning pinkish due to production of large conidiophores of conidia

**Ascomata:** Ostiolate perithecium, borne free or associated with stroma, erumpent pseudo-parenchymatous superficial, scattered to casepilose, fleshy, ovoid to globose, blue to voilet.

**Ascus:** Clavate, uni-tunicate with undifferentiated apex non-myeloid ellipsoidal to broadly clavate 8- spored.

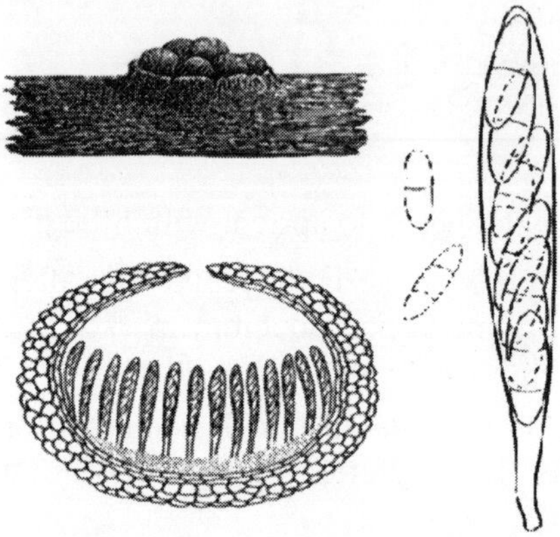

a. Cluster of perithecia on erumpent stroma. b. Section through perithecium with asci. c. Asci with ascospores. d. Mature ascospores

**Ascospores:** Pharagmosporous, ellipsoid, fusoid, pyriform, straight or curved, smooth, hyaline, slightly yellowish

**Mycotoxins:** Moniliformin, nivalenol

**Diagnostic features**: Colonies pink to violet, chlamydospores absent. Perithecium dark blue, spherical, asci cylindrical.

### 30. *Gliocladium* Corda

It has about 20 species described but *G.roseum* and *G.virens* are most common.

**Colonies**: White or some shades of salmon pink or Rose Spreading, white

**Conidiophores:** Erect on aerial or submerged hyphae or hyphal ropes two types

i. **primary conidiophores** divergent verticellate branched phialides in whorl of 3-4/conidial heads are discrete

ii. **Secondary conidiophores** formed particularly in the colony center, diversity penicellate, phialides in closely appressed whorts

**Conidia**: Conidia cohering in imbricate columns or irregular slimy masses, elongate, slightly asymmetrical, apex obliquely rounded, base slightly protruding. Hyaline or pinkish in mass, smooth walled.

**Telomorph:** *Nectria*

**i) *G. virens***

**Colony:** White, greenish or some shades of salmon pink or rose

**Hyphae:** Hyaline, branched, septate

**Reverse of colony:** Initially pale becoming dark with age

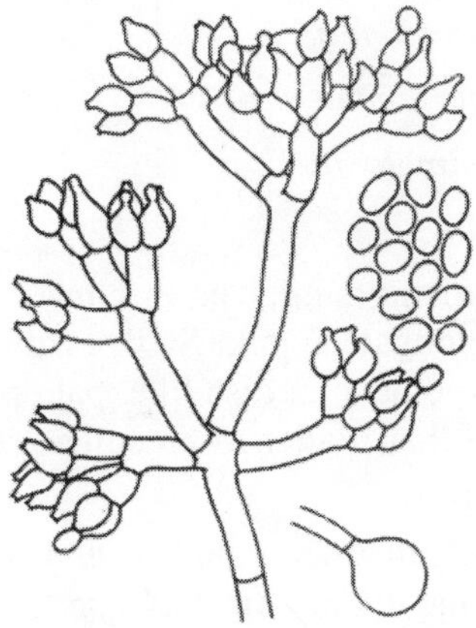

**Conidiophores:** Erect or suberect arising from the substratum or from aerial hyphae, hyaline, simple or branched at the apex in a dense fasciculate fashion bearing metulae and phialides, besides penicellate conidiophores primary, simply verticellate conidiophores may be present, ultimate sporogenous cells.

**Phialides:** Divergent, slightly or closely appressed bottle shaped, frequently more convex on one side

**Conidia:** Hyaline or pigmented, non-septate, ovoid or sometimes flattened on one side, conidial heads in slimy balls or held together in loose columns which are not distinguishable

**Mycotoxins:** Virens, viridin, gliorein, anratio gliocladium

**Diagnostic features:** Penicillate conidiophores, single celled conidia are produced in slime balls

## 31. *Gloeotinia* Wilson Noble and Gray

A true sclerotium does not develop. Sporodochia consist of a core of closely septate, branching hyphae with the terminal cells of each branch bearing 1-4 microconidiophores. Microconidiophores are septate, guttulate, hyaline, and penicillate. Microconidia (spermatia) are unicellular, uninucleate, ovoid, guttulate or biguttulate, hyaline. Macroconidiophores are short barrel-shaped cells, arise laterally on the hyphae. Macroconidia are budded from the apex of the macroconidiophores and arrange in clusters, smooth, unicellular, uninucleate, cylindrical to slightly cresentric with rounded ends, and usually biguttulate Apothecia. Apothecia are small, fleshy, and cup-shaped. Hymenium composed of fine, densely intertwining hyphae. The asci are cylindrical and clavate, ascospores are hyaline, smooth, elliptical, fusoid to broadly fusoid, and usually biguttulate. Paraphyses are fusiform, hyaline, non-septate.

**Anamorph:** *Endoconidium temulentum*

**i) *G. temulenta* (Prill &Delaer) Wilson**

**Synomy** =*Phialea temuteonta*

=*Ciboria temmlenta*

=*Stromatinia temmlenta*

=*Hymenoscyphis temmlenta*

=*Sclerotinia temmulenta*

=*Philea mucosa*

**Ascomata**: Ascomata arising singly or in cluster upto 7. Apothecium raised with the help of elongated stalk. Stalk cylindrical, smooth concolourous with sarcomata. Apothecium disc like, pale pinkish-cinnamon, darkening with age, concave then flattened receptical. Hymenium fleshy with asci and paraphyses.

**Ascus:** Ascus cylindrical, distinctly stalked and round above. Ascal pore not blue in melzers reagent, unitunicate, 8- spored arranged in single row.

**Ascospores**: Single celled, hyaline, ellipsoidal or in equilateral biguttulate

**Paraphyses**: Cylindrical, apex rounded, not enlarged

**Anamorph:** Produces indistinct conidia of two types. Macroconidia slime spores, unicellular, cylindrical, pink in mass. Microconidia are produced in a pink sporodochia developed on the substratum

**Mycotoxins:** Alkaloids

**Diagnostic features**: Ascomata, stalked apothecium with hymenium of paraphyses and cylindrical asci. Ascospore apex round with negative melzers reagent, single celled and biguttulate.

## 32. *Hormiscium* Kunze

**Colony:** Folicolous, black, descrete, mycelium composed of branched, septate

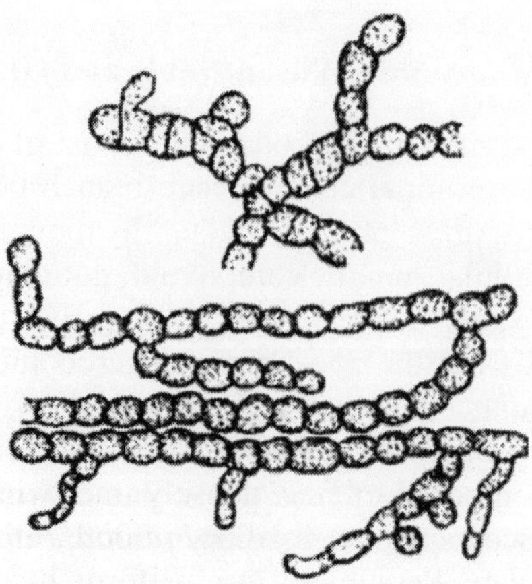

**Conidiophores:** Miconematous, entire mycelium develops into chains of rounded or oblong 1-celled conidia which separate with difficulty

**Conidia:** 1-celled, round or oblong

**Mycotoxins:** A species of *Hormiscium* is reported to produce Cytochalarin B

**Diagnostic features:** Entire mycelium transforming into chains of conidia. This has been merged with *Torula* and it is no more valid genus.

### 33. *Hyalodendron* Diddens

**Colony:** Spreading, white

**Conidiophores**: Erect, micronematous, mononematous, simple or branched, bearing one to a few conidia at the apex of the branches, conidia frequently in small clusters, becoming catenulate at the apex of the branches.

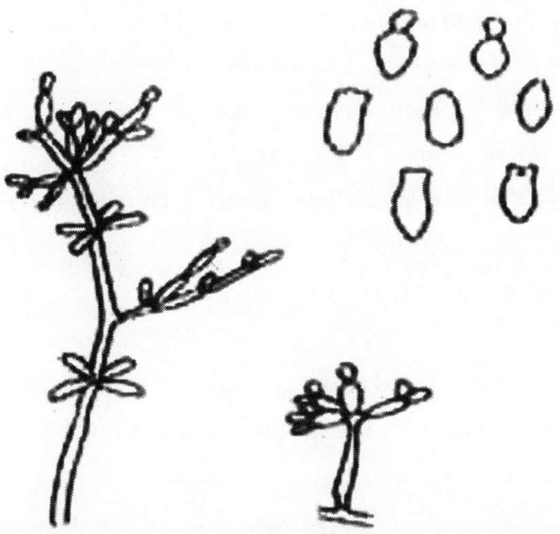

**Conidia:** Frequently in small clusters becoming catenulate by acropetalous formation of new conidia, chains often branched, 1-celled, hyaline, variable in shape, ovoid to cylindrical or oblong

**Teleomorph**: *Ophiostoma*

**Mycotoxins:** A species of Hyalodendrin is reported to produce Hyalodendrin, Hyalodendrin tetrasulfide

**Diagnostic Features**: Conidiophore erect, simple or branched bearing one to a few conidia, becoming catenulate at the apex of the branches.

### 34. *Lasiodiplodia* Ellis and Everh

*Lasiodiplodia* is a valid name while sphaeropsis, paradisiaca is the oldest name. The other known genera are *Hemidiplodia, Nematodiplodia, Paradiplodla, shizodiplodia, Synnemadiplodia* are The genus has about 25 species described.

**Colonies:** White in the beginning gradually change dark-brown and then to dark-chocolate-brown with age Pycnidia stromatic, separate or aggregated, carbonaceous thick walled, non ostiolate, dehiscence by irregular rupture.

**Conidiophores**: Absent.

**Conidia**: Born directly on the conidiamatous wall cells, holoblastic, determinate, acrogenous, hyaline when young later becoming dark-brown, 2-celled with longitudinal straiations from apex to base.

**Teleomorph:** *Bohyospharia rhodina*

**i) *L. theobromae* (Pat) Griffiths & Maubl Bull**

**Synonymys** = *Botryodiplodia ananarsae* (Sacc.) Petr. (1929)

=*B. elasticae* Petch (1906)

=*B. theobromae* Pat. (1892)

= *B. tubericola* (Ellis & Everh.) Petr (1923)

=*Chaetodiptochia grisea* Petch (1906)

= *Diplodia anonassae* Sacch. (1917)

=*D. theobromae* (Pat.) W. Nowell (1923)

=*D. tubericola* (Ellis & Everh.) Tanbenh (1915)

=*Lasidioplodia migra* Griffon & Maube

=*L.triflorae* B.B. Higgins (1916)

= *L. tubericola* Ellis & Everh. (1896)

=*L. triflorae* (B.B. Higgins) Zambett. (1955)

=*Macrophoma vestita* Prill & Decacr. (1894)

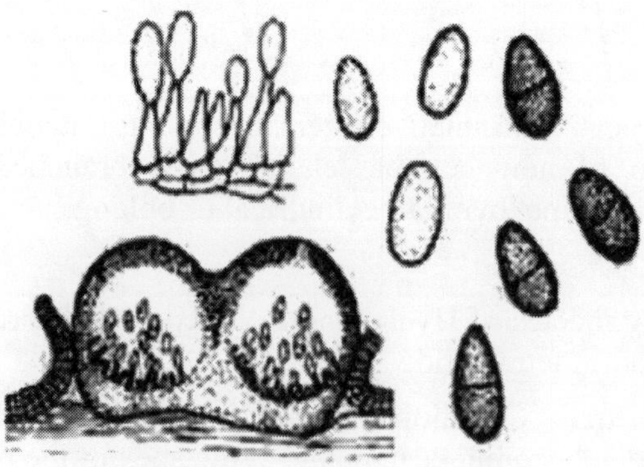

**Colony:** White begin to turn dark-brown hyphae, superficial, branched, septate, white turning dark-chocolate-brown with age. Pycnidium, eustromatic, immersed or superficial, separate or aggregated and confinement, globose, carbonaceous, dark-brown, thick walled, textura angularis, paler and thinner towards the conidiogenous region, often with dark-brown superficial hyphae over the surface, non-ostiolate, dehiscence by irregular rupture

**Conidiophores:** Absent

**Conidiogenous cells:** Directly on the conidiamatous wall cells, holoblastic, determinate, discrete, cylindrical, hyaline, smooth, with no percurrent or sympodial proliferation formed from cells lining the inner pycnidial walls

**Conidia:** Acrogenous, hyaline when young, later becoming dark-brown, medianly euseptate, thick walled, ellipsoid, base truncate with longitudinal straiations from apex to base

**Paraphyses:** Hyaline, cylindrical, septate

**Mycotoxins:** 4-hydroxy mellein, desmethyl lasiodiplodin

**Diagnostic features:** Pycnidium eustromatic, carbonaceous black, thick walled, non-ostiolate, holoblastic conidial cells, conidia hyaline one-celled when young but becoming 2-celled dark-brown strate at maturity.

## 35. *Metarrhizium* Sorok

Colonies white, spreading, hyphae hyaline branched, parasitic on insects. Conidiophores in low mounds, covered by conidia, erect, branched closely or loosely grouped, forming a sporulating layer or phialides borne singly, in pairs or in whorls. Conidia apical, produced in basipetal chains compacted in to columns long- ovoid to cylindrical with rounded ends, 1-celled, olive-green in mass.

### i) *M. anisopliae* (Metschn) Sorok

**Synonymys** = *Metarrhzium velutinum* Borowrka

= *Entomorphthora anisopliae* Metsch

= *Isaria destructor* Mtschn

= *M. album* Pelch

= *M. brunneum* Pelch

= *Mepothecium communae* Pidopl. Kiril

Parasitic on insects belonging Coleoptera (of families Elateridae and curculionidae)

**Colony:** Slow growing

**Conidiophores:** *Penicillium* type conidiophores are aggregated to form synnemata, aporogenous cells, flask shaped with thickened neck blows out a conidia are delimited at the base by completely closed septum

**Conidia:** Conidia single celled, hyaline, smooth walled

**Mycotoxins:** Cytochalasin C and D (Zygosporin A), Cyclodepsipeptides destruxin A, B, C or D, prolylvalyl anhydride

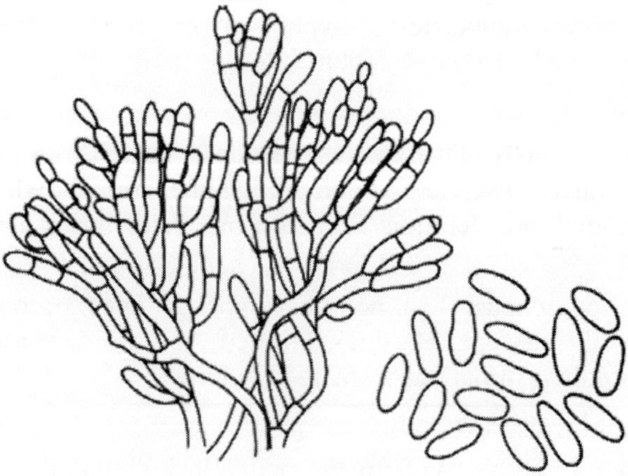

**Diagnostic features:** Parasitic or coleopteran group of insects, synnematous, phyalidic, single celled, hyaline, smooth walled.

### 36. *Microdochium* Syd.

The genus *Microdochium* has been re-described and amended to cover four species *Sporodochium*, erumpent, sub epidermal. Conidiogenous cells discrete, little differentiated, ampulliform producing a few apical conidiogenous denticles. Conidia blastoconidia, fusiform to falcate 1-2 celled hyaline smooth walled.

**i) *M. bolleji* (Spragne) de Hooge & Hermanides Nijhof**

**Synonymys** = *Gloeosporium bolleyi* Sprague
= *Aureobasidium bolly* (Sprague) V. Arx.

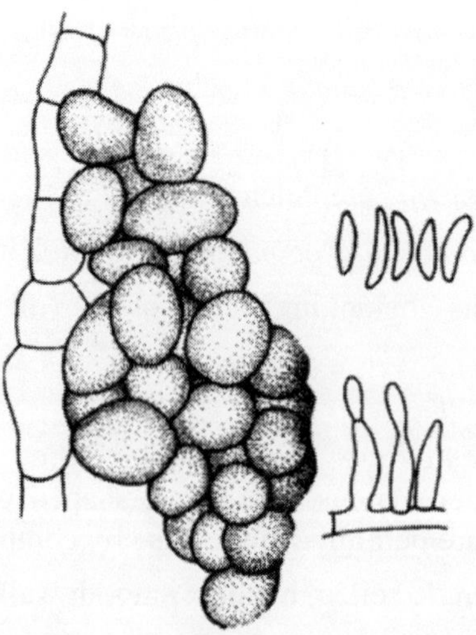

The genus *Microdochium* has been described and amended by Sutton (1960) to cover species growing on living and dead leaves of various plants. Sporodochium erumpent, sub-epidermal, conidiogenous cells discrete, little differentiated, ampulliform, producing a few apical conidioferous denticles, blastoconidia fusiform or falcate, 0-1 septate, hyaline, smooth walled.

**Colony:** Spreading, broadly moist, smooth, pink, red-brown to black releasing an orange pigment into the formation of clustered chlamydospores conidiogenous cells, lateral projections of intercalary cells discrete conidiogenous cells which arise singly or in sporodochial clusters, either directly on creeping hyphae or on globose subtending cells, ampulliform, forming apical clusters of conidia in an inconspicuously sympodial order

**Conidia:** Cresent shaped, one celled, hyaline, thin walled

**Mycotoxins:** Trichothecenes, T2 toxin

**Diagnostic features:** Erumpest sporodochium, conidiogenous cells discrete-little, differentiated, anmpuliform, produce conidia on denticles, blastoconidia fusiform or falcate, 0-1 septate, hyaline.

**ii)** ***M. dimerum*** **(Penz)**

**Synonymys** = *Fusarium aqueductum* var *dimerum* (Penz.) Raillo

= *F. dimerum Penz*

**Colony:** Slow growing, forming yeast like colonies, white to orange, appear wet.

**Conidiophore:** Monophialidic unbranched or loosely branched with phialides.

**Phialides:** Short, swollen.

**Conidium:** Falcate, lunate, mostly non-septate rarely upto 3-septate, more or less pedicillate, slightly pointed at both ends.

**Chlamydospores:** Intercalary single or in chains, globose, oval, hyaline, smooth walled

**Mycotoxins:** Trichothecenes

**Diagnostic features:** Wet colonies yeast like, white to orange, onophialidic conidiophore, phialides short and swollen, conidium falcate, non-septate pointed at both ends, chlamydospores intercalary single.

## 37. *Magnaparthe*

Colonies spreading, dark-brown, non stomatic perithecia, dark-brown, globose with long and cylindrical neck. Ascus unitunicate, thin walled, short, septate floating freely in perithecium. Ascospores long, fusiform, curved, 3-celled, slightly constricted at the septa, hyaline to yellowish brown.

**Anamorph:** *Nakatea*

**i)** ***M. grisea***

**Colony:** Spreading dark-brown producing conidiophores, conidia followed by sarcomata

**Ascomata:** Non-stomatic perithecia, dark brown, globose with long and cylindrical neck, slightly or not producing beyond the surface of the substratum

**Ascus:** Unitunicate, thin walled, short stipulate floating freely in the perithecium at maturity deliquent

**Ascospores:** Ascospores long, fusiform curved, 3-celled slightly constricted at the septa, hyaline to yellowish-brown at maturity, retiquiae wide and band like partially gelatinizing at maturity

**Anamorph:** *Pyricularia grisea*

**Mycotoxins:** Pyricularin

**Diagnostic features:** Colonies dark-brown, non-stomatic perithecia, globose, ascus unitunicate, ascospores long fusiform.

### 38. *Monographella* Petr.

*Monographella nivalis* is produces smooth walled perithecia on sterilized straw. Perithecia in culture are superficial at first white to pink later becoming grayish-black. Ascospores hyaline fusiform, 1-3 septate.

**i) *M. nivalis* (Schaffn.) E. Miill**

**Synonymys** = *Calomectria nivalis* (Schaffn)

= *Griosphaeria nivalis* (Schaffn) E. Miill

= *Micronectriella nivalis* (Schaffn) C. Booth Von Arx

= *Sphaerulina divergens* (Rehm.) Petr

= *Calonectria graminicola* (Berk & Br.) Sensu Wollenw.

**Anamorph**: *Gerlachia nivalis* (Ces. & Sacc.) W. Gams. & E. Miill

= *Fusarium nivale*

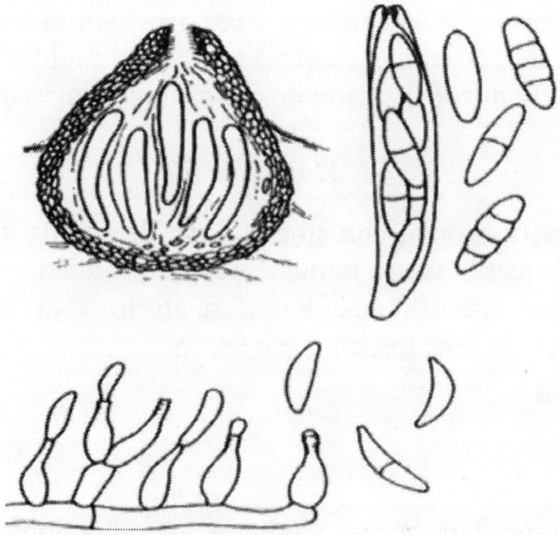

**Colony:** Slow growing, reaching more than 1cm diam, white to pale peach or apricot

**Mycelium:** Floccose or flatted, sporulating sparsely or pale -orange sporodochia

**Conidiogenous cells:** Short, doliform to obpyriform with apical annellate zone

**Conidia:** Broadly falcate with almost equal ends, base only slightly flattened 1-3 septate

**Chlamydospores:** Absent

**Ascocarp:** Homothallic produces perithecia on sterilized straw after 10-15 days at18°C, perithecia in culture superficial while at first then to pink later turning to grayish black. Ascospores hyaline, fusiform, 1-3 septate

**Mycotoxins:** Benetenotide

**Diagnostic features:** Sporodochia pale-orange, conidiogenous cells short and oliform to obpyriform showing an apical annelated zone, conidia broadly falcate, chlamydospores absent. Homothallic perithecia in culture superficial, at first white to pinkish later grayish-black, ascospores fusiform hyaline 1-3 sepatate.

## 39. *Myrothecium* Tode ex Fries

The genus *Myrothecium* includes those imperfect molds in which the fructification is a sporodochium of hyaline, compacted, irregularly branched conidiophores bearing unicellular slimy phialospores which are green to black in mass. Two other genera, *Metarrhizium* (Metsch.) Sorokin and *Dendrodochium* Bonorden, include species in which unicellular spores, green to black in mass, are borne on branched conidiophores grouped in sporodochia. *Metarrhizium,* isolates can be distinguished from those of *Myrothecium* by their dry spores which adhere in tall columns, but green-spored *Dendrodochium* spp., including *D. toxicum* Pid. and Bilai, are very similar to *Myrothecium,* and Tulloch (1972) considers that it is necessary to extend the present concept of *Myrothecium* to include them.

Three of the 13 species included in *Myrothecium* by Tulloch (1972) have been reported to produce material toxic to mammals. *M. roridum M. verrucaria* and *M. leucotrichum* (synonyms: *M. jollymannii* Preston, *M. indicum* Rama Rao). *M. leucotrichum* is distinguished from the other two toxin-producing species by the hyaline, thin-walled, several-celled setae which surround its sporodochia. *M. verrucaria* by fusiform spores, 6.5 to 8.0 µm in length, and *M. roridum* by rod-shaped or narrowly ellipsoid spores, 5.5 to 7.0 µm in length.

### Key to the species

1. Conidiophores in synnemata or sporodochium ...................................... *M. masonii* (1)

1a. Conidiophores single ............................................................................................ 2

2. Conidia not more than 4 µm long on agarics (Fig. 14a)........................ *M.inundatum* (2)

2a. Conidia more than 4 µm ........................................................................................ 3

3. Setae present conspicuous mostly non-septate (Fig. 14b) ..................... *M.gramineum* 4
3a. Setae – septate ........................................................................ *M. leucotrichum* 5
4. Conidia with longitudinal striations ....................................................... *M. cinctum* (4)
4a. Conidia without longitudinal striations ............................................................... 5
5. Conidia slightly curved, sporodochia stalked (Fig. 14c) ......... *Nectria bacteridoides* (5)
5a. Conidia straight, sporodochia usually sessile ........................................................ 6
6. Conidia cylindrical or narrowly ellipsoidal ............................................................. 7
6a. Conidia navicular, limoniform or broadly ellipsoidal ............................................... 8
7. Conidia 5.7 x 1.5-2.5μm, usually on Coffea (Fig. 14d) ........................... *M. advena* (6)
7a. Conidia mostly 6-8 x 1.5-2.5 μm (Fig. 14e) .......................................... *M. roridum* (7)
7b. Conidia mostly 10-11 x 1.3 μm...................................................... *M. carmichaelii* (8)
8. Conidia 6-10 x 1.5-2 μm (Fig.14f) ..................................................... *M. verrucaria* (9)
8a. Conidia 12-17 x 7-9 μm ................................................................. *Nectria ralfsii* (10)

**i) *M. leucotrichum* (Pock) Tulloch**

**Synonymys** = *M. tollymannia*

=*M. indicum*

**Colony:** White with greenish colour mars at maturity flyphae hyaline, with smooth

**Reverse:** Pale -rosy –buff.

**Spore mars:** Wet, shiny black-olivaceous when young, margin white floccose

**Sporulation:** In concentric zones, diffuse or coalesced in to marginate olivaceous,-black to black sporodochia.

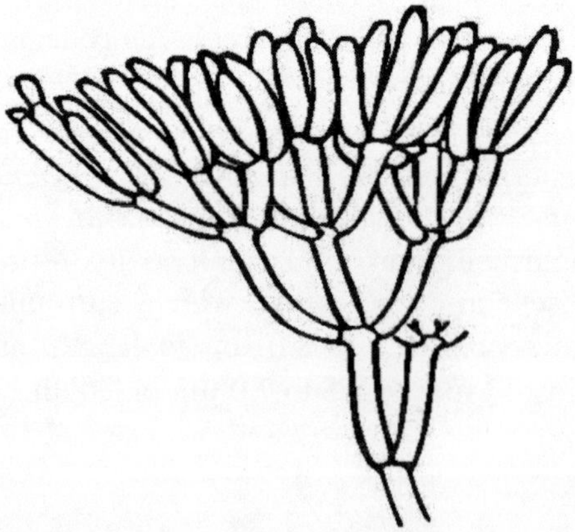

**Conidiophores:** Repeated branched closely interwoven, with mate branches bearing phialides hyaline or darkened at the nodes, septate

**Phialides:** Whorls of 3-6, closely produced in a dense paralle row, cylindrical or slightly clavate, hyaline occasionally with small collateral, percurrent.

**Conidia:** Cylindrical rounded at the apex often trinucleate at the base, pale – olovaceous –brown.

**Mycotoxins :** Roridins

**Diagnostic features:** Conidiophores single repeated branched, septate whorls conidia cylindrical rounded at the apex, non septate.

### ii) *M. roridum* Tode Fung

**Synonyms:** *Myrothecium advena*

**Colony:** White of with greenish colour mass at maturity.

**Hyphae:** Hyaline, thin, smooth, rarely branched spetate.

**Reverse:** Pale-rosy buff often streaked with pure yellow.

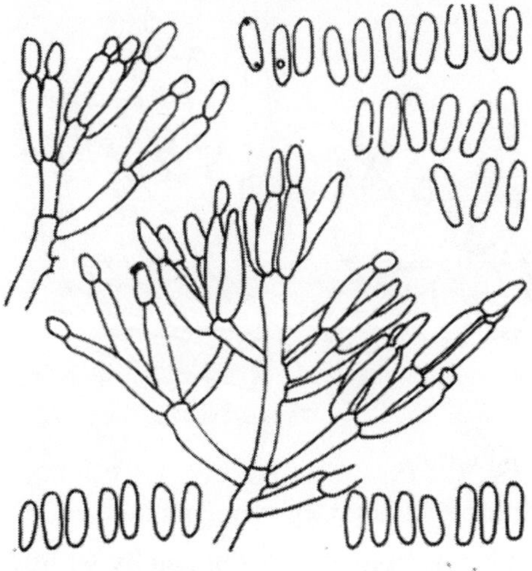

**Sporodochia:** Rarely synmematous

**Spore mass:** Flattened or convex, wet, shiny black-olivaceous when young margins pure white floccose.

**Sporulation:** In concentric zones, diffuse or coalesced into marginate olivaceous-block to black sporodochia.

**Stroma:** Loosely packed, cells elongate or isodiametric, hyaline forming a thin layer.

**Conidiophores:** Repeated branched closely interwoven, ultimate branches bearing phialides, hyaline or darkened at the nodes, spetate.

**Dark-sterile hyphae:** Occur occasionally between the conidiophores and phialides.

**Phialides:** Whorls of 3-6, closely packed in a dense parallel row, cylindrical or rarely slightly clavate, hyaline, sometimes darkened at the base, pore sometimes darkened and occasionally with small collarette percurrent.

**Conidia:** Rod shaped or narrowly ellipsoid, rarely slightly avoid, mostly with both ends rounded, occasionally one truncate, hyaline to dilute olivaceous, black in mass, guttules upto three.

**Mycotoxins:** Roridin A, B & C, verrucarin A & Z, Dehydroverrucarn.

**Diagnostic features:** Conidia rod shaped or narrowly ellipsoid, mostly both ends rounded.

**iii) *M. verrucaria* Alo. & schazein**

**Synonymys** = *Metarrhizium glutinosum* S A. pope
= *Gliocaladium fimbriatum* cikl man & Abbot
= *Peziza verruearia* Ab. Schwein.

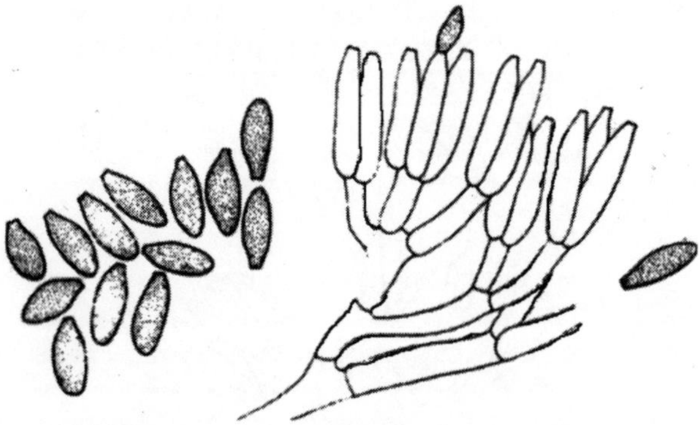

**Colony:** Spreading white.

**Conidiomata:** Sporodochia rarely synnemata

**Spore mass:** Wet black, convex, surrounded by white floccose margin.

**Reverse of colony:** Rosy buff

**Sporulation:** Diffuse or coalesced into pale olivaceam to black sporodochia with a white margin, occasionally stipitate.

**Stroma:** Rarely well developed sodiametric or elongated hyaline cells.

**Marginal hyphae:** Airling, some branched with a blunt tip, usually verrucose, sometimes smooth hyaline septate.

**Conidiophores :** Ansing from Knot of basal hyphae in culture or from thin stromatic layer, branched repeatedly usually forming 2-4 branches at each level, the ultimate branches bearing phialides, hyalive, septate.

**Phialides :** Three to six in a whorl, closely packed in a dense parallel layer, cylindrical sometimes slightly tapered towards apex, hyaline, dancered around the pore.

**Conidia :** Broadly fusiform, are end pointed, other protruding and truncate, clear in erythrosine and $NH_4$ solution with fan-tailed appendoxe on the pointed end.

**Mycotoxins :** Verrucavins A, B, J & K, Roridin, A4 Trichoverrol - A B, Trichoverrin – A, B.

**Diagnostic features:** Conidia broadly fusiform one end pointed, other protruding and truncate Fan-tailed appendoxe on the pointed end.

## 40. *Nigrospora* Zimmermann

**Synonymys** = *Basisporium* Molliard

= *Dichotomella* sacc.

**Colony:** First white with small shining black conidia easily visible under low power directing microscope, later brown or black when sporulation is abundant.

**Mycelium:** All immersed or partly superficial

**Stroma:** None

**Setae and hyphopodia:** Absent.

**Conidiophore:** Micronematous or semi-macronematous branched, flexuous, colourless to brown, smooth.

**Conidia:** Solitary with a violent discharge mechanism, acrogenous, simple, spherical or broadly ellipsoidal, compressed dorsiventrally, black, shining, smooth, 0-septate

**Key to species:**

1. Conidia 10-16 (Monthly 12-14) μm diam. *Nigrospora* state of *Khuskia oryzae*.................... 2

1a. Conidia 14-20 (monthly 16-18) μm diameter .................... *N. sphaerica*

2. Conidia 17-24 (Monthly 20-22) μm .................... *N. sacchari*

2a. Conidia 25-30 .................... *N. panici*

### i) *Nigrospora* state of *Khuskia oryzae* Huds

**Colony:** Spreading, grey cottony, reverse olivaceous -black

**Mycelium:** All immersed or partly superficial

**Stroma:** None

**Setae and hyphopodia:** Absent

**Conidiophores:** Micronematous or semi-macronematous, branched, flexous, colourless or brown, smooth

**Conidiogenous cells:** Hyaline, 8-12 μm diameter

**Conidia:** Solitary, acrogenous, simple, sphaerical or broadly ellipsoidal, compressed dorsiventrally, black shinning, smooth, non-septate

**Teleomorph:** *Khuskia* (Family Amphisphaeriaceae) is a manotypic genus (*K. oryzae*) is known only on the natural substrate of *Vetiveria zizinoides* and *Panicum maximum*.

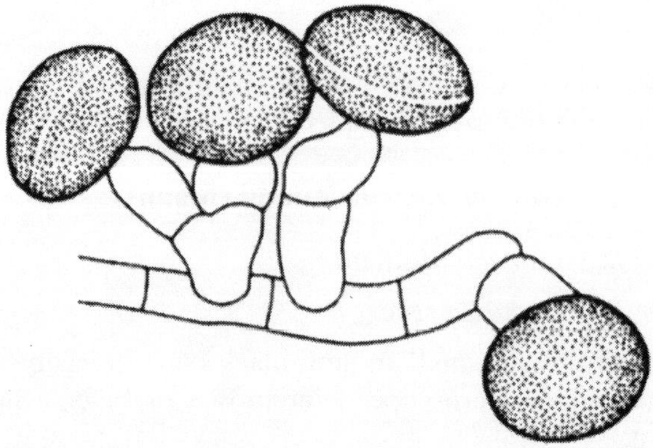

**Perithecia:** Perithecia are formed in elongate clusters of 1-7 below epidermis becoming erumpet at later stage with papillate ostiole.

**Ascus:** Ascus clavate, short-stalked uni tunicate without any apical differentiation.

**Ascospores**: Ascospores hyaline somewhat allantoid with distal end more rounded than the basal usually becoming unequally 2-celled on germination.

**Mycotoxins:** 6 allyl-5, 6 dihydro-5hydroxypyran-2-one 4,5 dihydroxyoct-6 exoic acid, 1,4 lactone

**Diagnostic features:** Black, smooth, non-septate with voilent discharge mechanism.

**ii)** ***N. sphaerica*** **(Sacc.) Mason**

**Colony**: Spreading, grey, cottony

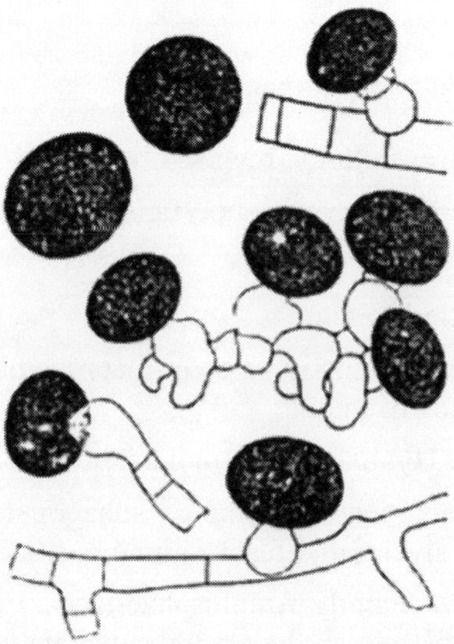

**Mycelium**: All immersed orpartly superficial

**Stroma:** None

**Setae and hypopodia**: Absent

**Conidiophores:** Macronematous or semimacronematous, branched, flexious, colourless or brown, smooth

**Conidiogenous cells**: Hyaline, 8-12 μm diameter

**Conidia:** Solitary, acrogenous, simple, spherical, black shinning, smooth, aseptate

**Teleomorph**: None

**Mycotoxins:** 6 allyl-5, 6 dihydro-5hydroxypyran-2-one 4,5 dihydroxyoct-6 exoic acid, 1,4 lactone

**Diagnosticfeatures:** Conidia spherical, 10-14μm in diameter *N. sacchari* (speg) Mason mostly on Saccharum but also other substratum

***Khuskia oryzae*** **Hudson**

**Ascomata**: Perithecia formed in clusters of 3-7 in uniseriate rows or irregularly arranged crumpent, globose or ovoid, papillate ostiolate

**Ascus**: Shortly stalked, unitunicate clavate, 8-spored

**Paraphyses**: Interspersed with asci, filamentous septate

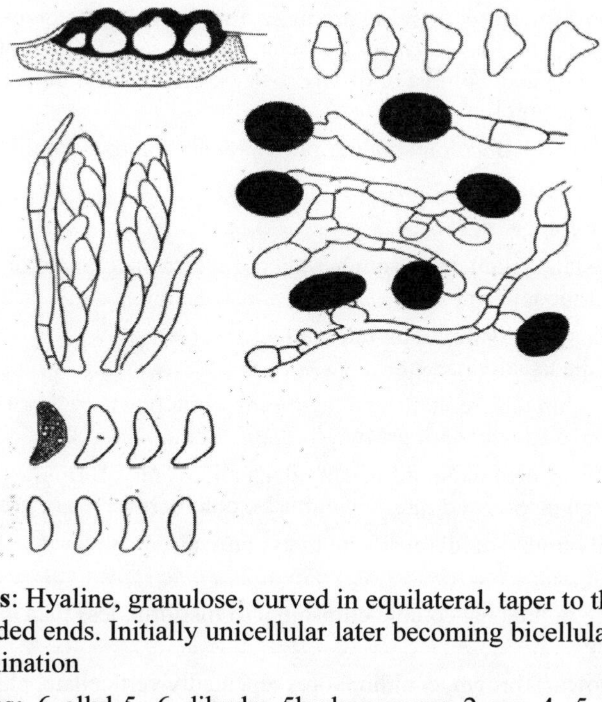

**Ascospores**: Hyaline, granulose, curved in equilateral, taper to the base and has rounded ends. Initially unicellular later becoming bicellular at the time of germination

**Mycotoxins:** 6 allyl-5, 6 dihydro-5hydroxypyran-2-one 4, 5 dihydroxyoct-6 exoic acid, 1, 4 lactone

**Diagnostic features**: Perithecia crumpent with papillate ostiolate ascus shortly stalked, unitunicate clavate, 8-spored, paraphyses septate, ascospores hyaline, single celled, bicelled during its germination

## 41. *Paecilomyces* Bainer

**Colony:** Growing fast

**Hyphae:** Funicoluse, olivaceous to brown.

**Conidiophore:** Mononematous, macronematous, simple or branched, septate, hyaline to pigmented, arise directly on aerial hyphae.

**Conidiogenous cells:** Phialides, borne singly in pairs or in verticals or on the penicellate heads, sometimes irregular arranged.

**Phialides:** Awl shaped, divergent, base swollen with slender long neck.

**Conidia:** Catenate tending to slip or slime down, hyaline or darkly pigmented smooth or roughened, sometimes spherical ovoid to fusiform, sometimes spherical

**Chlamydospores:** Aleuro spores or chlamydospores are produced sometimes Aleuro spores

**Teleomorph**: *Byssochlamys*

### Key to species of *Paecilomyces*

1. Colonies yellow-brown; odour sweet aromatic; usually thermotolerant or thermophilic (Sect. *Paeciolomyces*) Ascomycota absent: Conidiophores repeatedly branched; conidia of different sizes; more or less ellipsoidal; chlamydospores present

1a. Colonies of other bright colours; odour not sweet aromatic; mesophillic .................... 2

2. Colonies vinaceous to violet .................................................................... 4

2a. Colonies other colours .................................................................... 3

3. Conidiophores stalks pigmented, rough-walled; reverse uncoloured or purple; aleurioconidia absent .................................................... *P.lilacians*

3a. Conidiophores stalks hyaline, smooth-walled; reverse yellow; aleurioconidia usually present .................................................... *P. marquandii*

4. Conidia finely echinulate; colonies white to pink, thin, powdery, synnemata absent: reverse usually dark green .................................................... *P. carneus*

4a. Conidia smooth-walled; colonies usually deep and woolly, forming synnemata on insects and rarely in cultures; colony reverse not green .................. 5

5. Colonies usually pink; conidia fusiform to cylindrical. 3-4x1-2µm .................................................... *P. fumosoroseus*

5a. Colonies white to luteous; conidia ellipsoidal to fusiform, less than 3.5 µm long .................................................... *P. farinosus*

5b. Colonies yellowish–brown, conidiophores repeatedly verticellate, phialides distributed irregularly along the fertile hyphae, conidia strongly ellipsoidal, smooth walled, size variable.................................................... *P.varioti*

**i) *P. lilacinus* (Thom)**

**Synonymys** = *Penicillium lilacinus* Thom

**Teleomorph:** Absent

**Colony**: Vinaaceous brown, spreading reverse uncoloured

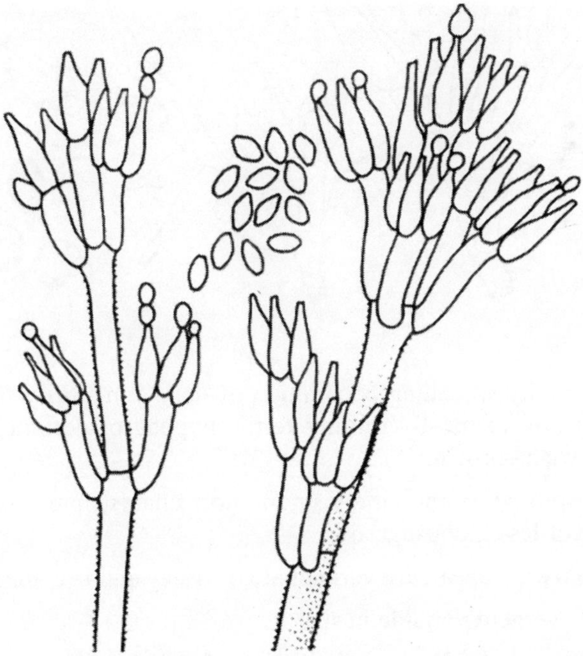

**Conidiophores**: Erect mostly arising solitarily from the horizontal mycelium, rarely synnematous yellow to purple rough walled with densly clustered metulae and conidiophores.

**Conidium**: Ellipsoidal to fusiform, smooth walled, slightly rough walled, aleuroconidia absent.

**Mycotoxins:** Leucinosttins

**Diagnostic features:** Colonies vinaceous brown to voilet, conidiophores stalks pigmented, rough walled reverse uncoloured or purple aleuroconidia absent, conidia ellipsoidal to fusiform.

**ii) *P. vareoti* Bainier**

**Synonymys** = *Monilia Formosa* Sakug
= *Penicillium divericatum* Thom
= *P. varoti (Bainier)* Sacc
= *P. viniferum* Sakug
= *Spircara diverticata* (Thom). J.C. Glman

**Teleomorph:** Absent

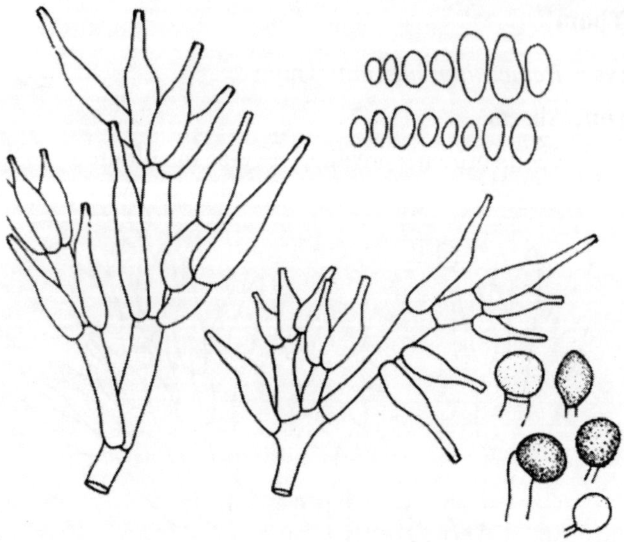

**Colony:** Broadly spreading becoming powdery at maturity with aerial growth consisting of mostly trailing fertile hyphae of definite ropes of hyphae yellowish –brown.

**Chlamydospores:** Borne singly or in short chains, inter colony or terminal, more or less globose, brown

**Conidiophores:** Repeatedly verticellate or freely and irregularly branched.

**Metulae:** Divergent variable in size

**Phialide:** Irregularly distributed along the fertile hyphae along with acuminate conidium bearing tubes usually bent away from the cries of the cell, and widely divergent at the apex bearing long tangled chains of conidia.

**Conidia:** Strongly elliptical, yellowish to brown, smooth walled uneual in size with in the same colony

**Mycotoxins:** Ustic acid, dehydroustic acid, 4, 6 dihydroxy-3-methoxy phthalic acid, variotin

**Diagnostic features:** Colonies yellowish–brown, chlamydospores repeatedly verticellate phialides distributed irregularly along fertile hyphae, conidia strongly ellipsoid, smooth walled.

### 42. *Papulaspora* Peuss

**Colony:** Spreading, white

**Vegetative structures:** Asexual reproductive structures are absent but produce compact clustures of small cells or bulbish which are sclerotium like and serve to multiply the fungus.

**Mycotoxins:** A species of *Papulospora* is reported to produce Papulosporin, tascal, diketopiperazines

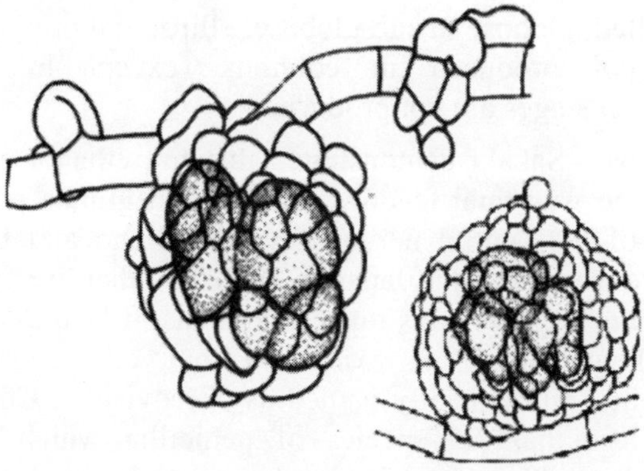

**Diagnostic features:** Asexual reproductive bodies are absent. But produce clusters of cells or bulbish which are sclerotium like and serve to multiply the fungus.

### 43. *Penicillium* Link ex Fr.

Though genus *Penicillium* was described by Link (1809) *P. expansum* as type but it was validated only by Gray in 1821. But Thom and Raper (1949) described *P. crustosum* as type material as per article 13 of code of Botanical Nomenclature. Later Hawksworth and Pitt (1976) redesignated *P. expanum* as neo type. The generic name *Penicillium* is based on imperfect fungus and consequently it represents a genus of hyphomycetes. Raper and Thom (1949) included other fungi with sexual state in *Penicillium* and modified the generic concept.

The term conidiophore as used by Raper and Thom (1949) is replaced by stipe. The branched part of the conidiophore above the stipe is referred to as penicillus. The term metulae is maintained for the penultimate branches which bear the whorl of phialides. All cells between metulae and stipe are referred to as branches. The type of branching is designated as simple, not branched, monoverticillate, one stage branched (biverticellate asymmetrical as in *P. citrinum* or species) or (biverticellate asymmetrical as in *P. italicum* or *P. expansum*) two stage branched or three to more stage branched (Terverticillate as in *P. griseofulvum*). The conidiophores can be mononematous (single as in *P. chrysogenum*) or distinctly synnematous (coremial) (as in *P. claviforme*).

The conidiophores may arise from the substrate (velvety), from simple aerial hyphae (= lanose), from frostose fascicles of aerial hyphae (funiculose) or from erect fascicles = synnematous (fasciculate). In several species the conidiophores are both mononematous and loosely synnematous as in *P. expansum*. In *P. claviforme* and *P. concentricum* the synnemata are differentiated into a distinct stalk and an enlarged head composed of the sporulating structures. Species mostly grow well on czapek's agar and sporulate with in 5 days. Odour mostly strong and pungent, aromatic in some species. Conidiophores mononematous or synnematous often with one of these features predominating depending on the species. Usually 2-3 stage branched, hyaline, smooth or rough walled. Phialides usually consisting of a cylindrical basal part and a short neck in some species different with a shorter or longer neck. Conidia smooth or rough walled, globose to sub-globose, ellipsoidal or cylindrical, hyaline to greenish, usually not produced in columns (except in *P.claviforme* and *P.concetricum*). Perfect stages are not produced.

In 1891, in Japan, Sakaki demonstrated that hot ethanol extract from moldy unpolished yellow rice was fatal to dogs, rabbits and guinea pigs with symptoms indirectly paralysis of the central nervous system (Ueno and Ueno, 1972). As a result, the yellow rice was banned in Japan in 1910. Another incident that moldy corn infested by *P. puberulum* extract was toxic to animals in Nebraska in 1913. In 1940 Miyake *et al*. (1940) have isolated *P. toxicarium* from yellowed rice which produced highly toxic metabolite, subsequently named as citreoviridin. Cole and Cox (1981) have recorded no less than 85 species of penicillia which are known to be mycotoxigenic and about 42 mycotoxins are known to be produced by one or other species of Penicillia. The literature on this aspect has accumulated mainly on the chemistry and toxicology and most of the literature lacks specificity with misidentification of species of *Penicillium*. For instance, citrinin was reported to be produced by at least 22 species which may be attributed to misidentification. Misidentifications, even by experts have also been common. Of 12 *Penicillium* species reported in the literature to produce tremorogenic toxin, only two were found to be accurate (Pitt, 1979). Pitt (1979) have developed a classification of species of *Penicillium* based on broader range of standardized characters that used previously. Frisvad (1985) tried to characterize species of *Penicillium* on the basis of secondary metabolites they produce. Bridge *et al*. (1986) tried to provide an integrated classification of *Penicillium* based on morphological, ultrastructural, biochemical and enzymological characteristics. It is hoped that a multidisciplinary approach to the classification of filamentous fungi will certainly improve the understanding of different groups of fungi. Classification based on secondary metabolites (Frisvad, 1986) and electrophoretic finger printing of certain isoezymes (Cruickshank and Pitt, 1987) was also proposed. Pitt (1988) and Samson and Pitt (1990) have developed more convenient guide to the identification of toxigenic penicillia. Despite of the enormous amount of interest in this genus, its classification remains very confused largely due to the variability of individual strains in culture because of their limited potential for gene exchange and because of the inherent instability of character expressions in some strains.

*Penicillium* is a large genus with 150 species recognized which probably is the most conservative. The *Penicillium* is divided into sub-genera based on number and arrangement of phialides (elements producing conidia) and metulae and rami (elements supporting phialides) on the main stalk cells (stipes). Pitt (1989) classified *Penicillum* into four subgenera *Aspergilloides* in which phialides are borne directly on the stipe without intervening supporting elements, while furcatum and biverticillatum in which phialides are supported by metulae. On the other hand, in *Penicillium* both metulae and rami are usually present. Majority of important mycotoxigenic and food spoilage species are found in the subgenus *Penicillium*.

General procedures for enumeration of food-borne fungi are effective for enumerating all common *Penicillium* species. However, some *Penicillium* species grow weakly on media like PDA. Dichloran-Rose Bengal – chlorophenicol (DRBC) agar or Dichloran + 18% glycerol (DG 18) agar is generally recommended (Pit and Hocking, 1985).

Pitt (1979) gave comprehensive account of taxonomy of genus *Penicillium.* Good computer assisted key for identification of species of *Penicillium* is provided by Pitt (1988), while Samson *et al.* (1995) and Pitt (1996) classified food-borne species of *Penicillium* based on secondary metabolites. Correct identification of species of *Penicillium* can be made by observations under standard conditions such as media, incubation period and temperature, in addition to microscopic morphology, gross physiological features including colony diameter, nature of conidiophore metulae and sterigmata, colour of conidia and colony pigments.

About 14 species of *Penicillium* are reported to be principal mycotoxigenic. Majority of them produce single toxin, while some species such as *P. expansum* (citrinin and patulin), *P. crustosum* (CPA, penetrim A and roquesfortin), *P. roqueforti* (Patulin, PR toxin, Roquefortine) and *P. aurantiogriseum* (penicillic acid, vermcosidin, viomellein and xanthomegnin) produce more than one mycotoxin. Similarly toxins like penetrim A and patulin are elaborated by four species, while CPA by seven species. Toxins like citrinin and patulin are simple molecules, while others like penetrims are very complex. Thus, *Penicillium* toxins are of great complex and diverse. Only ochratoxin A is carcinogenic, while citreovirdin, citrinin, patulin and penetrim A exhibit wide range of toxicity, from acute to chronic. The toxins produced by *Penicillium* can be broadly divided into 2 groups. They affect liver and kidney and are asymptomatic or cause generalized debility in humans or animals. While others are neurotoxins which are characterized by sustained trembling. With the low levels of *Penicillium* in nature animal health is affected resulting reduced fertility, increased susceptability to infectious diseases, reduced feed conversion and ill thrift. Many species produce several compounds which are known to be toxic and the possibility of synergy also cannot be ignored. The potential involvement of *Penicillium* toxins in low level diseases in humans undoubtedly warrants continued research. Species of *Penicillium* reported to produce one or other mycotoxins are précised in Table 3.16.

**Table 3.16.** Mycotoxins produced by species of *Penicillium*

| | |
|---|---|
| *Penicillium aethiopicum* | Viridicatum toxin, tryptoquialanins |
| *P. albocoremium* | Roquefortine C |
| *P. allii* | Roquefortine C |
| *P. atramentosum* | Roquefortine C, rugulovasine A, B |
| *P. aurantiogriseum* | Penicillic acid, verrucosidin |
| *P. bialowiezense* | Unknown (Frisvad, unpublished). |
| *P. brevicompactum* | Mycophenolic acid, botryodiplodin |
| *P. camemberti* | Cyclopiazonic acid |
| *P. carneum* | Patulin, penetrim A, mycophenolic acid, penicillic acid, isofumigaclavine A |
| *P. caseifulvum* | Rugulovasine A |
| *P. cavernicola* | Territrems, tremorogenic mycotoxins |

| | |
|---|---|
| *P. chrysogenum* | PR-toxin, Roquefortine C, Secalonic acids |
| *P.citrinum* | Citrinin, quinolactcin, anthraquines |
| *P. clavigerum* | Penitrem A, Patulin, Cyclopiazonic acid, viomellein |
| *P. commune* | Cyclopiazonic acid |
| *P. concentricum* | As in other coprophilic species, many antibiotically active extrolites are found, including patulin, patulidin, barceloneic acid, meleagrin |
| *P. confertum* | Asteltoxin |
| *P. coprobium* | Patulin, roquefortine C |
| *P. coprophilum* | Roquefortine C, alternariol |
| *P. crustosum* | Penitrem A, roquefortine C |
| *P. cyclopium* | xanthomegnin, viomellein, vioxanthin |
| *P. digitatum* | Some metabolites were shown to be toxic in the chick embryo test |
| *P. dipodomyicola* | α-cyclopiazonic acid |
| *P. dipodomyis* | Deoxybrevinamide, Penicillins |
| *P. discolor* | Chaetoglobosin A, B,C |
| *P. echinulatum* | Territrems |
| *P. expansum* | Patulin, chaetoglobosin C, roquefortine C, citrinin, communesin A, B |
| *P. flavigenum* | Penitrem A, roquefortine C |
| *P. formosanum* | Patulin |
| *P. freii* | Viomellein |
| *P. gladioli* | Patulin |
| *P. glandicola* | Patulin, penetrim A, roquefortine C |
| *P. griseofulvum* | Patulin, cyclopiazonic acid, roquefortine C |
| *P. hirsutum* | It is not known whether roquefortine C will be produced in vegetable |
| *P. hordei* | Roquefortine C |
| *P. italicum* | Italinic acid, Brevinamide |
| *P.islandicum* | Citrinin, luteoskyrin, islandi toxin |
| *P. marinum* | Patulin, communesins |
| *P. melanoconidium* | Penicillic acid, verrucosidin, xanthomegnin, viomellein, vioxanthin |
| *P. mononematosum* | Fumitremorgin A, B, verrucologen, isochromantoxin |
| *P. nalgiovense* | The toxicity of nalgiovensin, nalgiolaxin, dipodazin, diaporthins are unknown |
| *P. neoechinulatum* | Penicillic acid |
| *P. nordicum* | Ochratoxin A, verruculone |
| *P. olsonii* | Verrucolone, brevienes, asperhenamata |

| | |
|---|---|
| *P. palitans* | Cyclopiazonic acid, fumigaclavine |
| *P. paneum* | Botryodiploidin, patulin, roquefortine C |
| *P. persicinum* | Roquefortine C |
| *P. polonicum* | penicillic acid, verrucosidin, nephrotoxic glycopeptides |
| *P. radicicola* | It is not known whether this species can produce citrinin, penicillic acid, terrestic acid or roquefortine C |
| *P. roqueforti* | PR-toxin, roquefortine C, isofumigaclavine A, B. Mycophenolic acid |
| *P. sclerotigenum* | Patulin, roquefortine C |
| *P. solitum* | Unknown |
| *P. thymicola* | Unknown |
| *P. tricolor* | Xanthomegnin, viomellein, vioxanthin, asteltoxin |
| *P. tulipae* | penitrem A, roquefortine C and penicillic acid |
| *P. ulaiense* | Unknown |
| *P. venetum* | Roquefortine C |
| *P. verrucosum* | Verrucolone, Ochratoxin A, Citrinin, Verrucin A & B |
| *P. viridicatum* | Penicillic acid, Xanthomegnin,viomellein, vioxanthin, Brevianamide A, Viridic acid, Viridamine |
| *P. vulpinum* | Patulin, roquefortine C |

In assessing the claim of toxigenicity of *Penicillium,* several points need to be kept in mind.

1. Many of the identifications reported in the literature are incorrect and thus has resulted in confusion.
2. A number of genuinely toxic compounds are produced by species not usually occurring in foods or feeds. Such toxins are of little practical importance.
3. Some compounds often cited as mycotoxins in the literature are of low toxicity.
4. Growth of moulds does not always mean production of toxin

**Key to mycotoxigenic species of *Penicillium***

**I)** Producing macroscopic colonies in 7 days, stipes bearing predominately terminal, terverticillate penicilli, metulae borne on well defined rami, phialides borne on metulae, phialides ampulliform, cylindrical to acerose.......................................................................... **Subgenus *Penicillium***

**i)** Compact multiramulate penicilli with long stipes and velutinous colonies ..................................................................**Section *Coronata***

Colony reverse colour cream to yellow, conidiophore multiramulate penicilli, conidia ellipsoid, finely roughened.

**a)** Regularly produces terminal penicilli which are multi-ramulate.......................................................... **Series:** ***Olsonii*** 1 (2)

**1)** Short-broad-penicellate conidiophores, apically inflatended metulae, finely roughened ellipsoidal conidia. Ehrlich reaction yellow ............. ***P. brevicompactum*** (1)

**2)** Colony reverse colour cream to yellow, conidiophore multi ramulate penicilli, conidia ellipsoid, finely roughened, Ehrlich reaction yellow. ....................... ***P. olsonii*** (2)

**ii)** Low velutinous colonies growing fast on almost all substrates, rough walled stipes, large smooth walled globose conidia. High tolerance to propionic acid, acetic acid, lactic acid and other acids ............................................................ **Section *Roqueforti* 1 (2)**

**a)** Grow poorly at low –water activity ($a_w$) but grow well at low temperature and caeses to grow at 37°C ........... **Series *Roqueforti* (2)**

**1)** Good growth on media with 0.5% acetic acid or 1% propionic acid, black-green reverse on CYA and YES media, conidia globose smooth walled. Ehrlich reaction voilet or yellow ...................................***P. roqueforti*** (3)

**iii)** Phialides short broad, phialides with broad collula and divaricate penicillus structures ......................... **Section and Series Chrysogena (3)**

**1)** Fast growth on YES medium with yellow reverse and strong sporulation. Relatively short phialides with short broad collula, globose to subglobose to broadly ellipsoidal smooth walled conidia. Ehrlich reaction negative .......................................................***P. chrysogenum*** (4)

**2**) Slow growth on YES medium with orange reverse on YES, smooth walled, dark-green conidia. Ehrlich reaction negative...............................................***P.nalgiovense*** (5)

**a)** Conidiophore terverticillate, colonies fasciculate, yellow to curry-yellow reverse, produce ellipsoidal conidia. Grow fast (29-35mm) in 7days on CYA at 25°C ......................**Series *Aethiopica***

**1)** Colonies sulcate with golden yellow reverse at 37°C, conidia smooth walled ellipsoid. Ehrlich reaction negative ........................................................***P. aethiopicum*** (6)

**iv)** Strongly fasciculate to coremiform colonies, conidiophores with smooth stipes, and terverticillate to quarterverticillate conidiophore ....................................................... **Section *Penicillium* (5)**

**a)** Grow rapidly and produce stipes with smooth wall ...................................................................**Series Expansa** 5 (1)

**1**. Terverticillate conidophores, phialides cylindrical, tapering to a distinct collulum, metulae cylindrical, conidia smooth walled ellipsoidal. Ehrlich reaction strong voilet..................................................... ***P. expansum*** (7)

**b)** Colonies with surface texture velutinous or fasciculate, conidia enmasse blue-green, grey or rarely olive. Colonies on CYA at 25°C net exceeding 30mm diameter ............ **Series *Utricicolae* 2 (1)**

**1)** Highly branched conidiophores, very short phialides, grey conidia with only small element of green, smooth walled and ellipsoid. Ehrlich reaction voilet ............................................................***P.griseofulvum*** (8)

**c)** Terverticillate penicilli, conidiophore two stage branched, grow rapidly in CYA and MEA in 7days at 25°C. Conidia cylindrical to pyriform ............................................ **Series *Italica*** 3 (1)

**1)** Poor growth and no acid production on CREA medium, crustose on YES medium cylindrical to ellipsoidal smooth walled conidia ........................................... ***P.italicum*** (9)

**v)** Conidiophore and conidial structures are irregular and exceptionally large, biverticillate, divaricate, conidia olive-green, large ellipsoidal ................................. **Section and Series *Digitata***

**1)** No growth on Czapek agar and CYAS, large cylindrical to ellipsoidal smooth walled olive-green conidia. Ehrlich reaction negative.................................. ***P.digitatum*** (10)

**vi)** Fasciculate colony texture and grow fast, rough walled conidiophore stipes, dark-green rough walled conidia .................................................. **Section and Series *Viridicata* (7)**

**1)** Dark–brown halo on Raulin-Thom agar, velulinous weakly fasciculate, smooth walled conidia. Ehrlich reaction pink.......................................... ***P.aurantiogriseum*** (11)

**2)** Colonies on CYA green, blue-green on MEA, conidia blue-green, smooth walled. Ehrlich reaction pink to red or yellow-brown ......................................... ***P.cyclopium*** **(12)**

**3)** Stipes rough walled, yellow-brown reverse on CYA and YES, smooth walled conidia. Ehrlich reaction negative ............................................................... ***P.tricolor*** (13)

**vii)** Deep fasciculate to coremiform colonies conidiophore with strongly roughened walls globose, smooth walled conidia ............................................................**Series *Corymbifera*** (2) 10

**1)** Weakly fasciculate, dark–brown reverse, garlic foot, globose smooth walled conidia. Ehrlich reaction pink to red ..............................................................................***P.allii*** (14)

**a)** Slow growing on nitrite or creatine morphological similar to viridicatum .....................................................**Series *Verrucosa*** 3 (11)

**1)** Slow growth, no growth at 30°C, Red-brown reverse on YES, smooth walled globose to sub-globose conidia. Ehrlich reaction negative ..............................***P.verrucosum*** (15)

**2)** Good growth on UNO and nitrate, poor growth and no acid globose to subglobose. Ehrlich reaction yellow-green ............................................................ ***P. nordicum*** (16)

**a)** Colonies floccose, conidia en massive persistently white or pale-grey green **Series Camemberti** 4 (12)

**1)** Floccose colonies, good growth on CREA, weak and restricted growth on Czapek (1000 ppm propionic acid) smooth walled conidia. Ehrlich reaction weak voilet .......................................................... ***P.caseifulvum*** (17)

**2)** Colonies velulinous to floccose, good growth on CREA, Reverse colour on YES cream coloured to yellow, sub-globose to ellipsoidal smooth walled conidia. Ehrlich reaction strong voilet ............................ ***P.commune*** (18)

**3)** Colonies crustose on CYA and MEA after 7-10days, high growth rate, good growth on CREA and YES. Strong sporulation and high growth rate. Ehrlich reaction negative ........................................................ ***P.crustosum*** (19)

**4)** Conidia smooth walled, globose, white floccose colonies, poor sporulation. Reverse colour yellow, smooth walled, globose to sub-globose conidia. Ehrlich reaction pink .................................... ***P.camemberti*** (20)

**a)** Rough or thick walled green conidia. Conidiophore stipe roughwalled ...........................................................**Series-Solita** 5 (1)

**1)** Brilliant red diffusible colour on YES agar, velutinous to fasciculate colonies, conidia rough walled, globose to sub-globose, Ehrlich reaction weak voilet........... ***P.discolor*** (21)

**II)** Penicilli characteristically biverticillate, verticels always terminal, metulae numerous in symmetrical appressed or divergent verticels, phialides typically acerose with collula conical tapering at uniform rate to narrow apical pores, conidia ellipsoidal to fusiform................................................................. **Subgenus *Biverticellium* (2)**

**i)** Surface of colonies on CYA and MEA at 25°C velutinous, floccose or funiculose ............ **Section *Simplicinum***

**a)** Colonies on MEA exceeding 25mm diameter in 7days ..........................................................................**Series *Islandica***

**1)** Colonies at 37°C exceeding 1o mm diameter in 7days, stipes non-verticicelate, exceeding 30mm in length, conidial chains in columns or irregular. Colony pigmentation on CYA at 25°C red ................. ***P.islandicum*** (22)

**III)** Penicilli predominately biverticillate consisting verticils of 2-5, metulae terminally with or without subterminal or intercalary metulae, phialides ampulliform ......................................................**Subgenus *Furcatum*** (3)

**i)** Penicillia predominately produced as terminal verticils of metulae ........................................................ **Section *Furcatum***

**a)** Colonies on CYA not exceeding 4 mm diameter, exceeding 35 mm diameter in 7 days at 25°C. Conidia of various shapes ..............................................................................**Series *Citrina***

**1)** Colonies on MEA not exceeding 25mm in diameter in 7 days. Mycelium white or grey orange (apricot), conidiophore stipe walls smooth, if sclerotia produced then not black conidia not exceeding 3-5 mm diameter ........................................................................ ***P. citrinum*** (23)

## Key to species of *Penicillium* based on conidiophores branching and conidial characters ( Raper andThom ,1949)

I. Monoverticellate (Aspergilloides)

Lacking branches, bearing a terminal whorl of phialides at the apex of the stipe e.g. *P. frequentans*

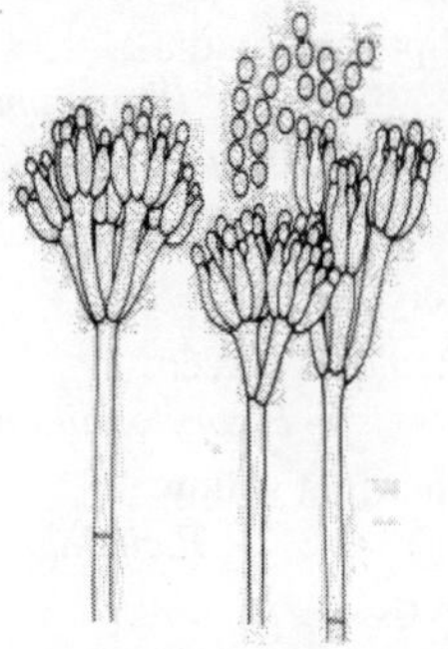

Ascospores species

- Cleistothecium present — *P.jaranicum* series -(*P.brefeldianum*)
- Sclerotia present Pink — *P.thomii*,
  Yellow to orange — *P.lapidosum*

Sclerotia or Cleistothecium present

Vesiculate stipes – Aspergilloides
Non-vesiculate stipes – Exilicaulis

Cleistothecia or sclerotia absent (lacking)

i) Quick growth
- i. Long conidiophores and globose conidia
  - Velvety: Smooth conidia – *P.frequantans*
  - Rough conidia – *P.purpurescen*
  - Floccose, rough conidia – *P.spinulosa*
- ii. Short conidiophores, elliptical smooth conidia – *P.spinulosum*

ii) Restricted: growth

Vesiculate

- i. Long conidiophores elliptical rough conidia – Strictly monoverticellate, – *P.lividum*
- ii. Moderate conidiophores – elliptical, smooth conidia, strictly monoverticellate –*P. implicatium*
- iii. Short conidiophores – elliptical to subglobose, smooth to finely rough conidia,occasionally branched –*P.fellutanum*

iii) Slow growth, short conidiophores globose conidia and floccose

Non- Vesiculate

- i. Conidia echinulate + No. pigment – *P.restrictum*
- ii. Conidia finely rough + vinaceous pigment – *P.vinaceum*
- iii. Conidia smooth + bright yellow pigment – *P.citreoviride*

II. Biverticellate – One stage branching

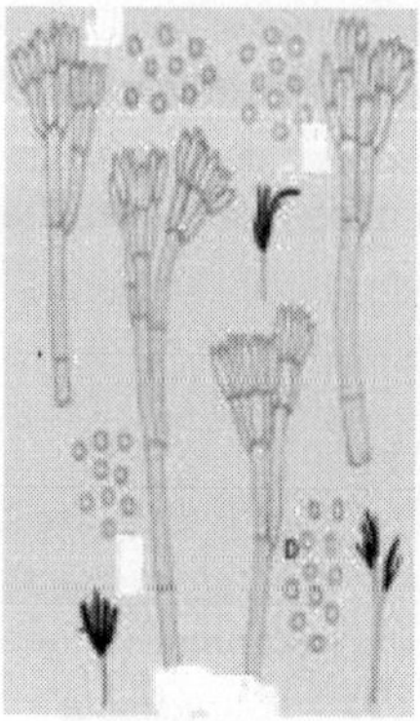

**a. Biverticellate – asymmetrica** (Furcatum) one stage branching but arranged symmetrically around the *P.citrinum*

i. Large globose conidia – Megaspora series (Rare)

ii. Conidia otherwise:

Spreading colonies – Oxalic series

Restricted colonies – Citrina series

**Oxalic series**

i. Conidia rough – colonies floccose, stipes rough, conidia slightly ellipticalphialides abruptly tapered – *P.simplicisum* (in Divarticata)

ii. Conidia smooth:

a. Spherical conidia – *P. novae*-zelandiae and others.

b. Colonies velvety large elliptical Conidia

Good growth on Czapek – *P.oxalium*

Poor growth on Czapek – *P.digitatum*

**Citrina series**:

Colonies restricted and velvety

i) 0-3 unequal metulae often dark- green, black reverse – *P.corylophilum*

ii) 2-4 equal metulae often bright yellow reverse – *P.citrinum*

iii) 5 or more metulae colourless reverse. – *P.paseilli*

iv) 3-5 or more compressed metulae, bright colour in reverse – *P.herquei* & others

**b. Biverticellate – symmetrica**

One stage branching but arrange symmetrically around the axis and with lancortate

eg. - *P.funiculosum*

Talaromyces (Ascosporic)

a. Cleistothecia yellow:

i. Elliptical spinulose ascospores

Spreading growth – club shaped perithecial initials –*T.flavus*

Restricted growth –

Irregular perithecial initials –*T.wortmannii*

Helically coiled initials, smallish ascospores – *T.helicus*

ii. Elliptical ascospores with transverse bands –*T. luteus*

iii. White to cream cleistothecia and elliptical spinulose conidia – *T. trachyspermus*

iv. White to pinkish brown growth, thermotolerant –*T. thermophilus*

**c. Biverticellate (non-ascosporic)**

i) Producing coremia -*P.declaueii*

ii) Not producing coremia

Funnel shaped heads lanceolate phialides – other Biverticellate symm.sp.

Compact asymmetrically arranged – Furcatum series

**Non- coremial species**

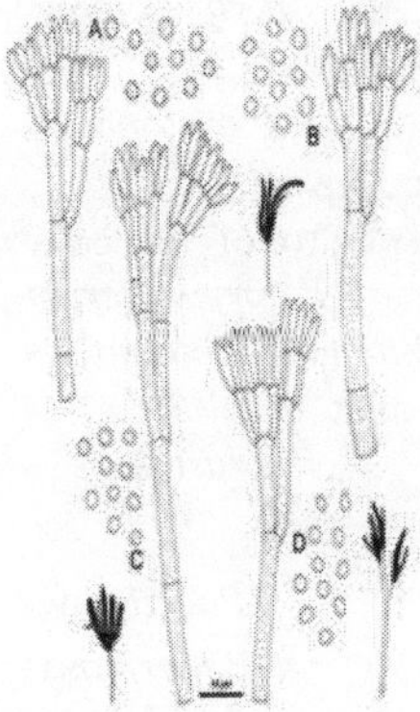


i. Short Conidiophores mostly from trailing hyphae or ropes of hyphae.

i. Globose conidia – Rough conidia, broad, compact penicilli –*P.verruculosum* and others

ii. Elliptical conidia:

- Very ropy colonies, *P.funiculosum* slightly rough conidia - *P.funiculosum*
- Conidia in conical mass -*P.piecum*
- Smooth, large hairy conidia, yellow – orange mycelium - *P.islandicum*

ii. Long conidiophores arising mostly from substram:

Rough conidia –

- Globose - *P.aculeatum*
- Elliptical:

i) No purple – red pigment, yellow or orange and restricted growth – *P. rugulosum*

- Red colours in reverse – *P.pururogenum*
- Yellow to brown colours in reverse with yellow aerial mycelium – *P.variable*

ii) Purple – red pigment narrow funnel shaped heads – *P.purpurogenum*

III. Terverticellata (*Penicillium*)

Two or more stage branching usually symmetric eg. *P.expansum*

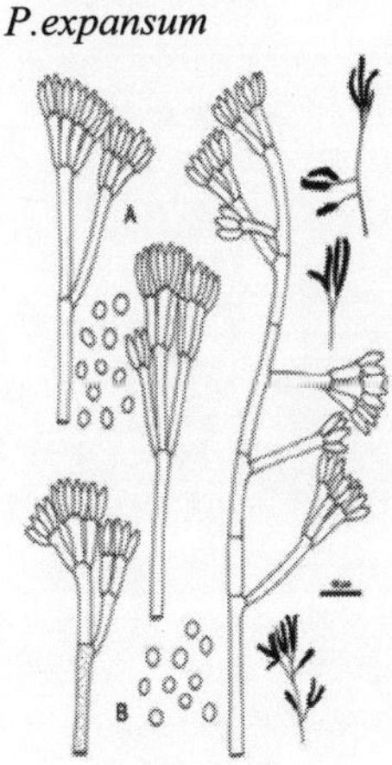


i. Species with smooth stipes:

- Globose conidia – spreading colonies loose penicilli – *P.chrysogenum*
- Restricted colonies compact penicilli - *P.brevicompactum*

Poor growth on czapeks agar – *P.italicum*

ii. Species with rough stipes:

-Cylindrical conidia– Grey rots of citrus (seldom elsewhere) -*P. digitatum*

-Blue rots of apples (in soil or elsewhere) – *P.expansum*

-Grey restricted growth spreading heads, very short phialides – *P.griseofulvum*

Floccose colonies
Funiculose colonies
Velvety colonies

iii. *Fasciculate* colonies *Fasciculate*

-White or very pale –green – *P.camemberti*

-Not green (Geosmithia series) – *P.pallidum*

-Spreading velvety poor growth on czapecty almost black – green – *P.roquefortii*

Reverse………………………………….. ***Fasciculata***

- Conidia rough – *P.echinulatum*
- Conidia smooth or nearly so –Producing pink

*P.viridicatum*: conidia at 20$^0$C – *P.gladioli*

series Coremia with pink stalks – *P.clariforme*

Not as above with fasciculate colonies smooth globose conidia rough stipes -*P.visidicatum*

- Restricted yellow – green colonies – *P.viridicatum*
- More spreading very green and blue-green – *P.cyclopium*
- Blue – green colonies with deep-wine-red drops on surface causing rot of lilaceans bultrs – *P.corymbiferum* (*P.hirsutum)*
- Conidia forming crusts on malt agar – *P. crustosum*

Key to common species of *Divaricata*

i. No green colonies – -*Paecilomyces*

ii. Green colour: Irregularly branched eg. -*P.nigricans*

IV. Divaricata

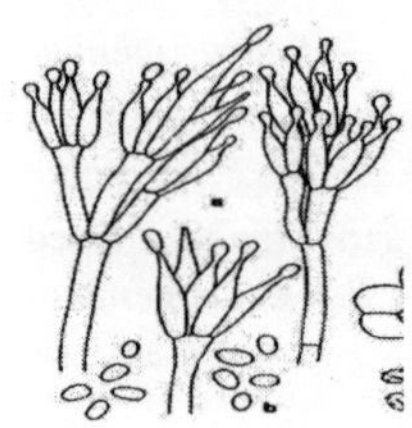

Growth spreading elliptical conidia, abruptly tapered phialides with long tube:

-*Delicate* conidiophores almost smooth -*P.janthinellum*

- Fragmentary heads –*P.ochrochloron*
- Robust rough conidiophores - *P.smplic sienus* (*P.simplic sinum)*

Growth restricted globose conidia:

- Smooth conidiophores rough conidia – *P.nigricans*
- Rough conidiophore with smooth conidia – *P.canescans*

---

## Description of species of *Penicillium*

### i) *P. aethiopicum* Frisvad

**Subgenus:** *Penicillium*

**Section:** Chrysogena

**Series:** Aethopica

**Teleomorph:** None

**Colony texture:** Sulcate on CYA

**Colony colour CYA:** Dull-green

**Exudates droplets on CYA:** Copious, clear

**Reverse colour on CYA:** Golden yellow

**Reverse colour on YES:** Yellow to curry-yellow

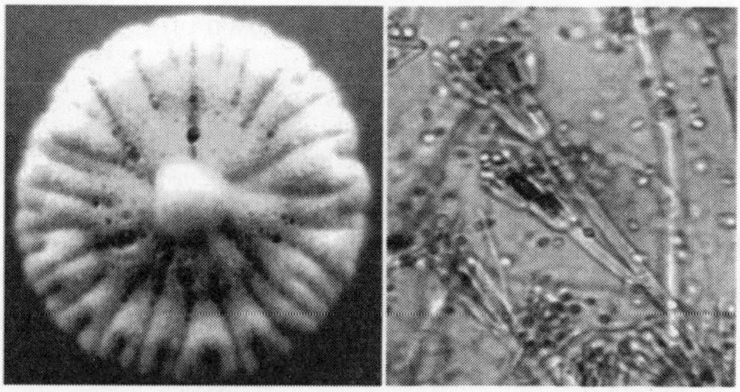

**Diffusible colour:** Occasional; pale-orange

**Ehrlich reaction:** No reaction

**Sclerotia:** None

**Stipes:** Smooth to rough walled

**Synnemata or fasciculation:** Weakly fasciculate

**Rami:** 15-25 nm

**Metulae:** 12-17 µm

**Phialides:** Short collula

**Conidia:** Smooth walled, ellipsoidal in long columns

**Mycotoxins:** Viridicatum, tryptoquialanins

**Diagnostic features:** Markedly sulcate colonies with a golden-yellow reverse, conidia ellipsoidal, smooth walled, growth at 37ºC.

**ii) *P. allii* Vincent and Pitt**

**Subgenus:** *Penicillium*

**Section:** Viridicata

**Series:** Corymbifera

**Teleomorph:** None

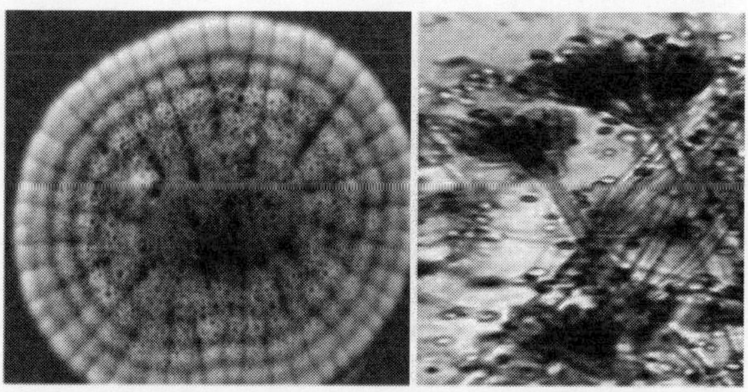

**Colony texture on CYA:** Granular to weakly fasciculate only sulcate in centre of colony

**Colony colour on CYA:** Dull-green

**Exudate droplets:** Small clear, yellow or brown on CYA

**Reverse colour on CYA:** Dark-brown

**Diffusible colour on CYA:** Amber to brown

**Reverse colour on YES:** Yellow-brown to dark-yellow-brown

**Ehrlich reaction:** Pink to red

**Conidiophore:** Terverticillate, some irregular structures

**Phialides:** Cylindrical

**Metulae:** Cylindrical

**Rami:** Cylindrical

**Stipe:** Short, rough walled

**Synnemata or fasciculation:** Weakly fasciculate especially on OAT

**Sclerotia:** None

**Mycotoxins:** Roquefortine C

**Diagnostic features:** Smooth walled conidia, fulvic acid or analoges, a dark-brown reverse, garlic rot.

iii) ***P. aurantiogriseum*** **Diereky**

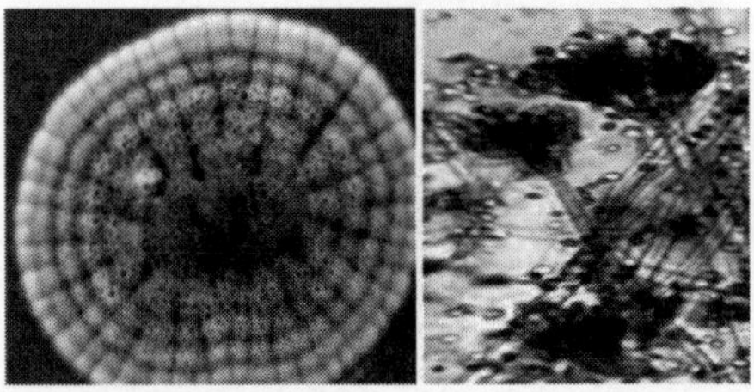

**Subgenus:** *Penicillium*

**Section**: Viridicata

**Series:** Viridicata

**Teleomorph**: None

**Colony texture:** Velutinous to weakly fasciculate

**Colony colour on CYA**: Blue-green to grey-green to dark dull-green

**Exudates droplets on CYA**: Occasionally present, pale or light-brown

**Reverse colour on CYA**: Curry-yellow to warm-orange to reddish-brown

**Reverse colour on YES**: Yellow

**Diffusible colour on YES**: Often present, orange to red-brown

**Ehrlich reaction**: Pink

**Conidiophores**: Terverticillate few biverticellate and quarter verticellate

**Conidia**: Cylindrical with collula

**Metulae stipe**: From sub-surface hyphae, walls smooth to finely roughened.

**Synnemata or fasciculation**: Weak fasciculation

**Sclerotia**: None

**Mycotoxins:** Penicillic acid, verrucosidin

**Diagnostic features**: Anacin, aurantiamin. Conidia smooth walled, dark-brown halo on Raulin Thom agar

**iv) *P. brevicompactum***

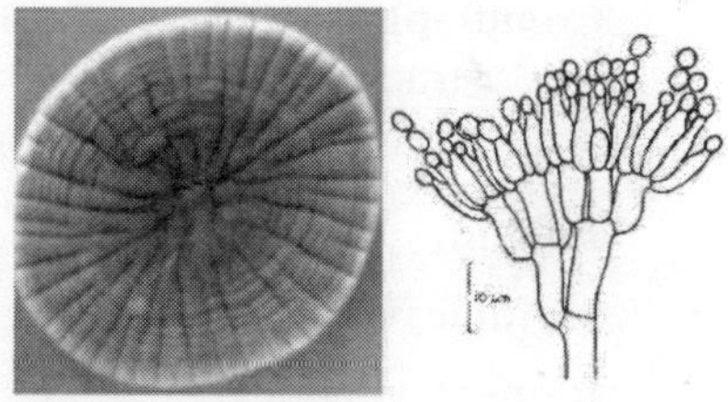

**Subgenus**: *Penicillum*

**Section:** Coronata

**Series:** Olsoni

**Teleomorph:** None

**Colony texture:** Velulinous

**Colony colour on CYA:** Dull-green to grey-green

**Exudates droplets on CYA:** Pale to yellow or reddish-brown

**Reverse colour on CYA**: Beige to yellowish-cream

**Reverse colour on YES:** Cream coloured to beige in few strains a conspicuous red, reverse and diffusible pigment is produced.

**Diffusible colour on CYA**: Pale or light-brown

**Ehrlich reaction**: No reaction or yellow

**Conidiophore:** Long, appressed, terverticillate

**Conidia:** Finely roughened, ellipsoidal

**Phialides:** Cylindrical with gradually tapering collula

**Metulae:** Cylindrical apically inflated

**Rami:** Cylindrical, smooth walled

**Stipes:** Smooth walled

**Synnemata or fasciculation:** None

**Sclerotia:** None

**Mycotoxins:** Mycophenolic acid, botryodiplodin

**Diagnostic features**: Finely roughened, ellipsoidal conidia, short broad penicilli, apically inflated metulae, Mycophenolic acid.

**v) *P. camemberti* Thom**

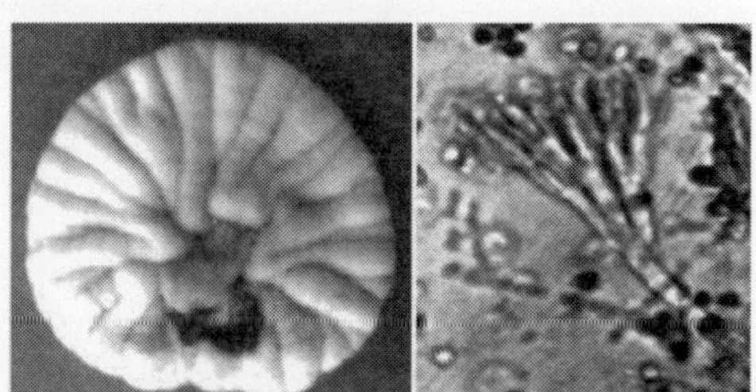

**Subgenus**: *Penicillium*

**Section:** Viridicata

**Series:** Camemberti

**Teleomorph**: None

**Colony texture:** Floccose

**Colony colour on CYA:** White rarely grey-green

**Exudates droplets**: None or rarely grey-green

**Reverse colour CYA:** Pale-cream coloured or yellowish

**Reverse colour YES:** Yellow

**Diffusible colour:** None

**Ehrlich reaction:** Voilet (No reaction in few old cultures)

**Conidiophores**: Terverticillate to quarter vertiallate, sometimes irregular structures.

**Conidia**: Smooth walled, globose to subglobose

**Phialides:** Cylindrical with long wide collula

**Metulae**: Cylindrical

**Rame:** Cylindrical

**Synnemata or fasciculation**: None

**Sclerotia:** None

**Mycotoxins:** Cyclopiozonic acid

**Diagnostic features:** White, floccose colonies, poor sporulation, cyclopiazonic acid.

**vi) *P. caseifulvum* Lund**

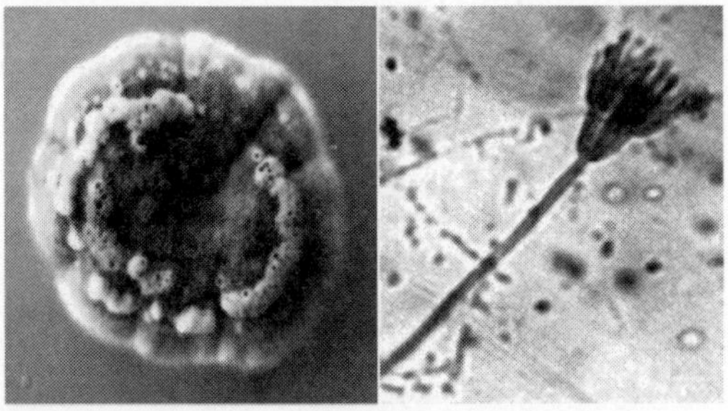

**Subgenus**: *Penicillium*

**Section**: Viridicata

**Series**: Camemberti.

**Teleomorph:**None

**Colony texture**: Floccose

**Colony colour on CYA:** Greyish-green to grayish-blue-green (Furquoise)

**Exudate droplets on CYA**: Small, clear droplets at colony margin.

**Reverse colour on CYA**: Creamish yellow to brown-yellow

**Reverse colour on YES:** Vividly orange, rarely only cream yellow.

**Diffusible colour on CYA**: None

**Ehrlich reaction:** Violet (weak reaction)

**Conidiophore:** Born on aerial hyphae, terverticillate

**Conidia**: Smooth walled, subglobose to broadly ellipsoidal.

**Phialides**: Cylindrical with short narrow collula

**Metulae:** Cylindrical

**Rami:** Cylindrical

**Stipes:** 300-800 μm

**Synnemata or fasciculation**: None

**Sclerotia**: None

**Mycotoxins:** Regulovasine

**Diagnostic features**: Smooth walled conidia, floccose colonies, good growth on CREA weak but consistent, growth on CZP regulovasine.

**vii) *P. chrysogenum* Thom**

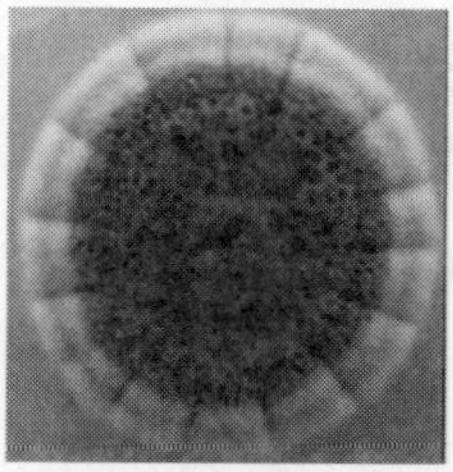

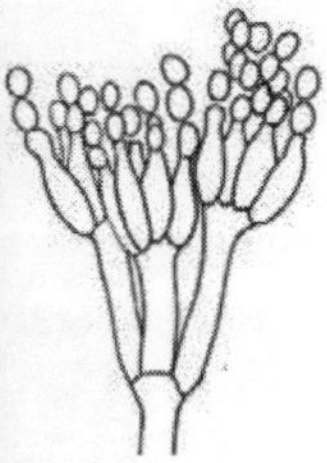

**Subgenus**: *Penicillium*

**Section:** Chryscgenea

**Series**: Chrysogena

**Teleomorph:** None

**Colony texture**: Floccose to velutinous

**Colony colour on CYA**: Blue-green to green

**Exudates droplets on CYA**: Often present, copious, yellow

**Reverse of colony on CYA:** Cream, yellow rarely brown

**Reverse of colony on YES**: Citrine yellow

**Diffusible colour:** Yellow pigment often produced

**Ehrlich reaction**: No reaction or yellow reaction

**Conidiophores**: Bi-, ter- and quarter verticellate both appressed and divergent ram borne from aerial and sub-surface hyphae.

**Conidia:** Smooth-walled, globose to sub-globose to broadly ellipsoidal.

**Phialides**: Cylindrical with short broad collula

**Metulae:** Cylindrical

**Rami**: Cylindrical

**Stipes**: 200-300 μm X 3-4 μm

**Synnemata or fasciculation:** None

**Sclerotia**: None

**Mycotoxin:** PR toxin, roquefortine C, Secalonic acid.

**Diagnostic features**: Globose to sub globose to broadly ellipsoidal, smooth walled conidia, relatively short phialides with short broad collula, high growth rate on YES with a yellow reverse and strong sporulation.

**viii) *P. citrinum* Thom**

**Synonyms** = *P. brandicum*

= *P.botryogeum*

= *P. saratorys*

= *P. stekii*

**Subgenus:** *Aspergilloides*

**Section:** Citrina

**Series: Citrina**

**Teleomorph:** Absent

**Colony texture on CYA:** Velutinous, centrally, floccose

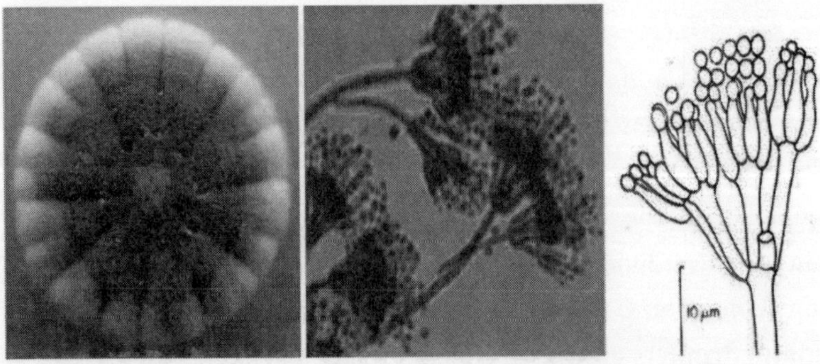


**Colony colour on CYA:** White to grayish orange or apricot.

**Exudates droplets on CYA**: Pale –yellow to reddish-brown

**Reverse colour on CYA**: Yellow, yellowish-brown, reddish-brown

**Reverse colour YES:** Pack, dull-brown, yellow brown or occasionally olive

**Diffusible colour on CYA:** Shades on yellow soluble pigment production

**Ehrlich reaction:** Negative

**Conidiophores:** Borne from surface or subsurface hyphae

**Conidia**: Spherical to subsphaerical, smooth or finely roughened.

**Phialides:** Verticals of 8-12, ampulliform

**Metulae:** Commonly spathulate, terminally vesiculate in loosely packed whorls of 3-4

**Stipes:** Smooth, characteristically terminating in a well defined verticals of 3-5 divergent metulae.

**Synnemata or fasciculation:** None

**Sclerotia:** None

**Mycotoxins:** Citrinin, quinolactcin, anthraquines.

**Diagnostic features:** Growth on CYA when incubated at 37°C 2-12 mm in diam, reverse on CYA in shades of yellow. Soluble pigment production on CYA and YES. Conidiophore heads symmetrical with flask shaped sterigmata and loosely packed metulae.

**ix) *P. commune* Thom.**

**Subgenus**: Penicillium

**Section:** Viridicata

**Series:** Camemberti.

**Teleomorph**: None

**Colony texture on CYA:** Velutinous to floccose or weakly fasciculate.

**Colony colour on CYA:** Blue-green to green

**Exudates droplet on CYA:** Often present, clear

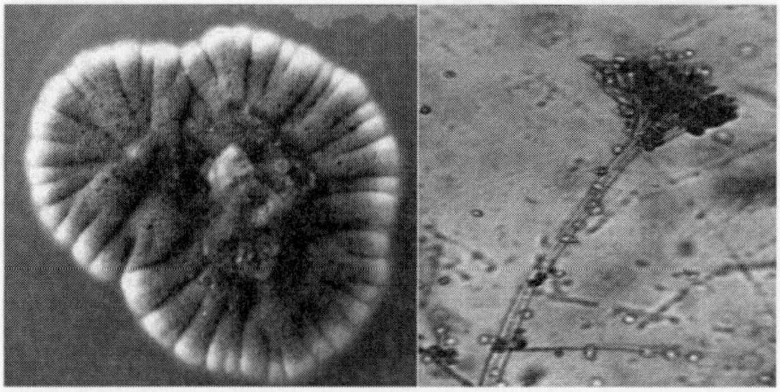

**Reverse colour on CYA:** Cream coloured to beige or cream yellow.

**Reverse colour on YES:** Cream coloured to yellow

**Diffusible colour on CYA**: None

**Ehrlich reaction:** Strong violet reactions

**Conidiophores:** Terverticillate, appressed elements born from subsurface hyphae.

**Conidia:** Smooth-walled, globose to subglobose (rarely), ellipsoidal.

**Phialides**: Cylindrical, tapering to a distinct collulum

**Metulae**: Cylindrical

**Rami**: Cylindrical

**Stripes**: Rough walled

**Synnemata or fasciculation:** Weakly fasciculate

**Sclerotia:** None

**Mycotoxins:** Cyclopiazonic acid.

**Diagnostic features**: Sub globose to ellipsoidal, smooth walled conidia, good growth on CREA cyclopiazonic acid, regulovarine A & B.

**x) *P. crustosum* Thom**

**Subgenus:** *Penicillium*

**Section:** Visidicata

**Series:** Camemberti.

**Teleomorph:** None

**Colony texture:** Velutinous to weakly fasciculate becoming crustose.

**Colony colour on CYA:** Dull-green to grey-green or blue-green at the colony margin

**Exudates droplets on CYA:** Copious, clear or brown.

**Reverse colour on CYA:** Cream coloured to yellow-brown.

**Reverse colour on YES**: Strongly yellow

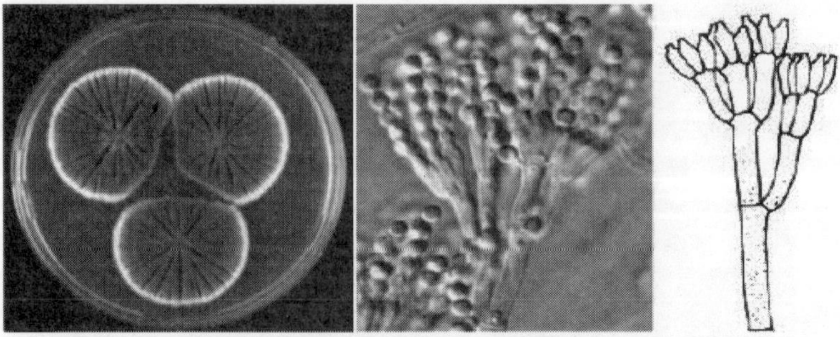

**Diffusible colour**: Pale-brown or none

**Ehrlich reaction:** None

**Conidiophores**: Terverticillate, appressed elements borne from subsurface hyphae

**Conidia**: Smooth walled, globose to subglobose.

**Phialides:** Cylindrical tapering to a distinct collulum

**Metulae**: Cylindrical

**Rami:** Cylindrical.

**Stipes**: Rough walled.

**Synnemata or fasciculation**: Weakly fasciculate

**Sclerotia:** None

**Mycotoxins:** Penetrim A and roquefortine C

**Diagnostic features:** Smooth walled conidia, crustose on CYA and MEA after 10days high growth rate, good growth on CREA, strong sporulation and high growth rate on YES agar.

**xi) *P. cyclopium* Westling**

**Subgenus:** *Penicillium*

**Section**: Visidicatum

**Series**: Viridicata

**Teleomorph**: None

**Colony texture:** Velutinous to weakly fasciculate

**Colony colour on CYA:** Blue-green to green, blue-green on MEA

**Exudates droplets on CYA**: copious, clear or light-yellow

**Reverse colour on CYA:** Yellow to orange to red-brown, occasionally only cream yellow

**Reverse colour on YES**: Strongly yellow also yellow obverse (poor sporulation on YES)

**Diffusible colour:** None or pink or red-brown

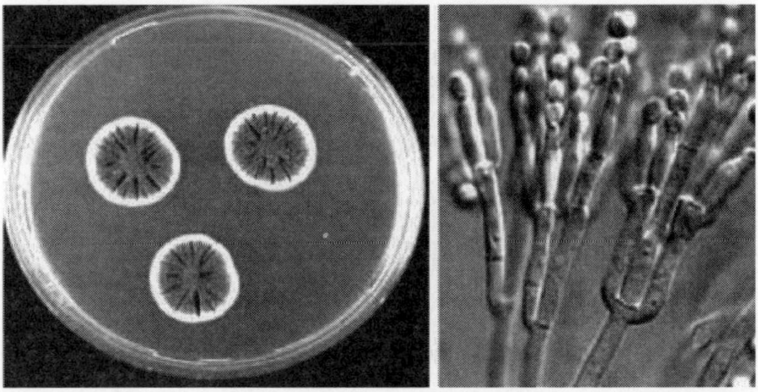

**Ehrlich reaction:** Terverticillate, appressed elements borne from sub-surface hyphae

**Conidia:** Smooth walled, globose to sub-globose

**Phialides**: Cylindrical, tapering to a distinct collulum

**Metulae**: Cylindrical

**Stipes**: Rough walled

**Synnemata or fasciculation**: Weakly fasciculate

**Sclerotia:** None

**Mycotoxins:** Xanthomegnin, viomellenin and vioxanthin.

**Diagnostic features**: Blue-green smooth walled conidia, green on CYA and blue-green on MEA, xanthomegnin, viomellein, vioxanthin, penicillic acid.

**xii) *P. digitatum* (Pers: Fr.)**

**Subgenus**: *Penicillium*

**Section**: Digitata

**Series:** Digitata

**Teleomorph**: None

**Colony texture**: Velutinous

**Colony colour on CYA:** Olive-green

**Exudate droplets on CYA**: Only in few isolates

**Reverse colour on CYA:** Grayish olive-beige

**Reverse colour on YES**: Cream-yellow with light brown centre

**Diffusible colour:** None

**Ehrlich reaction:** None

**Conidiophores:** Terverticillate, appressed elements, borne from sub surface or aerial hyphae.

**Conidia:** Smooth walled, ellipsoidal to cylindrical

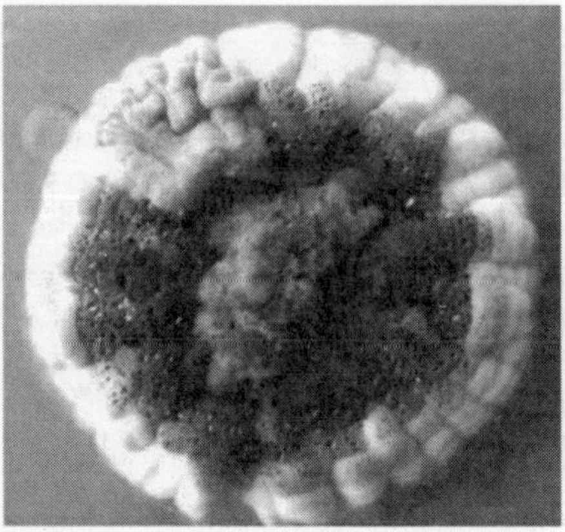
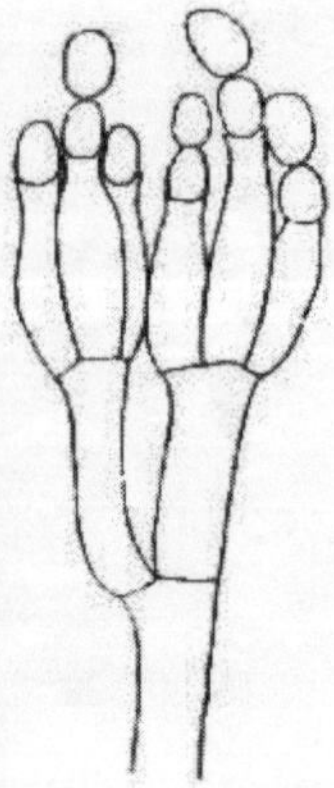

**Phialides:** Cylindrical tapering to a distinct collulum

**Metulae:** Cylindrical

**Rami:** Cylindrical

**Stipes:** Rough walled

**Synnemata or fasciculation**: None

**Sclerotia**: None

**Mycotoxins:** Some metabolites were shown to be toxic in the chick embryo test.

**Diagnostic features:** Large cylindrical to ellipriodal, smooth walled, olive-green conidia, no growth on Czapek agar & CYAS.

**xiii) *P. discolor* Frisvad & Samson.**

**Subgenus:** *Penicillium*

**Section:** Viridicata

**Series:** Solita

**Teleomorph**: None

**Colony texture on CYA**: Velulinous to fasciculate.

**Colony colour on CYA:** Dark-green

**Exudate droplets on CYA:** Clear

**Reverse colour on CYA:** Cream yellow

**Reverse colour on YES:** Orange turning to deep red with age.

**Diffusible colour:** Brillant-red on YES

**Ehrlich reaction:** Weak, violet

**Conidiophores:** Terverticillate, appressed elements borne from subsurface hyphae

**Conidia:** Rough walled, globose to subglobose

**Phialides:** Cylindrical, tapering to a distinct collulum

**Metulae:** Cylindrical

**Stipes:** Rough walled

**Synnemata or fasciculation**: Fasciculate on MEA and oat.

**Sclerotia**: None

**Mycotoxins:** Chaetoglobosin A, B and C

**Diagnostic features**: Dark-green, rough walled conidia, brilliant-red, diffusible colour on YES agar, Chaetoglobosin A, B and C.

**xiv) *P. expansum* Link**

**Subgenus:** *Penicillium*

**Sections**: Penicillium

**Series:** Expansa

**Teleomorph:** Absent

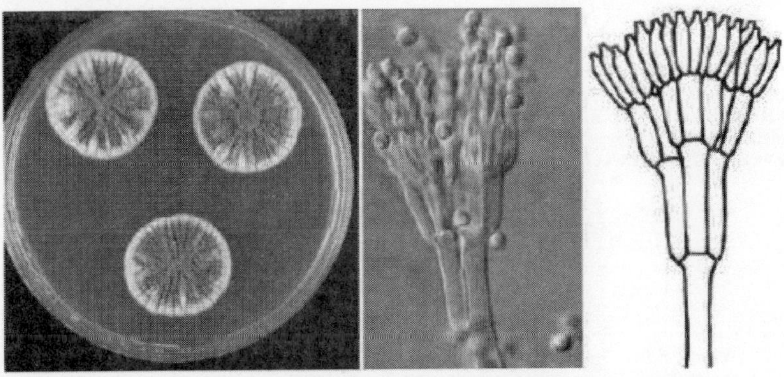

**Colony texture**: Floccose to weakly fasciculate

**Colony colour on CYA**: Blue-green to green

**Exudate droplets on CYA:** Copious, clear

**Reverse colour on CYA:** Cream to yellow with brown centre, orange – brown or dark-brown

**Reverse colour on YES**: Cream yellow or orange

**Diffusible colour on CYA**: Common, rosy-brown

**Ehrlich reaction**: Strong, violet

**Conidiophores**: Terverticillate, appressed elements borne sub-surface and aerial hyphae

**Conidia:** Smooth walled, ellipsoidal

**Phialides:** Cylindrical tapering to a distinct collulum

**Metulae:** Cylindrical

**Rami**: Cylindrical

**Stipes:** Smooth walled, occasionally rough walled at lower part.

**Synnemata or fasciculation:** Capitulate synnemata produced in fresh strains or after cold storage on MEA or OAT

**Sclerotia:** None

**Mycotoxins:** Patulin, chaetoglobosin C, roquefortine C, citrinin and communesin A& B.

**Diagnostic features:** Ellipsoid smooth-walled conidia, expan solids, patulin, chaetoglobosin commonsense A and B.

**xv) *P. griseofulvum* Diercka**

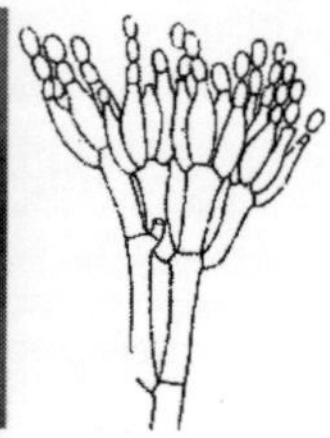

**Sub genus:** *Penicillium*

**Section**: Penicillium

**Series:** Urticicolae

**Teleomorph:** Absent

**Colony texture**: Velulinam to fasciculate

**Colony colour on CYA:** Grey to grey-green

**Exudates droplets on CYA**: Present, clear to yellow

**Reverse colour:** Cream, beige or brown

**Diffusible colour**: Weak pinkish-brown

**Ehrlich reaction**: Voilet

**Conidiophores**: Divergent, undulate or sinuate, terverticillate to quinverticillate

**Conidia:** Smooth walled, broadly ellipsoidal

**Phialides**: Flark shaped tapering to a distinct collulum

**Metulae:** Cylindrical

**Rami:** Cylindrical

**Stipes:** Smooth walled

**Synnemata or fasciculation**: Weakly fasciculate

**Sclerotia:** None

**Mycotoxins:** Patulin, cyclopiazonic acid, roquefortine C.

**Diagnostic features:** Ellipsoidal smooth walled conidia, very short phialides, grey conidia with only small element of green, very branched conidiophores, Griseofaluin, CPA.

**xvi) *P. islandicum* Sopp**

**Subgenus:** *Biverticelium*

**Section**: Simplicium

**Series:** Islandica

**Teleomorph:** Absent

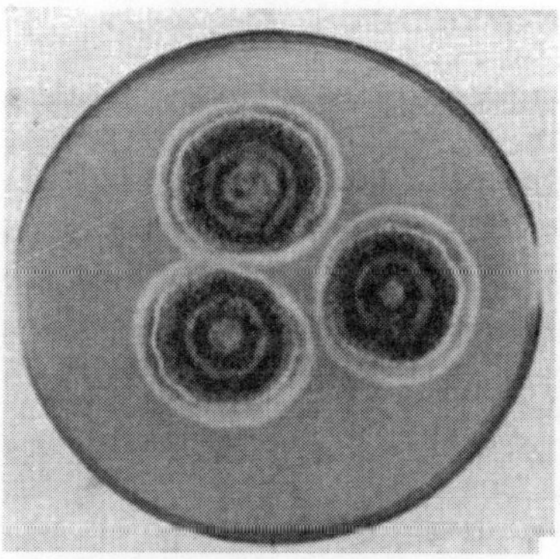

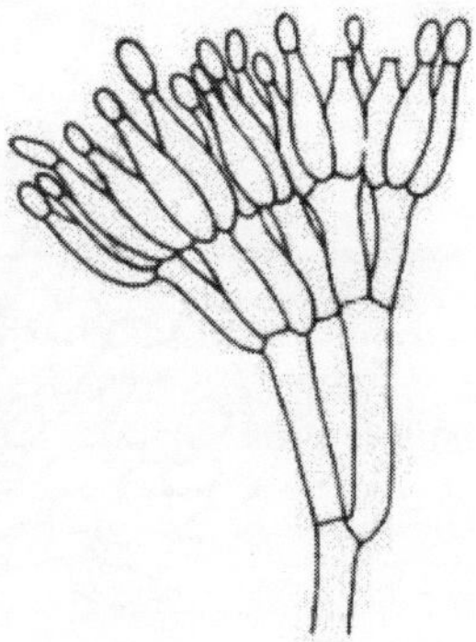

**Colony texture on CYA**: Floccose

**Colony colour on CYA:** Dark-green or bottle- green

**Exudate droplets on CYA:** Clear to pale-yellow

**Reverse colour on CYA:** Orange to rust-brown

**Reverse colour on YES:** Orange to rust brown, burnt sienna or henna sometimes reddish brown

**Diffusible colour on CYA**: Not produced

**Conidiophores:** Borne on aerial mycelium, bearing terminal penicilli, typically biverticellate occasionally with appressed ranus, non-vesiculate.

**Stipe:** Smooth walled

**Conidia:** Broadly ellipsoidal to sub spheroidal, smooth, heavy walled.

**Phialides:** 5-8 per metulae, acerose

**Rami:** Usually borne singly

**Metulae:** Closely approved verticals of 4-7 with abruptly narrowing collula

**Synnemata or fasciculation:** Absent

**Sclerotia:** None

**Mycotoxins:** Citrinin, luteoskyrin, islandi toxin.

**Diagnostic features:** Conidiophore heads symmetrical with lanculate sterigmata and closely packed metulae, colony on CZ yellow-brown to olove-brown pigment not diffusing into surrounding agar, conidia smooth walled.

**xvii) *P. italicum* Wehmer**

**Subgenus**: *Penicillium*

**Section**: Penicillium

**Series**: Italicum

**Teleomorph:** Absent

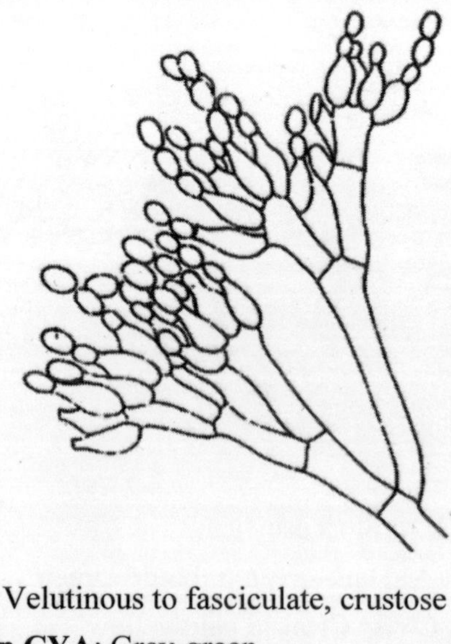

**Colony texture:** Velutinous to fasciculate, crustose

**Colony colour on CYA**: Grey-green

**Exudates droplets on CYA:** None or small clear droplets

**Reverse colour on CYA**: Pale, brownish-orange or red-brown

**Reverse colour on YES:** Red-brown

**Diffusible colour on CYA**: None

**Ehrlich reaction:** None

**Conidiophores**: Terverticillate, appressed elements, born from subsurface hyphae

**Conidia:** Smooth walled, ellipsoidal to cylindrical

**Phialides:** Cylindrical tapering to a distinct collulum

**Metulae:** Cylindrical

**Stipes:** Smooth-walled

**Synnemata or fasciculation:** Fasciculate in the colony margins.

**Sclerotia:** Reported in fresh isolates, colourless to light-brown

**Mycotoxins**: None

**Diagnostic features**: Cylindrical to ellipsoidal smooth walled conidia, poor growth and no acid production on CREA, crustose on YES, deoxybrevianamide.

**xviii) *P. nalgiovense* Laxa**

**Subgenus**: *Penicillium*

**Section**: Chrysogena

**Series:** Chrysogena

**Teleomorph**: Absent

**Colony texture**: Velutinous

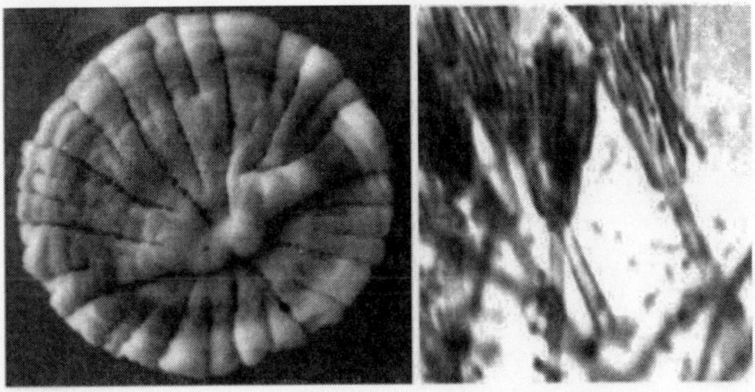

**Colony colour on CYA**: Dark-green (occasionally white)

**Exudates droplets on CYA**: Clear

**Reverse colour on CYA:** Cream to yellow to orange to dark orange-brown

**Reverse colour on YES:** Bright yellow to orange

**Diffusible colour:** Yellow-brown

**Ehrlich reaction:** None

**Conidiophores:** Bi, ter and quarter verticellate both divergent metulae and rami born from aerial and sub surface hyphae

**Conidia:** Smooth walled, globose to sub globose to broadly ellipsoidal

**Phialides:** Flask-shaped with short wide collulum

**Metulae:** Cylindrical

**Rami:** Cylindrical

**Synnemata or fasciculation:** None

**Sclerotia:** None

**Mycotoxins:** Toxicity of Nalgiovensin, nalgiolaxin is unknown

**Diagnostic features:** Smooth walled conidia, orange reverse on YES, divergent structures, dark-green conidia in fresh isolates.

**xix) *P. nordicum* Dragoni & Cantoni ex**

**Subgenus:** *Penicillum*

**Section:** Viridicata

**Series**: Verrucosa

**Teleomorph:** Absent

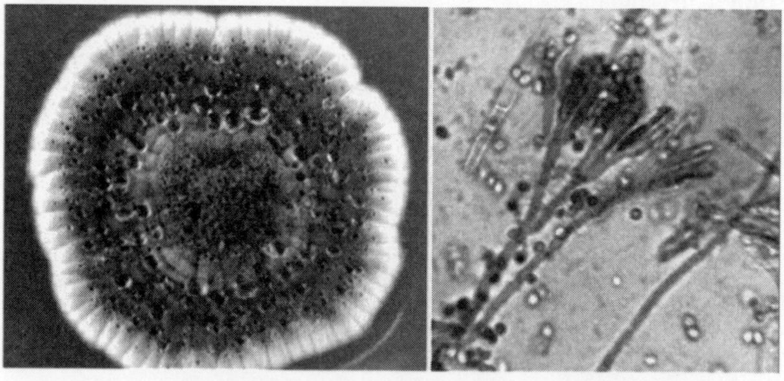

**Colony texture:** Velutinous to floccose, often weakly floccose

**Colony colour or CYA:** Green

**Exudates droplets on CYA:** Copious, clear to light-yellow

**Reverse colour on CYA:** Cream often with brown centre

**Reverse colour on YES:** Cream yellow

**Diffusible color on CYA:** None

**Ehrlich reaction:** Yellow-green reaction

**Conidiophores:** Terverticillate, appressed elements born from surface or sub surface hyphae

**Conidia:** Rough walled, globose to subglobose

**Phialides:** Cylindrical tapering to a distinct collulum

**Metulae:** Cylindrical

**Rami**: Cylindrical

**Stipes:** Rough walled

**Synnemata or fasciculation:** Weakly fasciculate

**Sclerotia**: None

**Mycotoxins**: Ochratoxin A, verruculone

**Diagnostic features:** Smooth walled conidia, good growth on UNO and nitrite, poor growth and no acid production on CREA, Rough conidiophore stipes.

### xx) *P. olsonii* Bai & Sartory

**Subgenus:** *Penicillium*

**Section:** Ceronata

**Series**: Olsonii

**Teleomorph**: Absent

**Colony texture**: Velutinous

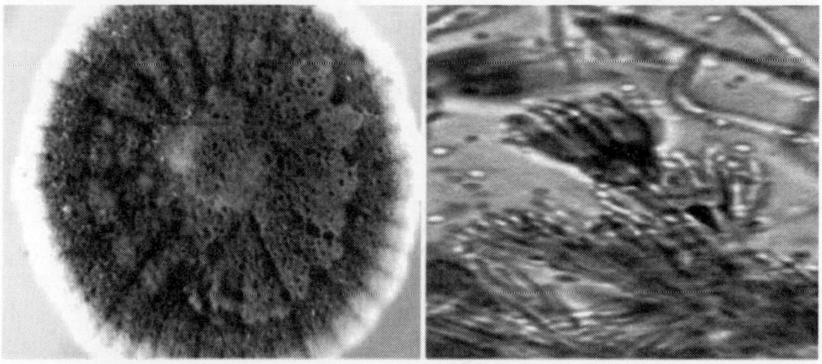

**Colony colour on CYA:** Greyish-green to dull-green

**Exudates droplets on CYA**: Clear to light-yellow

**Reverse colour:** Cream to light yellow

**Diffusible colour**: None

**Ehrlich reaction**: Yellow reaction

**Conidiophores**: Long, multi ronulate, appressed terverticillate

**Conidia:** Finely roughened, ellipsoidal

**Phialides:** Cylindrical with short collula

**Metulae:** Cylindrical but rapidly inflated

**Rami:**Cylindrical

**Stipes:** Smooth walled

**Synnemata or fasciculation**: None

**Sclerotia:** Occasionally large pale to light-yellow

**Mycotoxins:** Verrucolone, brevienes, asperhenamata

**Diagnostic features**: Ellipsoidal, finely roughened conidia, multi ramulate Penicillum verruculose.

**xxi) *P. roqueforti* Thom**

**Subgenus:** *Penicillium*

**Section:** Roqueforti

**Series:** Roqueforti

**Teleomorph**: Absent

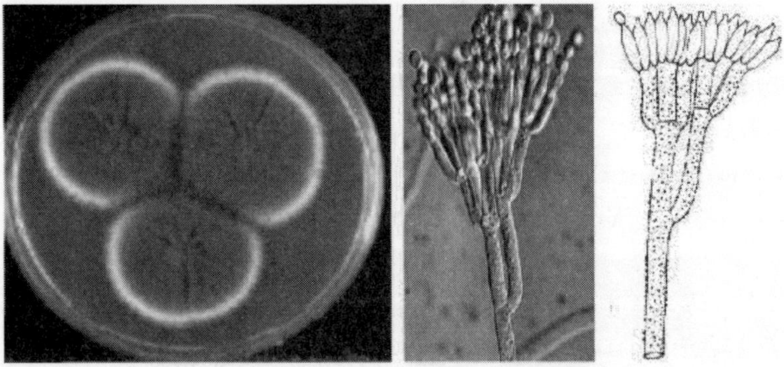

**Colony texture on CYA**: Velutinous

**Colony colour on CYA**: Green

**Reverse colour on CYA**: Blackish-green

**Reverse colour on YES:** Blackish-green

**Diffusible colour on CYA:** None

**Ehrlich reaction:** Voilet (Yellow in few old strains)

**Conidiophores:** Terverticillate, occasionally quarter verticellate, appressed elements borne from subsurface hyphae

**Conidia:** Smooth walled, globose

**Phialides**: Cylindrical with short collula

**Metulae:** Cylindrical

**Rami:** Cylindrical

**Stipes:** Rough walled

**Synnemata or fasciculation**: None

**Sclerotia:** Occasionally rudimentary sclerotic

**Mycotoxins**: P R toxin, Roquefortine, ISO furnigaclarine, mycophenolic acid.

**Diagnostic features**: Globose smooth walled conidia, good growth or medium with 0.5% acetic acid or 1% propionic acid, black-green reverse on CYA and YES.

**xxii) *P. tricolor* Frisvad**

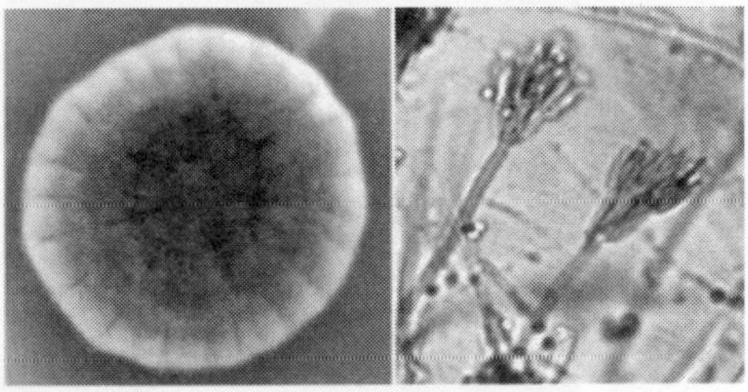

**Subgenus**: *Penicillium*

**Section**: Viridicata

**Series**: Viridicata

**Teleomorph**: Absent

**Colony texture on CYA**: Velutinous

**Colony colour on CYA:** Grey-green

**Exudates droplets on CYA**: Copious, brown

**Reverse colour on CYA**: Dark-yellow brown

**Reverse colour on YES**: Yellow-brown

**Diffusible colour on CYA**: Light yellow-brown

**Ehrlich reaction**: None

**Conidiophores**: Terverticillate, a pressed elements born from sub surface and aerial hyphae

**Conidia:** Smooth walled, globose to sub-globose

**Phialides:** Flask shaped tapering to a distinct collulum.

**Metulae:** Cylindrical apically sciollen

**Rami:** Cylindrical

**Stipes:** Rough walled

**Synnemata or fasciculation**: None or very weak fasciculation

**Sclerotia:** None

**Mycotoxins**: Xanthomegnin, viomellein, vioxanthin, asteltoxin

**Diagnostic features**: Smooth walled conidia, rough walled stipes, grey (green) conidia, yellow-brown reverse on CYA and YES.

**xxiii) *P. verrucosum* Dierckse**

**Subgenus:** *Penicillium*

**Section:** Viridicata

**Series**: Verrucosa

**Teleomorph**: Absent

**Colony texture on CYA**: Velutinous to floccose to weakly fasciculate

**Colony colour CYA:** Green

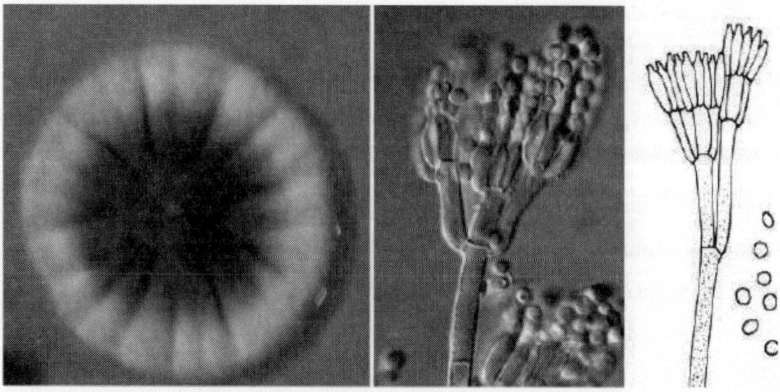

**Exudate droplets on CYA:** Copious, clear

**Reverse colour on YES:** Red-brown (Terracolta)

**Reverse colour on CYA**: Cream yellow often with brown entre

**Diffusible colour on CYA**: None

**Conidiophores:** Terverticillate, appressed elements borne from surface or sub-surface hyphae

**Conidia:** Rough walled, globose to sub-globose.

**Phialides:** Cylindrical to a distinct collulum

**Metulae:** Cylindrical

**Rami:** Cylindrical

**Stipes:** Rough walled

**Synnemata or fasciculation**: Weakly fasciculation

**Sclerotia:** None

**Mycotoxins**: Ochratoxin A. citrinin, verruculone, verrucins

**Diagnostic features**: Smooth walled conidia, red-brown reverse on YES, slow growing on all media, no growth at 30ºC.

## 44. *Periconia* Tode ex Fr

A hyphomycetes has been monographed for British islets by Mason and Ellis (1953) and for India by Raghuveer Rao and Dev Rao (1964). The type species *P. lichenoides* Tode ex Merat (1821) is now regarded as doubtful species and *P. byssoides* pers. ex Schw (1822) is perhaps most representative of the genus. It is characterized by hyaline and partly pigmented vegetative hyphae, conidiophores erect, mononematous, macronematous, brown, bearing an apical or sub apical head of dense branches with numerous short, usually monoblastic, conidiogenous cells which form short, usually monoblastic, conidiogenous cells which form short sometimes branched chains of blastoconidia. Conidia globose, brown, rough walled, formed in acropetal chain but maturing in basipetal sequence. Ellis (1996) described 27 species, while Raghuveer Rao and Dev Rao (1964) keyed out 28 species from India.

**i) *Periconia* spp.**

**Colony:** Effuse, small and compact, grey-brown, olivaceous-brown or black, hairy

**Hyphae:** Immersed, partly superficial

**Stroma:** Pseudoparenchymatous, mid to dark -brown

**Setae:** Absent

**Hyphopodia:** Absent

**Conidiophores:** Macronematous or semi macronematous, mononematous, brown mostly with stipe and sphaeroid head, branches present or absent

**Stipe:** Straight of flexous, pale to dark-brown, smooth, rarely verrucose, sometimes apex sterile or setiform

**Conidiogenous cells:** Monoblastic and polyblastic discrete on stipe and branches, determinate ellipsoidal, subsphaerical to spherical.

**Conidia:** Catenate, chains often branched arising at one or more points, subsphaerical to spherical, ellipsoidal, oblong or broadly cylindrical, pale to dark-brown, verrucose or echinulate, non-septate.

**Mycotoxins:** 2 chloro-8hydroxy-6 methoxy-3 methyl-3, 4 dihydroisocoumari

**Diagnostic features:** Brown, macronematous, conidiophore with stipe and sphaeroid head, conidiogenous cells, monoblastic or poly blastic, spherical conidiogenous cells which bear conidia in chain, conidia non-septate, verruculose or echinulate.

## 45. *Pestalotiopsis* Stey

**i) *Pestalotiopsis* sp.**

**Colony:** Compact or effuse, buff, grayish-brown, blackish-brown or black

**Hyphae:** Immersed, branched, septate, hyaline to pale-brown

**Conidiomata;** Acervulus, hyaline branched, septate, cylindrical or lageniform, formed from the upper cells of the pseudoparanchyma

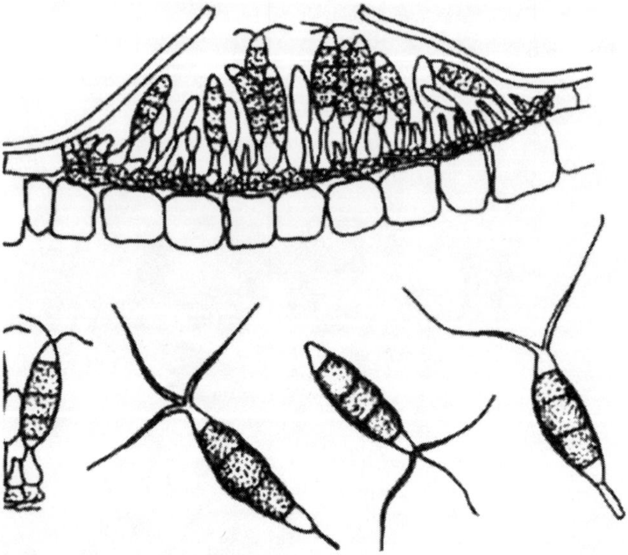

**Conidiogenous cells:** Holoblastic, annellidic, indeterminate, integrated, cylindrical, hyaline, smooth with several percurrent proliferation.

**Conidia:** Fusiform, straight or slightly curved, euseptate, base or rarely with branched appendage, apical cell, conic, hyaline with 2 or more apical, simple or branched, spathulate or espathulate appendages, median cells brown, sometimes versicoloured, thick walled, smooth or verrucose.

**Mycotoxins:** Species of pestalotopris in reported to produce. Ramulosin, hydroxy ramulosin, ambuic acid, pestalotiopsone

**Diagnostic features:** Colonies white producing black acervuli, conidiophore branched cylindrical or lageriform, holoblastic and annellate conidiogenous cells, conidia fusiform 5-celled, with spathulate appendages at the apex ovoid conical protruded base.

### 46. *Phoma* Sacc

**Synonymys** =*Plendomus*

=*Leptophoma*

= *Bakerophoma*

=*Sclerophemalla*

=*Polyopeus*

=*Dechterophoma*

=*Polyronellaea*

Freshly isolated strains grow rapidly on standard nutrient media at 20 to 25°C, producing dark mycelium and forming pycnidia within a few days of inoculation.

Old cultures may form pycnidia less freely. The black pycnidia are globose or ovoid with ostioles at the tips of small papillae. Conidiophores are inconspicuous. The very numerous pycnospores are ovoid, colorless, guttulate, and either unicellular or with a single septum. The presence of both one-and two-celled pycnospores has caused the mold to be assigned to both *Phoma* and *Ascochyta*. Macroscopically, cultures of *P. herbarum* var. *medicaginis* are brown-black, moist and somewhat wrinkled with little aerial mycelium.

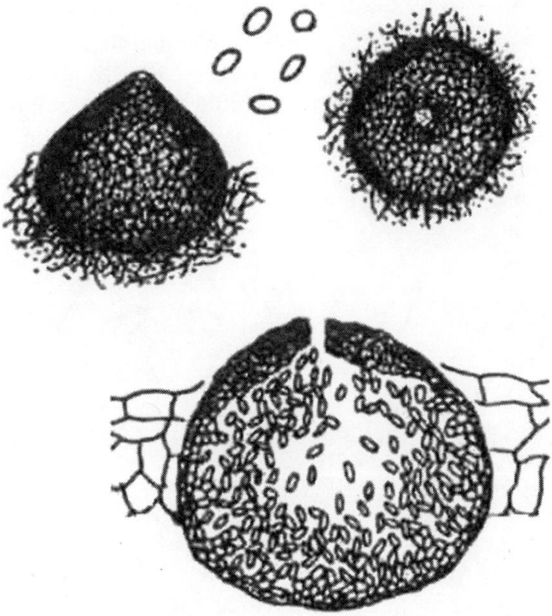

**i) *P. herbanum* var. *medicaginis***

**Colony:** Grow rapidly produce dark mycelium forming pycnidia

**Pycnidia:** Pycnidia globose, or ovoid with ostioles at the tips of small papillae

**Conidiophores:** In conspicuous produce numerous pycnidiospores.

**Conidia:** Ovoid, colourless, guttulate, either, unicellular, or bicelled, presence of both one celled and two celled conidia in the same pycnidium it is assigned to both ascochyta.

**Mycotoxins:** Brefeldin, cytochalasins B, phomazarin, terrein, emodin (*Phoma fovetfa, Cladosporium fulvum*), secalonic acid A (*Phoma terrestris*).

**Diagnostic features:** Pycnidia globose, ostioles at the tips of small papillae, conidia ovoid, hyaline biguttulate, 0 or 1 septate

## 47. *Phomopsis*

The genus Phomopsis produces black stromatic pycnidia on the stems, pods, and seeds of infected lupin plants. The first symptoms of infection are usually evident on the lower part of the stems, where yellow-brown, sunken lesions are formed. Black, stromatic pycnidia are produced in rows or irregularly scattered on these stem lesions. Dark-brown, water-soaked spots develop on infected pods; these pods then become dark-brown to black with pycnidia frequently produced in concentric rings. The interior of the pycnidial stroma consists of white pseudoparenchymatous cells surrounded by a black sclerotial plectenchyma. A flat fertile locule eventually develops in the upper part of the stroma under the black outermost layer. A long the inner wall of the locule, a dense pallisade of slender, simple or branched, tapering conidiophores are produced. Which produce two types of conidia: (1) Conidia are hyaline, one-celled, biguttulate, cylindrical with tapered ends, straight or very slightly curved. (2) β-conidia are produced in separate pycnidia on overwintered stems, and these are hyaline, filiform, and curved to strongly hooked conidia are extruded through the ostioles of mature pycnidia in cream-colored to salmon-pink droplets after 14 to 21 days. Only α-conidia have been observed on artificial media and they are similar in shape and size to those produced on the host.The ascigenous state of *P. leptostromiformis* is reported to be *Diaporthe lupini* Harkness.

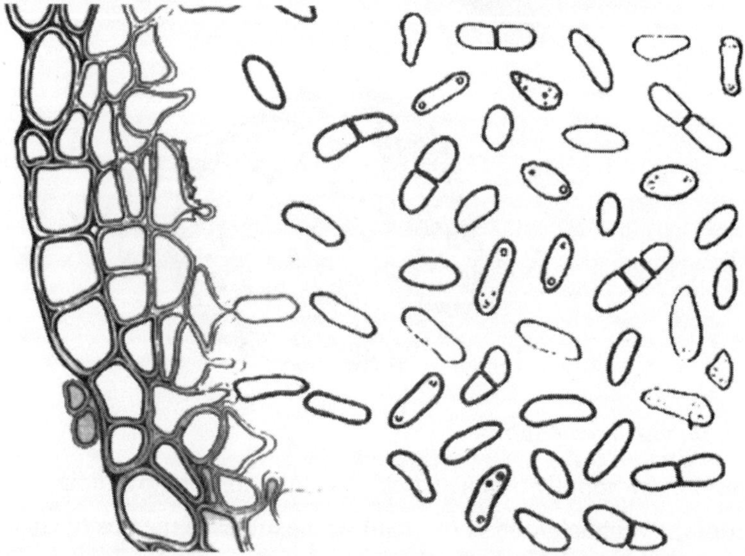

***P. leptostromiformis*:** A conspicuous, coarse white mycelium is visible on discoloured pods. Pycnidiam pseudoparenchymatous, surrounded by black sclerotia parenchyma. A flat fertile locule develops in upper part of the stroma which consists of pseudoparenchymatous cells surrounded by a black sclerotial plectenchyma. Under the black outermost layer along inner wall of the locule, a dense palisade of slender simple or branched, tapering.

α-conidia are hyaline, 1-celled, biguttalate, cylindrical with tapered ends, straight or slightly curved.

β-conidia are produced in separate pycnidia on clover wintered stems and these are hyaline, filiform and curved to strong hooked. Protopycnidia are at first white and gradually turn to become black stromatic pycnidia. The conidia are extruded through the oslioles of mature pycnidia in cream-coloured to salmon-pink droplets after 14-21 d. Only α conidia were observed in synthetic media.

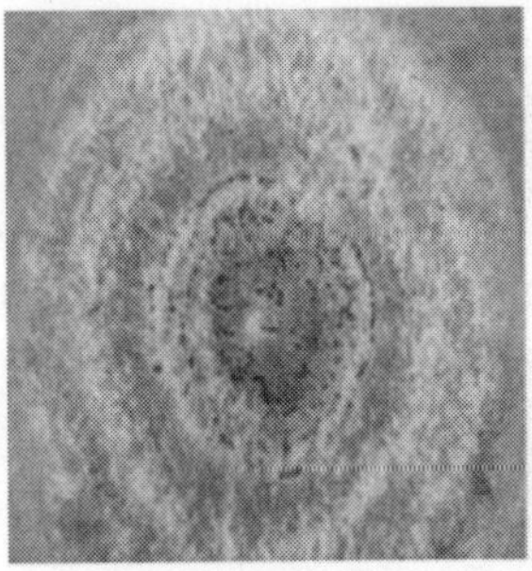

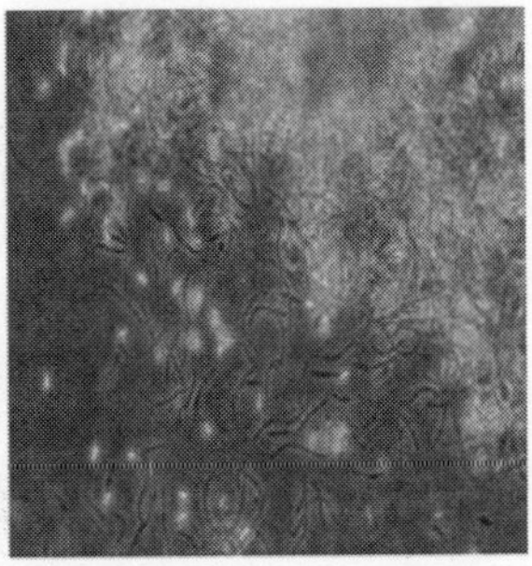

**Mycotoxins:** Lupinosin, cytochalasin H (*P. paspali*).

**Diagnostic features:** Stromatic pycnidial producing locules. Conidiophores are placed in a pallaside layer, simple or branched tapering conidia. α-conidia, 1-celled, biguttulate, cylindrical, straight or slightly curved. β-conidia are produced in separate pycnidium, hyaline, filiform, curved to shortly hooked.

## 48. *Piricauda* Bubak

**Colonies**: Effuse, thin, arachnoids, scarcely visible to the naked eye, with superficial, mycelium.

**Conidiophores:** Semimacronematous, mononematous, branched forming arched, sometimes anastomosing loop, slightly thicker than the ordinary hyphae.

**Conidiogenous Cells:** Monoblastic, integrated, terminal and intercalary, determinate, cylindrical or clavate, uniform, cicatrized, scars prominent. Erect, arising from the substratum, hyaline, simple, or branched at the apex, fasciculate, fascicles bearing metulae and phialides. Phialides divergent, slightly or closely adpressed, bottle shaped.

**Conidia:** Hyaline or pigmented, non-septate/ovoid or sometimes flattened on one side. Conidial heads are in shiny bulls or held together or in loose columns which are not distinguishable, solilitory, dry, pleuorogenous, broadly ellipsoidal, ovoid, spherical, sub-spherical or obpyriform, golden-brown, smooth or verrucose, uniform hyphae hyaline septate.

**Teleomorph:** *Nectria, Nectriopsis, Hypocrea* and *Ronmegueriella*.

### i) *P. coohinesis* (Subram.)

**Colony:** Effuse, thin, arachnoid, scarcely visible to the naked eye

**Mycelium:** Superficial

**Stroma:** None

**Setae and hyphopodia:** Absent

**Conidiophore:** Semi-macaronematous, mononematous, branched, forming arched, sometimes anastomosing loops, slightly thicker than ordinary hyphae

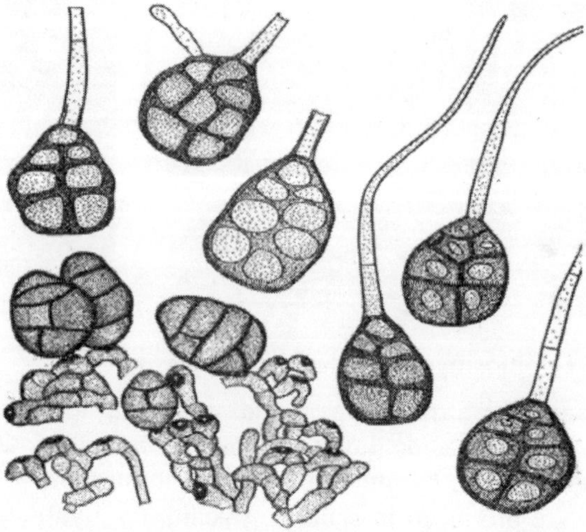

**Conidiogenous cells:** Monotretic, integrated, terminal and intercalary, determinate, cylindrical or clavate, uniform, cicatrized, scars prominent

**Conidia:** Solitary, dry, pleurogenous, broadly ellipsoidal, ovoid, spherical, subsphaerical or obpyriform often becoming rostrate, golden-brown, smooth or verruculose and muriform

**Mycotoxins:** Sporodesmin A

**Diagnostic features:** Conidiophore semi-macronematous, mononematous, cylindrical or clavate, cicatrized, scars prominent, broadly ellipsoidal, ovoid, golden-brown, verruculose and muriform

## 49. *Pithomyces* Berk. Ex. Br

*Pithomyces chartarum* is the causative agent of pithomycotoxicosis, commonly called facial eczema. They occurs in sheep and cattle grazing pasture in summer and fall in New Zealand and less frequently in Australia and South Africa. *P. chartarum* grows saprophytically on dead vegetable matter and is cosmopolitan in distribution. It has been reported more often from the tropical warm temperate zones than from the cool temperate zones. It occurs in sufficient numbers to cause animal disease virtually only in pasture and under rather restricted regimes of moisture and temperature.

### Key to species

Mature conidia mostly without longitudinal septa ........................................ 1

Mature conidia mostly with longitudinal septa ........................................ 5

1. conidia with 3 or more septa ........................................ 2

1a. conidia fewer than 3 septa ........................................ 4

2. conidia more than 11 μm thick ........................................ 3

2a. conidia less than 11 μm thick atro-olivaceus ........................................ 5

3. conidia 3-5 septate ........ *flavus* (1)
3a. conidia 6-septate elaeidi cola ........ 3
4. conidia 1 – septate ........ *cupiniae* (2)
4a. conidia 0 – septate ........ *africanus* (3)
5. conidia broadly ellipsoidal ........ 6
5a. Conidia variable in shape seldom ellipsoidal ........ 7
6. Conidia mostly with 3-transverse septa – ........ *chartarum* (4)
6a. Conidia mostly with 2 – transverse septa – ........ *maydicus* (5)
7. Conidia often clavate ........ 8
7a. Conidia not clavate ........ 9
8. Conidia with not more than 3 transverse septa ........ *sacchari* (6)
8a. Conidia often with 4-6 transverse septa ........ *cyanodontnis* (7)
9. Grows either an crataegus or palms ........ 10
9a. Grows on other than palms and crataegus ........ 11
10. Grows on crataegus ........ *quadrates* (8)
10a. Grows on palms ........ *pulvinalu* (9)
11. Conidia obturbinate or obgriform ........ *ellisii* (10)
11a. Conidia not obturbinate a obpyriform ........ 12
12. Conidia mostly more than 12 μm thick ........ *Karou* (11)
12a. Conidia 8μm or less thick ........ 13
13. Epispore cracking to form plates ........ *leprosus*– (12)
13a. Epispore not cracking ........ *graminicolo* – (13)

**i) *P. chartarum* (Berk & Curt.) M.B. Ellis**

**Synonymys** = *Sporidesmium chartarum* Berk. & Curt
= *S. echinulatum* Speg.
= *Scheleobrachea echinulata* (Speg.)
= *Sporidesmium bakeri* Syd
= *Pithomyces maydicus* (Sacc.) M.B. Ellis

*P. chartarum* is an imperfect mold which forms rough, dark, several-celled, barrel-shaped aleurospores, borne singly on short simple conidiophores. A small piece of denticle remains attached to the base of free spores. Septa in the spores are both transverse and longitudinal and their number none to five (usually three) transverse, none to three (usually two) longitudinal, distinguishes the spores of *P. chartarum* from other species.

*Scheleobrachia echinulata* Hughes is a synonymys of *P. charatum.*

**Colony:** Fast growing, at first hyaline, later becoming dark-grey

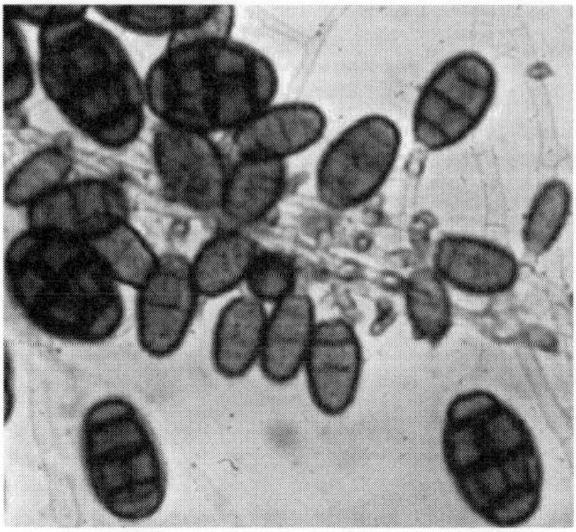

**Mycelium:** 1. Hyaline, sparsely septate often becoming verrucose.

2. The hyphal cell walls are reported to be dark-brown, thin walled, densely septate, smooth walled

3. Dark-brown, densely septate, smooth walled later becoming echinulate

**Conidiogenous cells:** Fertile hyphae sub hyaline to pale olive, conidiferous pegs globose to spherical

**Conidia:** Broadly ellipsoid, 3-4 transverse septate and mostly 2- longitudinal septa, often somewhat constricted at the septa

**Mycotoxins:** Sporodesmin A, F, G, H, E, sporodesmin C diacetate

**Diagnostic features:** Matures both with transverse and longitudinal septate.

### 50. *Pteroconium* Sacchario ex Grove

**Colonies:** Punctiform, scattered, black mycelium immersed stroma well developed erumpent, pseudoparenchymatous

**Conidiophores:** Basanxic, macronematous, mononematous, arising singly form lageniform ellipsoidal or doliform conidiophores another cells, usually unbranched straight or flexous cylindrical mid to dark-brown, septa, smooth or verruculose

**Conidiogenous cells:** Monoblastic or occasionally polyblastic, integrated, terminal and intercalary, cylindrical, denticulate

**Conidia:** Solitory, dry, acropleurogenous, simple, lenticular with a hyaline rim are germ slit, round, polygonal or dendric, pale to dark-brown, smooth, single celled

#### i) *P. pterospermum* (Cooke & Massee)

**Colony:** Sporodochia punctiform, scattered, black

**Mycelium:** Immersed

**Stroma:** Well developed, erumpent, brown or dark-brown, pseudoparenchymatous

**Setae and Hyphopodia:** Absent

**Conidiophores:** Basauxic, macronematous, mononematous, arising singly from lageniform, ellipsoid or doliform, conidiophore mother cells on the surface of the stroma, usually unbranched, straight flexuous, rather narrow, cylindrical, colourless or pale-brown except transverse septa, smooth or verruculose.

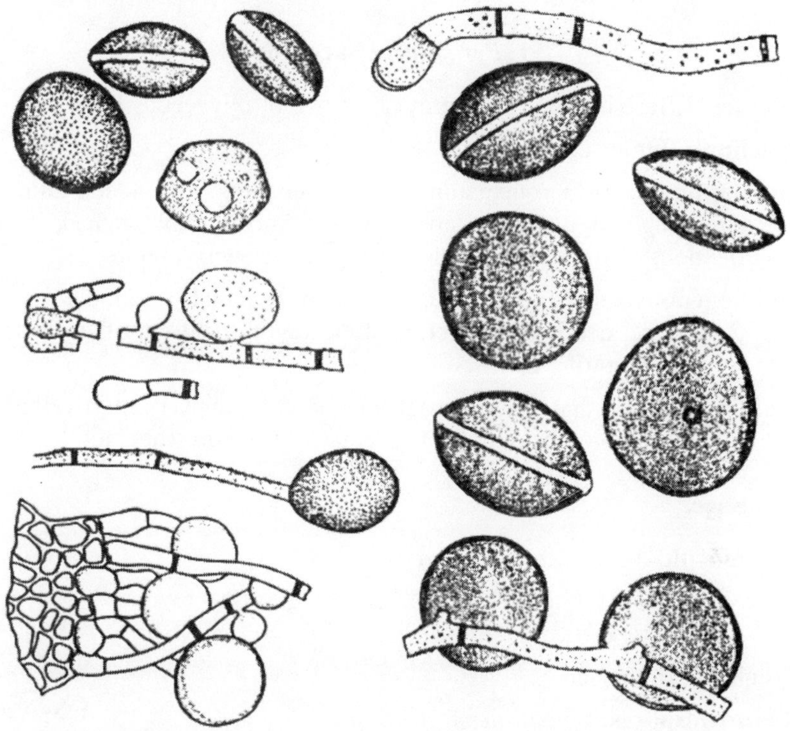

**Conidiogenous cells:** Monoblastic or occasionally polyblastic, integrated, terminal and intercalary, cylindrical, denticulate, denticles usually short, cylindrical

**Conidia:** Solitary, dry, acropleurogenous simple, lenticular with a hyaline rim or germ slit, round, polygonal or irregular in face view, sometimes lobed or dentate, pale to dark-brown, smooth, non-septate

**Key to species**

Conidia on natural substrate almost all lobed or dentate ................................ *pterospermum*

Conidia on natural substrate not lobed or dentate ............................*apiospora camptospora*

**ii) *Pteroconium* state of *Apiospora camptospora***

**Sporodochia:** Ellipsoidal or elongated, black

**Conidiophore:** Mother cells, 8-12 X 4-7 µm

**Conidia:** Lenticular, round or polygonal in face view, mid-dark-brown with a distinct hyaline rim or green slit.

**Mycotoxins:** 4-hydroxymellein, demethylviridin

**Diagnostic features:** Conidia lenticular, conidiphores basauxic, sporodochium elongated, black.

**Teleomorph:** *Apiospora comptospores*

## 51. *Pyricularia* Saccardo

**Colonies:** Effuse hairy, grey to grayish-brown or olivaceous-brown.

**Mycelium:** Immersed.

**Conidiophores:** Macronematous, mononematous, slender, thin walled emerging singly or in small groups, mostly unbranched, straight or flexuous, geniculate towards apex, pale –brown, smooth

**Conidiogenous cells:** Polyblastic, integrated, terminal, sympodial, cylindrical, geniculate, denticulate, each denticles cylindrical cut off by a septum to form a separating cell.

**Conidia:** Solitory, dry, acropleurogenous, simple, obpyriform, hyaline to pale-olivaceous-brown, smooth, 1-3 septate, hilum protruberant.

**i) *P. grisea* Sacc.**

**Teleomorph:** *Magnoparthe grisea*

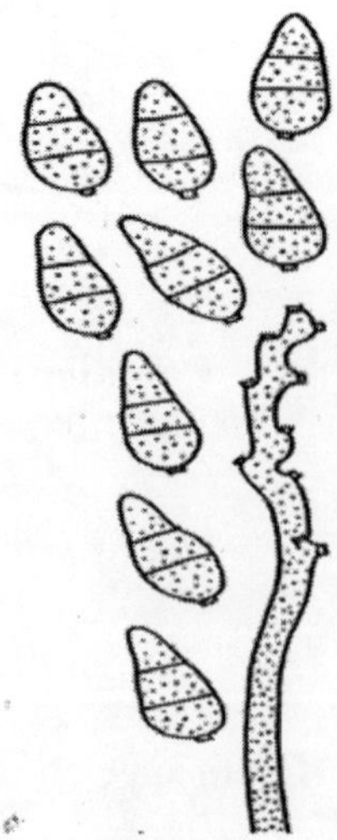

**Colony:** Effuse, thinly hairy, grey, grayish-brown, olivaceous-brown

**Mycelium:** Immersed

**Chlamydospores:** Sometimes formed in culture

**Stroma:** None

**Setae and hyphopodia:** Absent

**Conidiophores:** Macronematous, mononematous, slender, thin-walled, mostly unbranched, straight or flexuous, geniculate towards apex, pale-brown, smooth

**Conidiogenous cell:** Polyblastic, integrated, terminal, sympodial, cylindrical, geniculate, denticulate, thin walled, cut off as a rule by a septum to form a separating cell.

**Conidia:** Solitary, dry, acroleurogenous, simple, obpyriform, obtunicate or obclavate, hyaline to pale-olivaceous-brown, smooth, 1-3 septate, hilum often protruberant.

**Mycotoxins:** Pyricularin

**Diagnostic features:** Conidiophores macrone-matous, unbranched geniculate towards apex, conidiogenous cells polyblastic, sympodial, denticulate, conidia solitary, acroleurogenous, 1-3 septate with protruberant hilum.

**ii) *P. oryzae* Cav.**

**Colony:** Effuse, thinyhairy, grey, greenish-brown

**Mycelium**: Immersed

**Chlamidospores:** Absent

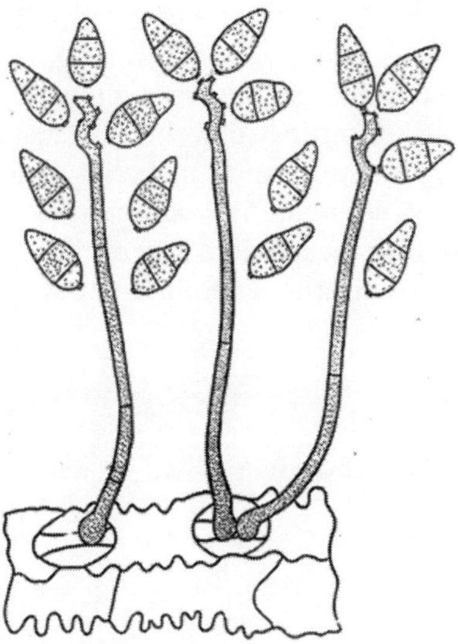

**Setae and hypopodia:** Absent

**Conidiophores:** Upto 130 μm long 3-4 μm thick, polyblastic integrated, terminal, sympodial, cylindrical, geniculate, denticulate, thin walled, cut off to form a separating cell

**Conidia:** Solitary dry acronematous, obpyriform, obtunicate or obclavate hyaline to pale pale-olivaceous brown, smooth, 1-3 septate, hylium protuberant.

**Mycotoxins:** Tenuzonic acid

**Diagnostic features:** Conidiophore geniculate towards apex, conidiogenous cells polyblastic, sympoidal, denticulate, conidia solitary, acropleurogenous, 1-3 septate with protuberant hylium.

**Key to species**

1. Conidia 17-28 X 6-9 .......... *P. grisea*

1a. Conidia 17-23 X 8-12 .......... *P. oryzae*

## 52. *Rosellinia* De Not

Colonies on oat meal agar, growing moderately fast with advancing zone appressed, even obscurely zonate, with white velvety to floccose mycelium over pale are third, with moist appressed mycelium over outer two-thirds, translucent to white or pale-pink, pale-vinaceous- pink, light-vinaceous cinnamon in reverse, exudates absent, slightly fragrant odor, no growth at 37 °C on oatmeal agar.

Colonies on V8 juice agar at 23°C, 12 hr photoperiod, growing slowly attaining 25-35mm in 14 days, submerged at the margin, mostly with appressed mycelium, with some floccose aerial mycelium towards inner third, zonate, translucent to pale-grayish, pink-vinaceous cinnamon, reverse translucent to cinnamon, Orange-cinnamon, cinnamon. or pale-reddish brown-russet Fawn color, odor fragrant.

Stromatic ascomata occurring singly but aggregated or fused in small clusters on a mycelial subiculum on decorticated wood; stromata hemispherical, papillate, smooth, shiny, black, subiculum a thin mycelial mat, pal-buff, or sometimes appearing only as a lightly colored, discoloration of the wood adjacent to the stomata: Ascus cylindrical with an amyloid apical plug; ascospores brownish-gray, broadly elliptical, to slightly reniform, smooth, without appendages or sheaths, with ventral germ slit.

### i) *R. necatrix* Prill

**Colony:** Spreading, blackish-brown, powdery with the production of conidia or conidiophores

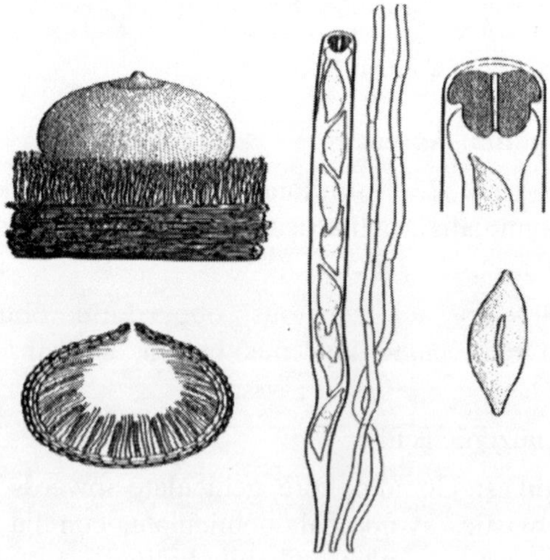

a. Exterior of perithecium seated on subculumn. b. Section through perithecium with asci. c. Paraphyses and ascus with ascospores. d. Apex of ascus. e. mature ascus

**Ascomata:** Single or gregarious borne on a felt like subculumn of dark hyphae, ostiolate globose or broadly or pyriform. Ostiolar neck papillate or comical

**Ascus:** Unitumicate, cylindrical, long talked with am amyloid special apparatus, liming inside of perithecial wall, 8-spored.

**Ascospores**: One celled, dark brown, ellipsoid, often laterally compressed with a long longitudinal germ slit.

**Anamorphs**: *Dematophora, Nodulosporiurn*

**Mycotoxins:** Cytochalasins-E

**Diagnostic features:** Ascomata borne on a felt like subiculum of dark hyphae, perithecium broadly obpyriform, ascus long stacked, ascospores one celled laterally composed with longitudinal germ slit

### 53. *Sclerotinia* Fuckel

Colonies spreading, reaching 4.0 cm in diameter in 5 days at 20ºC on OA, aractinoid white. Sclerotia tuber like or cushion shaped, black, 1-8mm or more diam consisting of a ring of dark, palisade like, thick walled cells, a cortex of 6.8 layers of hyaline, almost pseudoparenchymatous cells. It was a medulla of thick walled hyphae not embeded in a gelatinous matrix were some inter cellular spaces. Microconidia (Spermatia) of 2.5-3.2 µm diameter belonging to the genus *Myriococcum* Syd, are produce abundantly superficial sporodochia on flask shaped phialides, when nutrients on the substrate have been exhausted. The sclerotia mostly germinate to form apothecia (capogenic germination). Apothecia are produced in the field during spring and can also be obtained *in vitro* on nutrient deficient moist substrates after a period of cold storage. Apothecium primordium consists of irregularly shaped cells with thin walls lacking fibrous. Apothecial stalks are up to 3 cm long and 1-2 mm thick, they are concave, yellowish – brown, usually 3-8 mm diam. Asci cylindrical and with an amyloid pore. Ascospores ellipsoidal to almost fusiform, sometimes slightly asymmetrical, uninuecleate.

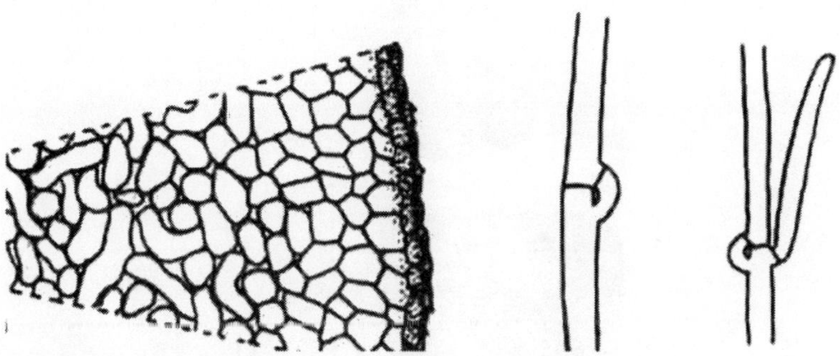

**i)** ***S. sclerotiorum*** **(Lib.) De Bary**

**Synonymys** = *Peziza sclerotiarum* Lib.

= *Sclerotimia libertiana* Fuckel

= *Whetzelimia sclerotiorum* (Lib.) Korf & Dumant.

**Anamorph:** *Sclerotium varium*

Sexual fruit bodies are absent. Sclerotia brown to black, globose to irregular, compact, mycelium usually light, parasitic principally on underground parts of the plants

**Colony:** Spreading, reaching 4.0 cm diam in 5 days at 20°C on OA, arachnoid, white.

**Sclerotia:** Tuber like or cushion shaped, black, 1-8mm or more in diam. Consists of a rind of dark palisade- like thick walled cells, a cortex of 6-8 cells and a medulla of thick walled hyphae not embedded in a gelatinous matrix and in which there are some inter cellular spaces. Micro-conidia (Spermata) 2.5-3.0µm diam. belonging to genus *Myrioconium* are produced abundantly in superficial sporodochia on flask shaped phialides when numbers in the substratum have been exhausted. The sclerotia mostly germainate to form apothecia (Corpogenic germination).

**Ascocarp:** Apothecia are produced in the field in spring and can also be obtained in nutrient deficient medium after a cold storage. Apothecial primordial irregular in shape with thin walls lacking fibrous layers. Apothecial stalks are 3 cm long and 1-2 mm thick. Discs are concave yellowish-brown, asci cylindrical with amyloid pore white ascospores were ellipsoidal to almost fusiform sometimes slightly asymmetrical and uninucleate

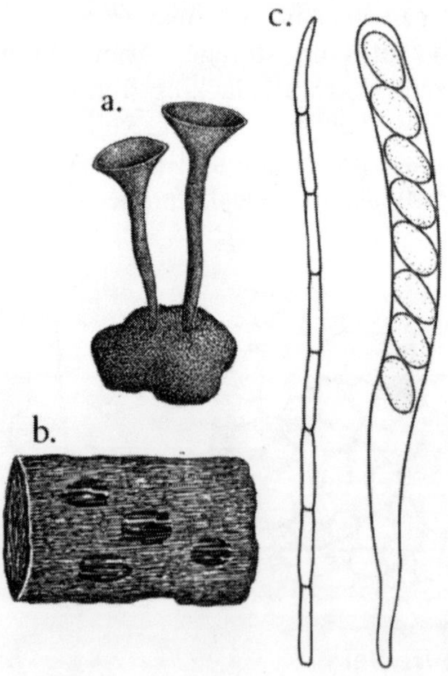


a. Apothecium arising from sclerotium.b. Black sclerotia embedded in dead stem of dill. c. paraphyses and ascus with ascospores. d. mature ascospore

**Apothecium:** 2-10 mm wide, cupulate or convex, cinnamon to umber colour. Sub-hymenium composed of textures intrica with hyphae brown walls, gelatinum, medullary excipulum composed of hyaline loosely interwoven textra intricata. Ectal excipulum composed of texturea prismatica with outer walls inflated to globose, with hyaline to type brown walls

**Ascus**: Unitunicate, cylindrical, tapering toward base, thin walled but slightly, thickened at apex, apical pore bluing in iodine, 8- spored

**Ascospores:** Hyaline, 1-celled, uniseriate, ellipsoid to somewhat flattened on one side, smooth biguttulate.

**Mycotoxins:** Sclerin, sclerolide

**Diagnostic features:** Sclerotia rubber like or cushion shaped, palisade like thick walled cells. Anamorph *Myrioconium*, ascocarp apothecium, cylindrical with amyloid pore, ascospores single celled, ellipsoid.

### 54. *Sepedonium* Link

**Colonies:** Spreading, white to pale–brown, effuse.

**Conidiophores:** Slender, typically verticellately branched.

**Conidia:** Hyaline, one-celled, ovoid, borne singly at the apex chlamydospores abundant large, spherical, apical, bright-yellow, tuberculate, 1-septate.

**Teleomorph:** *Hyphomyces*

#### i) *S. ampullosporum* Damon

**Colony:** Spreading, white to pale-brown, effuse

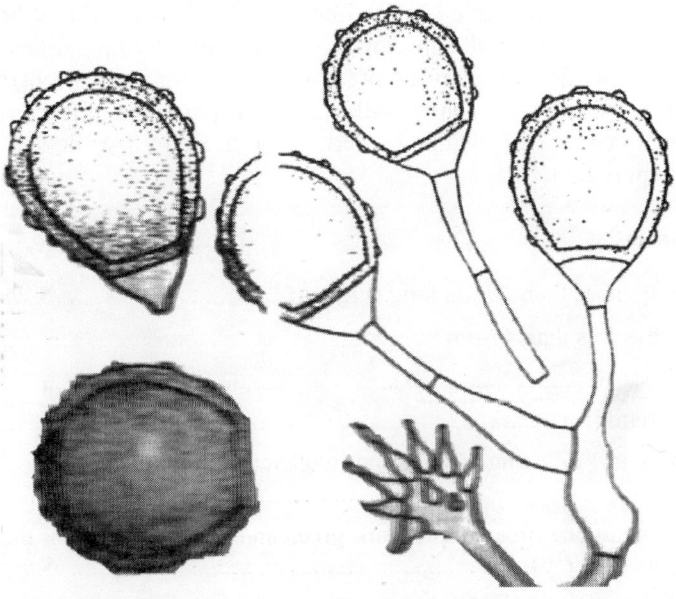

**Conidiophore:** Slender, typically verticellately branched, but branching sometimes sparse

**Conidia:** Hyaline, 1-celled, ovoid, bone singly at the apex (like *Verticellium*)

**Chlamydospores:** Abundant large, spherical apical, bright-yellow, tuberculate, hyperparasite

**Teleomorph:** *Hyphomyces*

**Mycotoxins:** Rugulosin, skyrin, sepedomin

**Diagnostic features:** Abundant, spherical, bright-yellow tuberculate chlamydospores

## 55. *Stachybotrys* Corda

*Stachybotrys* is a saprophytic fungi worldwide in to distribution. Soviet scientists have described this fungus under the name *S. alternans* Bonorden var. *jateli* with two different variants; one toxic and other nontoxic. While the literature of other parts of the worlds used the name *S. atra* Corda and in a few cases as *S. chartarum* (Ehrenberg ex Link) Hughes. *S. atra* is a typical cellulose-decomposing fungus that grows well on cellulose-rich substrates such as straw and hay, but it has also been isolated from cereal grains, plant debris, soil, peas, cotton, and sugar cane roots. It teleomorphic stage is *Melanopsamma*.

The aerial mycelium is colorless. The colonies on agar become black on standing, except for the peripheral zone, which remains grayish-white with abundant aerial mycelium. *S. atra* can grow at pH levels of 3.0 to 9.8, although the optimum pH values are 5.6 to 6.6. The optimum temperature for growth is 20 to 25°C, although the fungus grows within the range of 5 to 40°C.

Young hyphae are hyaline, become olive gray-brown, conidiophores are olive-brown or olive-gray, with a few septa, either simple or sympodially branched. The upper, usually darker part of the conidiophore bears a group of phialides (sterigmata). The phialides are hyaline or dark-colored, varying from 6 to 10 in number. The conidia, which are produced at the small tips of the phialides, differ greatly in size. Conidia are oval or rounded hyaline or dark-colored, smooth-walled or occasionally showing short spiny structures.

### Key to the species

1. Conidia mostly more than 20 μm long (Fig. 1a) ................................ *S. theobromae* (1)

1a. Conidia always less than 17 μm ................................ 2

2. Conidia reniform ................................ 3

2a. Conidia not reniform ................................ 4

3. Phialides brone in 1 or 2 rings around the swollen apex of the Conidiophore ................................ *S. oenanthes* (2)

3a. Phialides borne in one ring, simple, dark-green and verrucose apex of the conidiophore (Fig. b) ................................ *S.nephrospora* (3)

4. Conidia with dark longitudinal striations (Fig.1c) ................................ *S.cylindrospora* (4)

4a. Conidia without striations ................................ 5

5. Conidia when mature smooth ................................ 6

5a. Conidia when mature verrucose .......................................................................... 8

6. Conidia ellipsoidal or boat shaped truncate at the base................. *S. sanseveriae* (5)

6a. Conidia ellipsoidal not truncate at the base ............................................................ 7

7. Conidia dark-brown, 3.0-5.5 µm long.(Fig. 2 d) ................................ *S. parvispora* (6)

7a. Conidia greenish or grayish-brown 6.0-11.0 µm long
(Fig.2 e)................................................................ *Melanopsamma pomformis* (7)

8. Conidia often obliquely attenuated at the base, 8-12 x 4-6 µm
(Fig.2 f) ........................................................................................ *S. dichroa* (8)

8a. Conidia not obliquely attenuated at the base ............................................................ 9

9. Conidia 11-15 x 6-8 µm (Fig.2 g) ..................................................... *S. kampalensis* (9)

9a. Conidia 8-11 x 5-11 µm (Fig.2 h) ............................................................. *S. atra* (10)

9b. Conidia 6-8 x 4-5 µm (Fig.2 i) .......................................... *S. atravar microspora* (11)

**i) *S. atra* Corda**

**Synonymys**: = *Stachbotrys atrogrisea* Ellin & Everth
= *S. lobulata* Berk
= *S. scabra* Cooke & Harkin.
=*S .alternans*

**Colony:** Effuse, usually black or blackish-green

**Reverse:** Uncoloured

**Stroma:** None

**Setae and Hypopodia:** Absent

**Conidiophores:** First hyaline, soon becomes olivaceous-brown to black, rough or covered with granules especially towards the apex.

**Phialides:** Monophialidic, discrete, in groups at the apex of each stipe, determinate, clavate, ellipsoidal or broadly fusiform usually with a very small opening and no collateral

**Conidia:** Broadly ellipsoidal to sub spherical, dark -blackish – brown to black, verrucose.

**Mycotoxins:** Satratoxin C, D, E, F, G and H. verruculogen, verrucarin, trichoverrins.

**Diagnostic features:** Conidia verrucose, conidia not obliquely attenuated at the base

**ii) *S. chartarum* (Ehrenb) S.Hughes.**

**Synonymys** = *Chartarum* (Ehrenb) wallr.
= *Chartarum* Ehrenb.

**Colony:** Spreading covered with a dark powdery mass of conidia

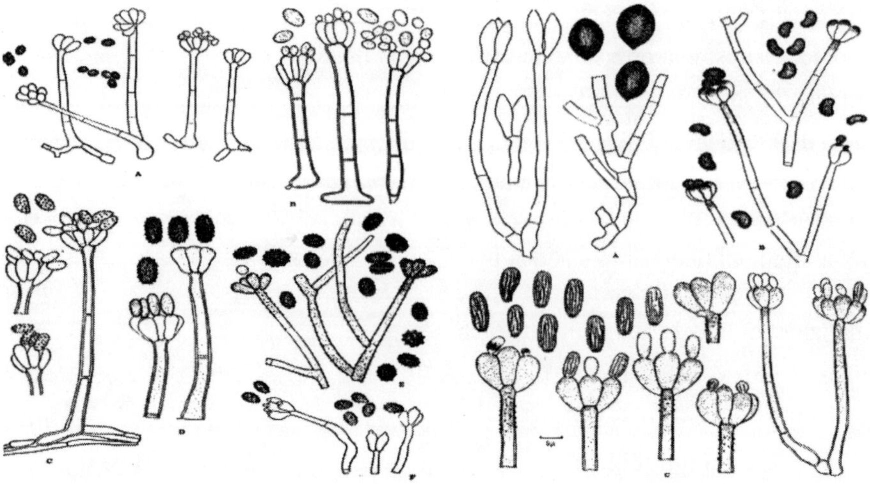

a) *S. parvispora* b) *S. Melanopsamma pomiformis* 2 a) *S. theobromae* b) *S. nephrospora* c) *S. cylindrospora* c) *S. dichroa* d) *S. kampalensis* e) *atra* var. *microspora*

**Reverse:** Uncoloured

**Conidiophores:** Simple or irregularly branched, hyaline at the base, dark-olivaceous roughed towards the apex bearing cluster of 4-10 phialides.

**Phialides:** Obovate to ellipsoidal, hyaline late becoming olivaceous.

**Conidia:** Aggregating into slimy masses, ellipsoidal at first hyaline at maturity dark olivaceous-grey, opaque, almost smooth walled to coarsely roughen with warts and ridges.

**Mycotoxins:** Satratoxin G and H, verruculogen, verrucarin, trichoverrins, trichoversols.

**Diagnostic features:** Conidia aggregating, slimy masses ellipsoidal, hyaline when growing become dark-grey at maturity, conidial wall with warts and ridges. Conidia of *S.lobulata* differ from *S. alternans* are wide except they elliptical, smooth and turn black on standing the fungus is non-toxic.

## 56. *Sphacilia* Lev.

**Colonies:** White, spreading, hyaline, branched from a stroma

**Sporodochium**: Stroma light, spreading

**Conidiophores:** Hyaline, simple, septate, bearing conidia terminally surrounded in compact palisade layer.

**Conidia:** Small, ovoid, hyaline, 1-celled produced in sugary honey dew.

**Telomorph:** *Claviceps*

### i) *S. purpurea*

**Colony:** Stromatic, different range colours such as purple black, blue, green, buff produce conidia in honey dew

**Conidiophores:** Produced in stromatic sporodochium, conidiophores hyaline simple, septate, arranged in a compact palaside manner bearing conidia terminally.

**Conidia:** Hyaline, small ovoid, 1-celled, produced in a surgery, honey dark

**Telopmorph:** *Claviceps*

**Mycotoxins:** Paspali, paspalicine, paspaline

**Diagnostic features:** Sporodochium on sclerotia developed in the spike, conidiophores simple, unbranched, arise in phialide layer. They produce conidia which is single celled, hyaline and has a slimy layer

## 57. *Talaromyces*

*Talaromyces* is a genus of family Trichocomaceae of Ascomycotina. Species in the genus form soft, cottony fruit bodies (ascocarps) with cell walls made of tightly interwoven hyphae. The fruit bodies are often yellowish or surrounded by yellowish granules.

### i) *T. avellaneus* C.R. Benfanin

**Colony:** Spreading, white to cream colour or yellow

**Ascocarp:** Cleistothecium, spherical or less commonly pedicillate, biseriate

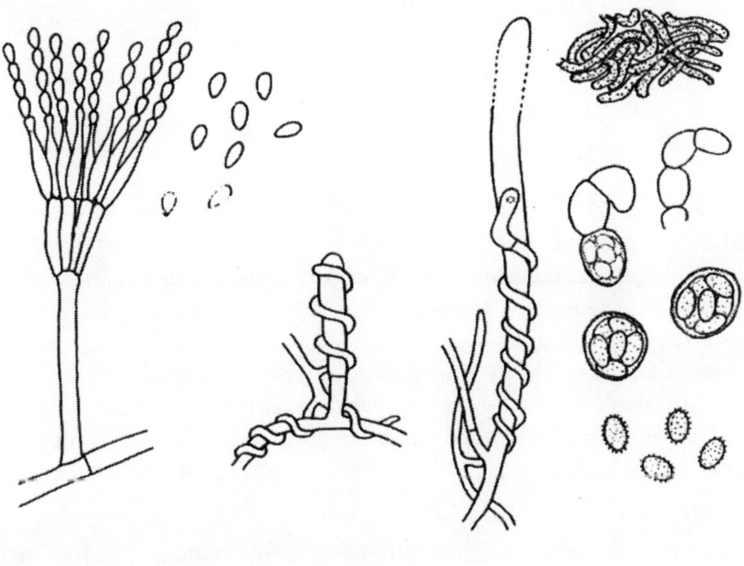

**Peridium**: Loosely interwoven hyphae

**Ascus:** Irregularly disposed, ellipsoidal to sphaerical unitunicate, deliquescing at maturity, usually borne as short chain, less commonly singly or irregularly, 8-ascospores

**Ascospores:** Irregularly arranged, single celled hyaline to yellow sphaerical to ellipsoidal, wall spinose

**Mycotoxins:** Anrthraquinone

**Diagnostic features:** Ascocarp, gymnothecium, asci spherical to ellipsoidal, hyaline to yellow, single celled spinose

**Anomorph**: *Penicillium*

### 58. *Thanatephorerus* Donk

Colonies fast growing usually filling a dish within 3 days, at first colourless submerged or with sime radiating aerial hyphae, rapidly becoming brown. Hyphae subhyaline to pale-brown, thick walled. Often constricted near the septa and where branching with conspicuous dolii-pore septum, lacking clamp connections, branched at wide angles. Sclerotia developing from irregular agglomeration of uniform multiform hyphae irregular. Basidial hymenia appearing on stems or leaves or above the soil. *Basidiocarp* resupinate, effuse, loosely attached to the substratum. Basidia arising in a symmetrical cymes or racemes from tufts of ascending hyphae, Basidia variable in shape, barrel shaped to cylindrical, sterigmata, straight and long. Basidiospores oblong to broadly ellipsoid, unilaterally flattened, apiculate at the base, hyaline, smooth walled, not amyloid, germinating by replication with secondary spores.

**i) *T .cucumeris* (Frank.) Donk**

**Synonymys** = *Hypochrus eucumeris*

= *Corticium solani*

= *Hypochrus solani*

= *Pellicularia filamentosa*

= *Hypochrus filamentosus*

= *Ceratobaridium filamentosum.*

Basidial hymenium appearing on stems or leaves just above the soil. Fructification resupinate, creamy, effuse, loosely attached to the substratum. Basidia arising asymmetrical cymes or racemes from tufts of ascending hyphae, variable in shape, barrel shaped to cylindrical sterigmata straight and as long as metabasidia. Basidiospores oblong to broadly ellipsoidal, unilateral falttend, apiculate at the base, hyaline, smooth walled, not amyloid, germinating by repetition with secondary ascospores. Basidiospores are mostly uninucleate, but sometimes up to 35% may be binuecleate.

**Anomomorph:** *Rhizoctonia solani* Khun.

**Colony:** Fast growing usually filling petridish with in 3-days, at first white submerged rapidly becoming brown.

**Mycelium**: Sub hyaline to plae –brown, thick walled often constricted near the septa, branched at wide angles, often anastomosing.

**Sclerotia:** Sclerotia developing from irregular agglomeration of uniform morlniform hyphae irregular in outline, solitary and about 1 mm in diameter sometimes confluent to conspicuous crusts, often in zones.

**Mycotoxins:** Slafranine, ferulic acid (*R. legguminicola*), 3 nitro-4-hydroxy phenyl acetic acid (*R. solani*), poly phynyl amyl anhydride, cytochalasin E (*Rosellina necatrix*).

**Diagnostic features:** No sexual fruit bodies are found. Hyaline with large cells septa of branch set off main hyphae, parasitic chiefly on roots and underground parts of the plant.

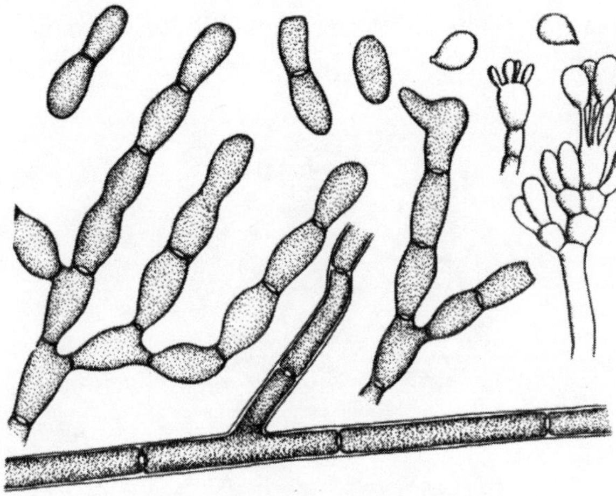

*Rhizoctinia solani*

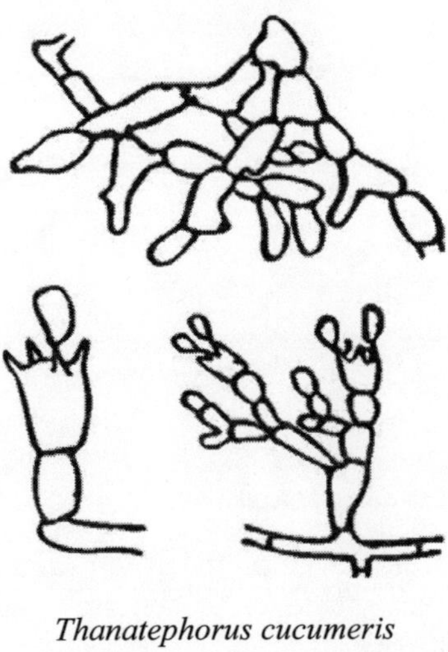

*Thanatephorus cucumeris*

## 59. *Thielavia* Zopf

**Ascomata:** A non-ostiolate, globose.

**Perithecium:** Glabrous, setose or tomentose, ascomal wall brown, composed of flattened cells (textura epidermoidea).

**Ascus:** In a fascicle or layered, ellipsoidal, saccate or cylindrical, rarely clavate, with a thin wall that is evanescent at maturity, 4-8-spored.

**Ascospores:** Single celled, dark-brown, fusiform, ellipsoidal, obovate or clavate, with single germ pore at one end

**Anamorph:** *Myceliophthora*

**i) *T. terricola* Glman& Abbof**

**Synonymys** = *Thielvdella*

**Colony:** Spreading, white appearing black, producing small perithecia

**Hyphae:** Hyaline, branched, spreading

**Ascocarp:** Perithecium developing from spiral ascogenia, non-ostiolate, globose, glabrous, setose or tomentose with a brown or pale wall of texture epidermoidea

**Ascus:** Fasciculate, clavate or ellipsoidal unitunicate, walls brown, without appendages, evanescent

**Ascospores:** Ellipsoidal or fusiform, apiculate, dark-olivaceous-brown, unicellular with one germ pore at the end which is not thickened

**Anamorph:** Rarely present, blastic or phialidic conidia produced in a few species

**Mycotoxins:** Malformina

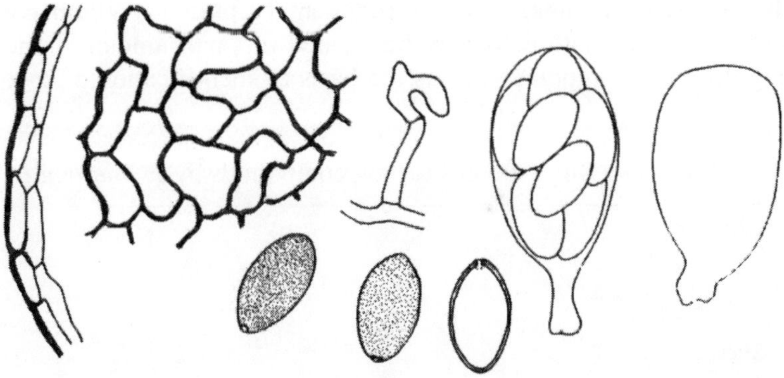

**Diagnostic features:** Colonies white with dark coloured ascospores, ascocarp wall glaberous, asci, ellipsoid or clavate, evercant, ascospores, ellipsoidal or fusiform anamorphoid with phialidic conidia.

### 60. *Torulopsis* Berlsee

Many wild types of yeast which sometimes cause problems in beverages and which appear in abundance on plating out many kinds of mouldy materials belong to *Torulopsis*. The cells are round or ovoid reproducing by multilateral budding capsules rare and never containing a starch like compound frequently fermentation. There are 22 species and one variety separated according to their fermentative abilities and ability to utilize nitrates (Table 3.17). Although they are common associates of mould, little is known as to have they contribute to mouldiness. Cells are round or ovoid reproducing by multilateral budding.

#### i) *T. anatomia* Zwellenberg

Growth in glucose-yeast extract-peptone water: After 3 days at 25°C the cells are spheroidal to short-ovoidal, (2-6) x (3-8μ), single, in pairs or in short chains. Elongate cells may occur. A ring may be present. Growth on glucose-yeast extract-peptone agar: after one month at 25°C the streak culture is whitish cream-colored.

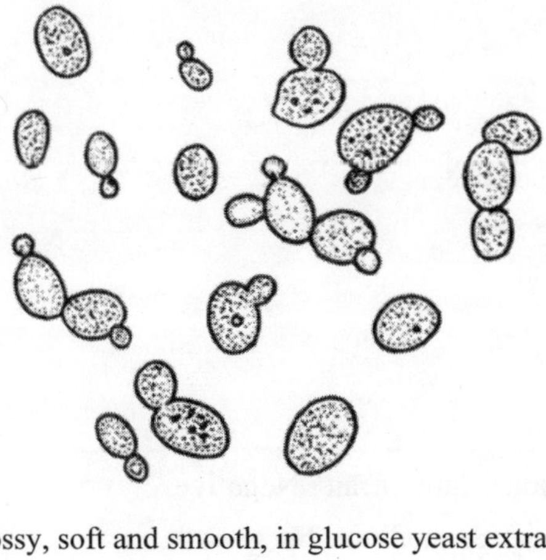

*T. anatomiae* glossy, soft and smooth, in glucose yeast extract peptone water.

**Cultures on corn meal agar:** A rudimentary psedomycelium consisting of branched chains of elongate cells may be formed in small amounts. The ability of assimilating different compounds forms the basis of identification of species (Table 3.17).

**Table. 3.17.** Ability of assimilation of carbon compounds by *T. anatomia*

| | |
|---|---|
| Glucose | + |
| Galactose | – |
| L- Sorbose | – |
| Sucrose | – |
| Maltose | – |
| Cellobiose | + |
| Trehalose | + |
| Lactose | – |
| Melibiose | – |
| Raffinose | – |
| Melezitose | – |
| Inulin | – |
| Soluble starch | – |
| D-xylose | – |
| L- Arabinose | – |
| D-Arabinose | – |
| D- Ribose | + (weak) |
| L- Rhamnose | + |
| Ethanol | + (weak) |
| Glycerol | – |
| Erythritol | – |
| Ribitol | – |
| Galactitol | – |
| D-Mannitol | + |
| D-Glucitol | – |
| α-Methyl-D-glucoide | + (weak) |
| Salicin | + |
| DL- Lactic acid | – |
| Succinic acid | – |
| Citric acid | – |
| Inositol | – |

Assimilation of potassium nitrate: Negative.

Growth in vitamin-free medium: No growth.

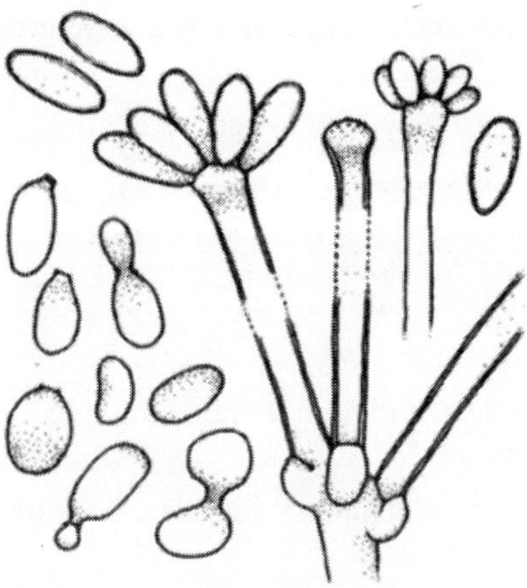

**Teleomorph:** *Filobasidium capsuligenum*

**Mycotoxins:** Torulopsin

**Diagnostic features:** Unicellular, spherical to short ovoid, in pairs or in short chains, elongated cells may also present, streak culture is whitish cream coloured ferments glucose, trehalose, cellibiose, mannitol and salicin. *Filobasidium* is its teleomorph

### 61. *Trichoderma* Pers. ex. Fr

**Synonymys** =*Aleurisma*

= *Pachyingbarium*

Colonies spread, repeated, branched, hyphae hyaline, septate, conidiophore in tufts with divergent often irregularly branched and flask shaped phialides. Conidiophores may end in a sterile appendage with phialides borne on lateral branches in some species. Conidia hyaline or usually green, non-septate, globose to spherical. Hyaline chlamydospores are usually present in the mycelium of older cultures. *Hypocrea* is the teleomorph of this genus.

Ishikawa *et al.* (1973) have reported production of trichodermin and trichoderminol by *T. hamatum*, *T. harzianum. Hypocrea* is the perfect stage of some species belonging to *Trichoderma* and *Trichothecium roseum.*

**Key to the species**

1. Conidiophore and branches short and stout often with sterile hyphal elongation bearing crowded phialides colonies white, whitish-green or green.......... 2

1a. Conidiophore and its branches strong and slender without sterile hyphal elongation, phialides not crowded. Colonies yellowish, bright-green to dark-green .......... 5

2. Sterile hyphal elongation absent, conidia lobose, hyaline.......... *T. piluliferum* (1)

2b. Sterile hyphal elongation present or modified or rarely absent, conidia not globose .......... 3

3. Conidia green, short, ellipsoidal, surrounded by a wide irregular veil........................................................................... *T. saturnisporum* (2)

3a. Conidia smooth walled or finely punctuated............................................. 4

4. Conidia hyaline, small, 2.4-3.8 x 1.8-2.2 μm.......................................... *T. polysporum* (3)

4b. Conidia green, small to large, 3.8-6.0 x 2.2-2.8 μm............................... *T. homatum* (4)

5. Conidia rough, 3.6-4.8 x 3.5-4.5 μm ..................................................... *T. viride* (6)

5a. Conidia smooth walled..................................................................................... 6

6. Conidiophore with complicated dendroid branching, phialides regularly disposed in no. 3 or more ....................................................................................... 7

6a. Conidiophore with simple branching system, phialides irregularly laterally disposed often arising singly .............................................................................. 9

7. Conidia ellipsoidal or oblong often appearing angular 3.0-4.8 x 1.9-2.8 μm ........................................................................................ *T. koningii* (6)

7a. Conidia shorter with a length: width ratio of less than 1.5 μm.................................... 8

8. Conidia obovoid with truncate ............................................................ *T. aureoviride* (7)

8a. Conidia shorter with a length: width ratio of less short obovoid......... *T. harzianum* (8)

9. Conidia subglobose to ovoid........................................................................ *T. reesei* (9)

9a. Conidia ellipsoidal............................................................................................ 10

10. Phialides usually distinctly attenuate at the base, conidia smaller, pale green 2.8-4.8 μm, mostly oblong ellipsoidal......................... *T. pseudokoningii* (10)

10a. Phialides usually only slightly attenuate at the base,conidia large, partly dark-green up to 7.0 μm long mostly ellipsoidal................. *T. longibrachiatum* (11)

**i) *T. harzianum* Rifai**

**Synonymys:** = *Sporotrichum narcissi* Tochinaii and Shimala

= *Trichoderma numbergii*

= *T. narcissi*

**Teleomorph:** *Hypocrea lixic*

**Colony:** Growing rapidly, smooth surfaced, watery white with sparse mycelial mat but soon develop aerial hyphae appearing granular or powdery with the formation of conidia rapidly turning to yellowish-green to dark-green producing tufts or pustules finzed by sterile white mycelium

**Reverse:** Colourless to dull- yellowish, buff or drab

**Exudates:** Colourless or amber or greenish-yellow, abundant in some isolates

**Odour:** Distinct or fairly earthy

Chlamydospores: Fairly abundant, intercalary or terminal on short branches, solitary contents subhyaline to pale-yellowish to brownish with age subglobose to ellipsoidal or pyriform, hyaline.

**Conidiophores:** Hyaline, smooth walled, straight or flexuous, much branched, macronematous form loose tufts, arising in distinct and continuous ring like zones outside the ring like zones. Branches mostly in groups of 2-3 and stand at right angled to the bearer or slightly bent towards the main branches, their length increases with the distance from the tip of main branch that gives a conical or pyramidal appearance.

**Phialides:** In flase verticils of upto 5 beneath the terminal phialides mostly in whorls of 3-4, usually arise at right angles to the bearer but occasionally bends towards the origin, short, ampulliform to lageriform, bulged at the middle, narrower at the base, attenuation abruptly into sharp pointed neck, terminal phialides which terminates into sharp pointed neck.

**Conidia:** Produced singly and successively, accumulate at the top of each phialides and form into globose conidial head, sub-globose or short obvoid often with truncate base, perfectly smooth walled but appear coarse in scanning electron microscope, pale-green, singly darker in mass.

**Mycotoxins:** Trichodermin

**Diagnostic features:** Colonies yellowish bright-green changing to dark-green, conidiophore with complicated dendroid branching, strong and slender without sterile, hyphal elongation, phialides irregular, laterally disposed often arising singly. conidia obvoid, short with width ratio less than 1.5.

**ii) *T. koningii* Oudem.**

**Colony:** Spreading, smooth changing to slightly hairy, color turning from watery white to greenish- white dull-dark-green and medium colour remaining or changed.

**Chlamydospores:** Terminal or intercalary

**Conidiophore:** Flexuous, much branched, formed in compact or loose tufts in ring- like zones with age these rings become less obvious because of the formation of the conidiophores on the aerial or creeping hyphae which are produced outside the primary conidial zones, main conidiophore giving out reversal side branches in groups of 2-3 and stand at right angles to the bearer, mostly paired, rarely verticellate, narrow at the base. Length of the side branches increasing with the distance from the apex of the main branch, hence the outline of the branching system appearing conical or pyramidal lateral branches also resemble more or less the main conidiophore branch.

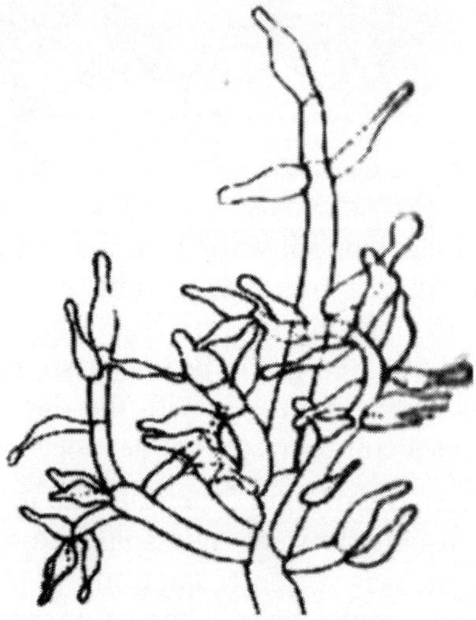

**Phialides:** Usually in false whorls of 2-3, solitary phialides uncommon, lageriform, more or less ampuliform and ninepin shaped, narrower at the base apices attenuate to conic.

**Conidia:** Sub cylindrical to narrow ellipsoid, occasionally ablong, rarely aboviod, ellipsoidal with distinct truncate based and rounded apex, smooth, pale–green to green, produced singly, accumulate at the apex of the phialides in to globose head.

**Mycotoxins:** Trichothecene

**Diagnostic features:** Colonies yellowish green changing to dark-green, conidiophores with complicated dendroid branching, phialides 3 or more irregularly disposed conidia ellipsoidal or oblong often appearing angular.

**iii) *T. polysporum* (Link ex Pers.)**

**Synonymys** = *Pachybasium hamatum* Sacchardo

= *Trichoderma sporulosum* (Link) S. Hughes

= *T. album* Preurs

= *Sporotrichum polysporum*

**Colony:** Growing slowly to moderate, mostly compact cushion shaped velvety pustules of conidiophores

**Reverse**: Uncolored

**Chlamydospores**: Terminal or intercalary

**Conidiophores:** Arise irregularly and form scattered compact tuff but close together on the surface of the colony branched oppositely or in groups of 3, rarely singly lower portion of main branch comprising fertile side branches whose length increases slightly with distance from the apex of main branch sometimes uppermost branches larger than the lowest branch. Each branch terminated in to a phialides or becomes swollen, apex of the main conidophores ends in sterile hyphal elongation, apex alternate.

Sterile hyphal elongation curved, flexous, or whip like sometimes coiled and many anastomose with each other. Sometimes the uppermost fertile branches are also terminated in to short hyphal elongation, septate hyaline.

**Phialides**: Mostly ampulliform (almost pear shaped) divergent, short, plump slightly wider above the middle than at the base and taper in to a short conical neck abruptly, stand at right angles to the bearer and arise in groups of up to 5 on the slightly swollen apices of the fertile branches, rarely arise singly.

**Conidia:** Produced singly and successively from the tips of phialides and form in to heads which unite with other heads produced by the neighboring phialides, hyaline, seen in mars, smooth ellipsoid often appear angula in conidial heads.

**Mycotoxins:** Trichothecene

**Diagnostic features:** Conidiophores and branches short and stout often with sterile hyphal elongation bearing crowded phialides colonies white turning whereas green to green conidia ellipsoidal, smooth walled, hyaline.

**iv) *T. viride* Pers**

**Synonymys=** *Pyrenium lingorum* (Tode)

=*Trichoderma lingorum* (Tode) Harz

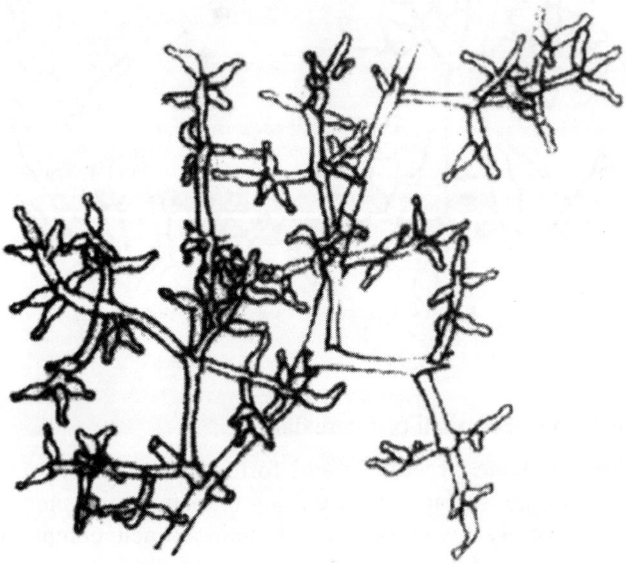

**Colony:** Growing rapidly, watery white becoming hairy from the formation of loose scanty aerial mycelium, floccose to arachnoid, somewhat whitish

**Reverse:** Uncolored

**Odour:** Coconut like in older colonies

**Chlamydospores:** Common

**Conidiophores:** Much branched, arise in compact or loose tuffs which often from broken or ring like zones ultimately a conifer like branching system is formed. All branches stand at wide angles to the bearer tip terminated by phialides.

**Phialides:** Form in false whorls beneath each terminal phialides generally not more than 2 or 3 phialides arise at right angles to the arise at right angles to the bearer occasionally arise singly or in opposite pairs along the branches, lagenform attienuated into long neck

**Conidia:** Globose or short ovoid or broadly ellipsoidal sometimes with distinct apiculus like base because of distinct minute roughening on their walls, bluish-green to dark-green.

**Mycotoxins**: Satratoxin F, G and H, trichotoxin, trichodermin, viridiol, viridin, gliotoxin

**Diagnostic features**: Colonies yellowish-bright-green to dark-green, conidiophores and its branches strong and slender without sterile hyphae elongation, phialides not crowded and conidia rough walled.

## 62. *Trichotheicum* Link ex. Gray

Colonies are fast growing First white later pinkish, culminate velvety, powdery form conidial formation zonate diurnal rhythm, conidiophores erect of sub-erect, single or in grooves, unbranched or sparingly branched, hyaline, septate, the tip bear basipetal imbricate, zigzag chains of conidia held together by mucilage conidia ellipsoidal two celled, thick walled hyaline with well marked truncate attachment point. This is monotypic genus, it produce trichothecine mycotoxin

### i) *T. roseum* Link ex Gray

**Synonymys:** = *Trichoderma roseum*

= *Cepahlothecium roseum*

**Colony**: Spreading, white later pinkish, powdery

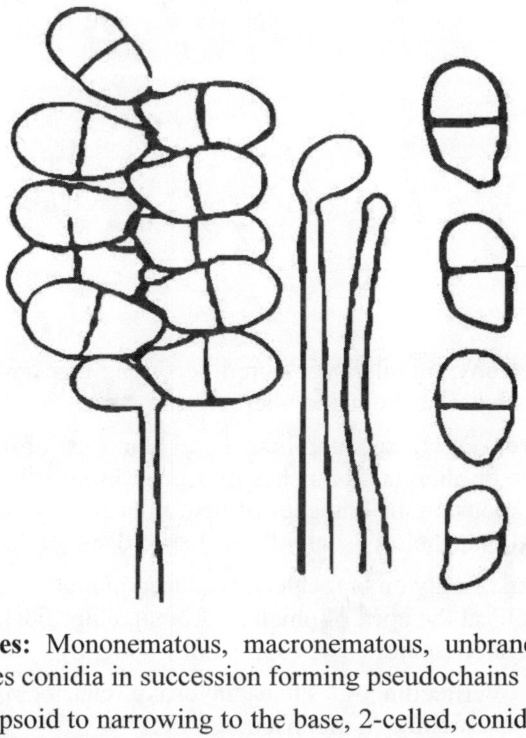

**Conidiophores:** Mononematous, macronematous, unbranched basauxigenic, produces conidia in succession forming pseudochains of conidia, septate

**Conidia:** Ellipsoid to narrowing to the base, 2-celled, conidia held in mucilage and gel separated due to drying thick walled, hyaline, smooth, well marked truncate attachment

**Mycotoxins:** Trichothecin, trichothecolono, crotocin and roseotoxin

**Diagnostic features:** Unbranched basauxigenic conidiophores hold conidia in mucilage 2-celled hyaline conidia.

### 63. *Verticillium* Nees ex. Link

**Synonyms** = *Acrostal ggmus*

= *Ascocylindrium*

= *Diheterospora phoconia*

**Colony:** Floccose, white, yellow, orange, orange-brown, dark

**Sterile hyphae:** Creeping, septate, branched, hyaline or slightly coloured

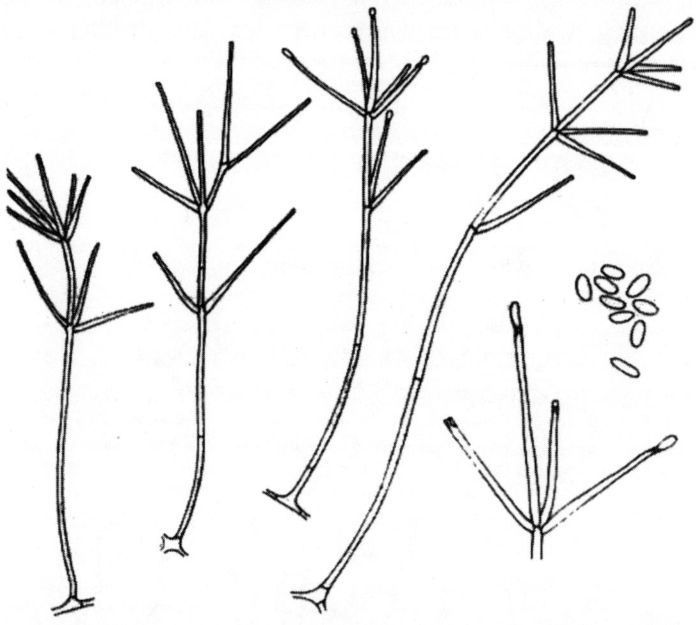

**Reverse of colony:** Similarly coloured becoming black with the formation of dark resting mycelium and microsclerotia

**Conidiophore:** Erect, septate, branched, branches of first order whorled, opposite or alternate, branches of second order whorled dichotomous or trichotomous on the branches of first order further branching is similar to terminal branchelets usually flask shaped distinctly pointed at the apex

**Conidia:** Borne singly on branchlets, produced in mucous and sliming down to form balls at the apex of phialides, round, ellipsoidal, ovate, inverted egg shaped, short-spindle-shaped, hyaline or slightly coloured

**Mycotoxins:** Verticellin A, 11 α-dihydroxy chaetocin, verucillin B, 13-hydroxy verticillin A, melinacinolin IV.

**Diagnostic features:** Conidiophore verticellate, pink spore balls appear on phialides, phialides single spores in large number.

**Key to the species:**

1 Colonies intensely orange to orange-brown due to the pigmented conidiophores and conidia; conidia ellipsoidal, 3.5-5 x 2.0-2.5 ...... ***Nectria inventa***

1a. Colonies white, yellowish or darkening ........ 2

2. Colonies darkening in reverse ........ 3

2a. Colonies remaining entirely white or yellowish ........ 7

3. Colonies producing discrete micro sclerotia but no chlamydospores: conidiophores hyaline; conidia 2.5-6(-8) x 1.4-3.2 ........ ***V. dahlia***

3a. Colonies usually not producing discrete micro sclerotia, but in some species dark resting mycelium and/or chlamydospores occur ........ 4

4. Only chlamydospores present ........ 5

4a. Dark torulose resting mycelium present ........ 6

5. Chlamydospores 5.5-8(-10) μm diam, usually produced singly; conidia mostly 1-celled, 4-8.5 x 1.5-2.5 μm ........ ***V. nigrescens***

5a. Chlamydospores 8.5-17 μm dia, commonly in chains; conidia often 2-celled, 4-10 (-12.5) x 2.5-3.5 μm ........ ***V. nubilum***

6. Chlamydospores present, 7.5-11 μm dia, in addition resting mycelim and some microsclerotia; reverse golden yellow in young cultures; conidia 3.5-10 x 1.5-3.5 μm ........ **(*V. tricorpus*)**

6a. Chlamydospores and microsclerotia absent; reverse white to cream in young cultures; conidia 3.5-10.5(-12.5) x 2-5 μm ........ ***V. albo-atrum***

7. Dictyochlamydospores present at least on SEA ........ 8

7a. Dictyochlamydospores always absent ........ 9

8. Dictyochlamydospores predominant; conidia short-cylindrical, 3-4 x 1.5-2.0 μm, cohering in chains ........ ***V. chlamydosporium***

8a. Dictyochlamydospores scanty; conidia subglobose, 2.9-3.3 x 1.9-2.2 μm, cohering in chains ........ ***V. catenulatum***

9. Well-differentiated, thick-walled conidiophores present with globose conidial heads; conidia slightly fusiform, straight or curved, 3.8-7.2 x 1.2-2.4 μm ........ ***V. fungicola***

9a. Phialides on little-differentiated, often prostrate conidiosphores; conidia commonly transversely attached to the phialides tip ........ 10

10. Conidia cylindrical with rounded tips ........ ***V. lecanii***

10a. Conidia with pointed tips ........ 11

11. Conidia straight, 3-10(-12) x 0.8-1.2 μm ........ ***V. lamellicola***

11a. Conidia ± curved, 5.8-8.0 x 1.5-1.8 μm ........ ***V. psallioate***

### 64. *Verticimonosporium* Matsushima

The genus *Verticimonosporium* was established with type species *V. diffractum.* it is characterized by macronematous conidiophores, hyaline, smooth, globose or subglobose conidia borne on verticellate conidiogenous cells. Matstushima (1975) published a second species *V.elliptium*, from Japan, which can be distinguished from the previously described type species *V.diffractum* by ellipsoidal conidia.

**Colonies:** Floccose, white reverse, hyphae hyaline, branched, septate, smooth and hyaline

**Conidiophores:** Macronematous, mononematous, simple or branched, easily had broken down at the fertile parts, straight, or flexuous, septate, smooth, hyaline often bearing verticellate conidiogenous cells at each swollen cells of the conidiophore. Conidiogenous cells obpyriform, smooth hyaline Conidia globose or subglobose or rarely ellipsoidal, smooth, hyaline.

***V. diffractum***

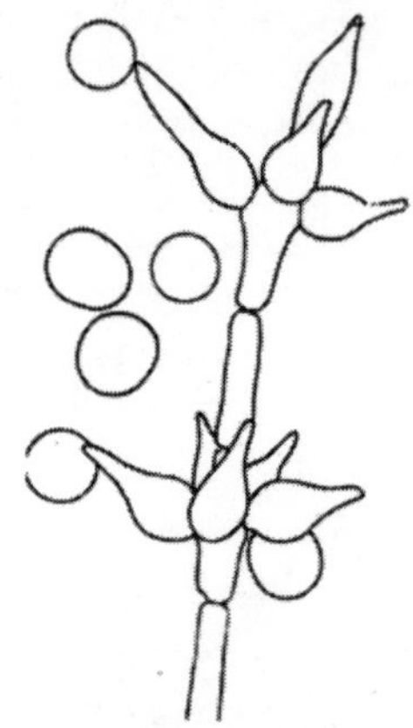

**Colony**: Floccose, white; reverse white, hyphae branched, septate, smooth and hyaline

**Conidiophores:** Conidiophores macronematous, mononematous, simple or branched, easily broken at the fertile parts, straight or flexuous, septate, smooth, hyaline, often bearing verticellate conidiogenous cells at each swollen cells of the conidiophore.

**Conidiophore**: Conidiogenous cells obpyriform, smooth hyaline

**Conidia:** Globose or sub-globose or rarely ellipsoidal, smooth, hyaline.

**Diagnostic features**: Colonies floccose, white reverse, conidiophores macronematous, easily broken down at the fertile parts, conidiogenous cells obpyriform, globose to subglobose, conidia smooth, hyaline.

## 65. *Zygosporium* Montgne

**Colony:** Effuse or compact, often thin, grey, brown, blackish brown or black.

**Stroma:** None

**Hyphopodia:** None

**Conidiophores:** Macronematous, mononematous, scattered, unbranched or banched, brown, smooth or minutely echinulate, stipe or branches bearing solitary dark-brown, curved, swollen vesicles often on a short or long stalks, upper part of stipe usually sterile, sometimes ending in a knob.

**Conidiogenous cells:** Usually monoblastic, discrete, determinate, ampulliform or ellipsoidal, often curved and tapering to a point, thin walled, colourless or pale, borne in pairs, three or four on the dark-brown vesicles.

**Conidia:** Solitory, acrogenous, simple, ellipsoidal spherical or subspherical, hyaline to brown, smooth verruculose or verrucose, non-septate. Nine species are included in this genus.

**Key to species**

1. Vesicle integrated, in chain ........ ***masonii*** (1)

1a. Vesicle discrete ........ 1

2. Vesicle short stalked borne laterally and singly just above the base of setiform type ........ 2

2a. Vesicle on usually long stalk arising at irregular intervals laterally on setiform stipes and also on athe mycelium ........ ***echinosporum*** (2)

2b. Vesicles on short or long status arising from the mycelium only........ 5

3. Conidia spherical ........ ***minus*** (3)

3a. Conidia ellipsoidal ........ 4

4. Conidia colourless or pale 7-12 x 4-7 μm ........ ***oocheoides*** (4)

4a. Conidia golden-brown 20-30 x 8-11 μm ........ ***geninatum*** (5)

5. Conidia spherical 4.5-6.0 μ diam. ........ ***giblum*** (6)

5a. Conidia ellipsoidal 5-10 x 4-6 μm ........ ***mycophilum*** (7)

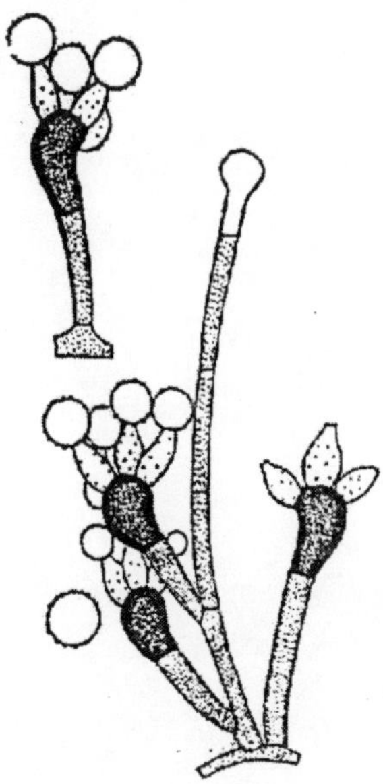

**i) *Z. masonii* Huges.**

**Colony:** Effuse or compact, brown, blackish-brown or black

**Conidiophores:** Long and thick with chains of upto 6 integrated vesicles and a hyaline, sterile apical region which terminates in small knob.

**Conidiogenous cells:** Oval to elongate bearing single blastoconidia

**Vesicles:** 7-12 μm long and 4-5 μm thick in broadest part.

**Conidia:** Ellipsoidal, hyaline

**Mycotoxins:** Zygosporin A, D, E, F, G.

**Diagnostic features:** Conidiophore with upto 6- integrated vesicles, hyaline, vesicular conidiogenous cells, non-celled, hyaline.

*Chapter 4*

# Molecular Detection of Mycotoxigenic Fungi and Mycotoxins

## 4.1. Introduction

The distribution of fungi among the various ecological niches of the biosphere seems to be infinite. Estimates suggest that out of 1.5 million fungal species, only less than a half have been described yet. This implies a backlog demand, which comes along with a rising importance of novel techniques for a rapid and unambiguous detection and identification of fungi to explore the fungal diversity as a coherent whole.

Most studies of natural fungal strains and populations focused on phenotypic differences in morphology and physiology and when possible, mating. However, these phenotypic features are often difficult to observe, quantify, standardize and analyze. Accurate identification of the fungal pathogens is in many cases a prerequisite for effective management of the diseases they cause and for ecological and population genetics studies. However, many fungal species are similar morphologically, and accurate species identification can be difficult. In recent years, there has been substanial progress in the development of innovative methods to analyze fungi and other organisms at the molecular level. Modern molecular techniques have been used for identifying fungi in a wide array of biological science disciplines. Molecular techniques, particularly the technology of the polymerase chain reaction, have revolutionized the molecular biology and the molecular diagnosis of fungi. The incorporation of molecular techniques into what has been traditionally considered as morphology-based taxonomy of fungi helps us in the differentiation of fungal species and varieties. Databases of genomes and genetic markers are used as source for molecular barcodes are being created and the fungal world is in progress to be unveiled with the help of and bioinformatic tools.

Molecular genetic markers have been employed for rapid identification of different kinds of fungi (Lieckfeldt and Seifert 2000; Njambere *et al.* 2008; Attanayake *et al.* 2009). The development of gene-specific primers for PCR amplification (White *et al.* 1990) has facilitated systematic studies and the detection and identification of fungi. The Internal Transcribed Spacer (ITS) region of nuclear ribosomal DNA has generally been considered as a convenient marker for molecular identification of fungi at species level because of its conserved as feature within species and multi-copy number per genome (Sanchez-Ballesteros *et al.,* 2000).

Toxins produced by fungi differ in structure as well as their role in toxicity and their mode of action (Knoche and Duvick, 1987). Toxins play diverse roles in disease, from impacting symptom expression and disease progress to being absolutely required for pathogenesis. Some toxins are directly kill the cells and promote for infection of other cells while others interfere with induction of defense responses or induce programmed cell death-mediated defense responses in order to generate necrosis required for pathogenesis (Wolpert *et al.*, 2002). All of the identified perylenequinone toxins are produced by members of the Ascomycota, the largest phylum within the fungal kingdom. The similarity of the fungal erylene-quinone structure to hypericin led to investigations of the fungal compounds as hotosensitizers (Daub, 1982); These play diverse roles as defense compounds in plants, pathogenesis determinants in fungi and as molecules responsible for photomovement of protozoans (Heitz and Downum, 1995).

The genes which are involved in secondary metabolite synthesis play prominent role in the formation of mycotoxins. In fungi, these genes are present in clusters and expression of one gene is dependent on the coexisting gene present in the cluster. Expression of all the genes is required for the formation of particular toxin molecule. These gene clusters are highly specific from one toxin chemotype to another toxin chemotype. First publication of a diagnostic primer pair was taken as time mark for each species depicted. In many cases, more systems were published later for the same fungus.

Schneider *et al.* (1997) developed a method for detection of *Rhizoctonia solani* isolates, pathogenic and non-pathogenic to tulips, using ITS rDNA sequences and they could further identify various anastomosis groups. Recent advancement in identifying fungal species using DNA markers is to develop DNA barcodes using species-specific oligonucleotides that are diagnostic of targeted species (Druzhinina *et al.,* 2005). Such specific DNA regions need to be explored for different groups of fungi. Henry *et al.* (2000) identified *Aspergillus* at species level and differentiated true pathogenic and opportunistic pathogens using ITS 1 and ITS 2, allowing for early diagnosis and screening of effective antifungal agents for patients.

The species of *Fusarium* are worldwide occurring plant pathogens and cause severe damage to numerous crop plants such as maize, wheat and barley (Mirete *et al.*, 2004, Nganje *et al.*, 2004, Youssef *et al.,* 2007, Li *et al.*, 2008). *Fusarium,* besides causing yield loss and kernel size (tombstone kernels), also produces numerous toxic metabolites while colonizing the plant and thereby heavily impairs the quality of grains used for the food and feed industry (Marcia McMullen *et al.,* 1997). The acute or chronic toxicity of *Fusarium* released compounds led to the introduction of national limits for mycotoxins in many nations or even to supranational application of regulations (European Community 2006).

The traditional method used to isolate and characterize fungi is the cultivation on particular medium and microscopic investigations. These conventional identific-ation methods are very time-consuming and have to be performed by skilled personnel to prevent incorrect identification and data interpretation. Moreover, conventional methods used for fungal detection are predominantly semi-quantitative. They determine the colony forming units from surface sterilized grains plated on particular solid medium (Krysinska-Traczyk *et al.,* 2007). To overcome the drawbacks of culture based identifications, rapid screening technologies based on DNA identification have been developed and are established for *Fusarium* species. In contrast to conventional detection methods, samples can be tested directly without any elaborate isolation and cultivation steps for a proper classification. These novel identification methods include PCR based technologies like DGGE, AFLP, or RFLP and also diagnostic microarrays. All these methods are relatively insensitive to microbial backgrounds and non-targeted organisms.

The broad application of DNA-based identification technologies increased the knowledge on diagnostic DNA fragments of *Fusarium* species: ITS or IGS sequences (Gagkaeva and Yli-Mattila, 2004, Konstantinova and Yli-Mattila, 2004, Mirete *et al.,* 2004, Yli-Mattila *et al.*, 2004 a, b, Jurado *et al.,* 2006, Kulik 2008), mitochondrial DNA (La′day *et al.,* 2004a, b), the b-tubulin encoding gene (Gagkaeva and Yli-Mattila, 2004; Mach *et al.* 2004; Yli-Mattila *et al.* 2004b), the translation elongation factor gene (Knutsen and Holst-Jensen 2004) and the calmodulin gene (Mule` *et al.* 2004) were sequenced from numerous *Fusarium* spp. to allow the design of highly specific PCR primers. On the other hand, the genes from biosynthesis pathways of mycotoxins (Niessen and Vogel, 1998; Bakan *et al.,* 2002; Lee *et al.,* 2001; Gonza′lez-Jae′n *et al.,* 2004; Nicholson *et al.,* 2004; Bezuidenhout *et al.,* 2006; Baird *et al.,* 2008) were studied accurately to distinguish producers and non-producers of certain toxins. The sequence information gained through different molecular taxonomic investigations and population studies forms a broad different techniques like DGGE, RFLP and AFLP were applied for the qualitative identification of targets, only the real-time PCR tended to be practicable for quantitative detection methods.

A reliable identification of species or strain does not always provide a straight forward and consistent approach to identify a toxin producing fungal strain. Delineating the molecular pathways involved in biosynthesis of toxin has proved to be highly valuable On one hand it helps to understand the dynamics involved in toxin produc-tion by a specific fungal species or strain and for growth and environmental conditions that are necessary for toxin production, while on the other hand, the metabolic pathway genes involved turn out to be more relevant and pertinent that can be exploited to identify the toxigenic potential of fungal species and their strains. Many mycotoxin biosynthetic pathway genes are present within gene clusters and some of these appear to have undergone horizontal transfer from one species to another. There are now present in several species (Niessen, 2007). Regions of homology within mycotoxin biosynthetic gene from different species can be used to develop primers to detect the presence of the relevant mycotoxigenic species. This strategy was successfully applied for aflatoxin producers (Shapira *et al.,* 1996), trichothecene-producing fungi (Paterson, 2006, Ramana *et al.,* 2011), fumonisin-producing *Fusarium* species (Gonzalez *et al.*, 2004) and also for producers of patulin (Paterson, 2006). The main advantage of pathway gene targeting is that genes are highly specific for a particular chemotype of the group which has the ability to produce mycotoxin or cause disease.

Over the past decade there has been great improvement in the development of nucleic acid based techniques and diagnostics of fungal pathogens. The benefits of these developments offer potential for rapid, specific and sensitive tests. The uses of fungal biosystematics and molecular techniques have paved the way for fungal phylogeny and evolution. Prior DNA sequence database of fungi is necessary for the design of pairs of oligonucleotide primers to confer the necessary degree of specificity. The genes which are involved in secondary metabolite synthesis are highly specific to the respective fungi. The molecular sequence analysis of such genes would clearly reveal the relationship among the fungal species based on their clustering in the phylodendrogram. Amplicons thus generated can be sequenced and the variability among such sequences can be exploited to design specific primer pairs for the target fungus. Designing of metabolic pathway gene specific primers and their use in multiplex PCR reactions for the detection of multiple toxigenic and pathogenic fungi from food crop or plant samples make PCR-based diagnostics a genuine option.

**DNA sequencing** approach can provide data for both high level phylogenetic analyses among species and for the analyses of genetic variation among strains within populations and species. Genes for phylogenetic analysis among species or higher taxonomic levels reside in the ribosomal RNA (rRNA) gene cluster, including the internal transcribed spacer (ITS) regions ITS1 and ITS2 and the inter genic spacer IGS, 5.8S rRNA, 18S rRNA, and 26S rRNA genes. These multi-copy genes are highly conserved within a species but can be variable among species. Protein coding and non-functional DNA fragments are usually more informative than the conserved rRNA genes for strains within a species. This is because rRNA genes are under strong concerted evolution pressure, whereas non-coding sequences and third-base substitutions are less constrained. With increasing accessibility and decreasing cost, multi locus DNA sequencing is becoming a main genetic data gathering tool in many studies (Xu *et al*., 2003).

### 4.2. Biosynthetic Gene Clusters

Secondary metabolites of fungi can be classified into three categories (i) Non ribosomal peptides and amino acid-derived compounds mediated by non-ribosomal peptide synthases (ii) Polyketides and fatty acid-derived compounds and which are synthesized by mediation of polyketide synthases (iii) Terpenes mediated by terpene synthases (Hoffmeister and Keller, 2007). Genes necessary for the synthesis and transport of these compounds are clustered in one genomic locus along with the key enzyme encoding genes (Osbourn, 2010). Accumulating data suggests that secondary metabolite biosynthetic genes are scattered throughout the genome and clustered in contrast to primary metabolite genes. The contiguous clustering of metabolic genes specific to one product has considerable impact on the regulation of these genes (Zhang *et al.,* 2004).

The genes which are involved in secondary metabolite synthesis play prominent role in the formation of mycotoxins. Expression of all the genes is required for the formation of particular toxin molecule. These gene clusters are highly specific for particular toxin chemotype.

### 4.3. Molecular Markers for fungal Complex Genetic Analyses

High genotypic diversity has been revealed within the species complex and even within a species (Gale *et al.,* 2002, Zeller *et al.,* 2003, Zeller *et al*., 2004, Karugia *et al*., 2009, Burlakoti *et al*., 2008, Schmale *et al*., 2006), using molecular markers. Molecular markers such as Sequence Characterized Amplified Regions (SCAR), Single Strand Conformational Polymorphism (SSCP), Randomly Amplified Polymorphic DNA (RAPD), Amplified Fragment Length Polymorphism (AFLP), Restriction Fragment Length Polymorphism (RFLP), Sequence Related Amplified Polymorphism (SRAP), Single Nucleotide Polymorphism (SNP) and Variable Number of Tandem Repeat (VNTR) have been used extensively to study genetic variability of the fungal complex. Recently, a high resolution, SNP-based multilocus genotyping assay was used to determine the diversity among the *Fusarium graminearum* complex and identified several novel species.

### 4.4. Polymerase Chain Reaction (PCR) in the detection of mycotoxigenic species

The PCR technology has spawned many procedures for typing strains and species, some of which have become standard methods for species and strain identification. PCR methods are easy to set up and have the advantage of requiring only minute amounts of starting material or template DNA. Although it is simple in concept, PCR methods have unraveled many, often overlooked complexities such as multi-ionic interactions, kinetic constants and enzymatic activities. These factors can repeatedly affect the reactants in a typically small PCR reaction volume over an extended time period. In contrast to hybridization, the PCR has found widespread use in microbial taxonomy and significant numbers of PCR-based assays have been developed for use with species associated with the production of mycotoxins (Parry and Nicholson, 1996). PCR can also be used in one- or two-step reverse-transcriptase. RT-PCR is used to detect genes that are being expressed rather than to detect the presence of the DNA that encodes a particular gene. Reverse transcriptase is used to generate a DNA copy (cDNA) from RNA (mRNA or total RNA) present in a sample. The cDNA is then subjected to PCR using primers designed for the target sequence. When designed to genes, characteristic of a particular species or to genes involved in mycotoxin biosynthesis and assayed in a real-time or competitive PCR format the mRNA level of target genes can be estimated in a manner similar to that for DNA (Alexander *et al.,* 1997). A number of approaches have been developed to detect a single species or strains of same species which have potential to produce mycotoxin. These methods can be broadly divided in to two.

**i. Universal primers** are being used to isolate specific DNA region followed by comparison of the sequence of target species with that of other species within databases. The coding portion of many fungal 18S, 5.8S and 28S rDNA genes is highly conserved and primers to these regions have been generated (Brown *et al.,* 2007). These allow the isolation of the Internal Transcribed Spacer Sequences (ITS-1 and ITS-2), which lie between the coding regions, from a wide range of fungi. The ITS region is amplified from the target fungus and sequenced to identify regions of DNA unique to the fungus of interest. Polymorphism within the ITS region is generally at the level of species, rather than among isolates of the same species. This makes it an ideal target for the development of species-specific PCR assays. For example the ITS-1 and -2 of *F. avenaceum* differed markedly from that of *F. graminearum* and *F. culmorum* and primers to this species did not cross-react with DNA from a range of other fungi. More highly conserved regions within the ITS is suitable for designing primers to differentiate one genus as applied to *Fusarium* (Nicholson *et al.,* 2004). When ITS region is of limited use, other characterized regions of the genome may be used. For example, galactose oxidase is produced by only a few fungal species and it can be used to produce PCR primers specific to *F. graminearum* based upon the sequence of the galactose oxidase (gaoA) gene (Carter *et al*., 2002).

**ii.** Another approach is to develop assays based upon polymorphisms detected within anonymous DNA regions. Two methods are commonly employed to detect such polymorphisms between fungal species. (1) Random Amplified Polymorphic DNA (RAPD) analysis, which has been used to study variability within and among species, many *Fusarium* species (*F. graminearum*, *F. cerealis* (*F. crookwellense*), *F. venenatum*, *F. torulosum*, *F. sambucinum* and *F. proliferatum*) (Carter *et al*., 2002). (2) The second method is the Amplified Fragment Length Polymorphism (AFLP) assay (Schmidt *et al.,* 2003). This method helped to develop a species-specific assay for *A. ochraceus* that did not cross-react with other *Aspergillus* or *Penicillium* species tested (O'Brian *et al.,* 2003).

The development of truly species specific PCR assays for mycotoxigenic fungi is dependent upon a robust taxonomic foundation. Primer sets should be tested against closely related species to ensure specificity. Unfortunately, it is not always correct from traditional, morphology-based taxonomies to determine which species are closely related. For example, PCR assays to *F. avenaceum* designed from ITS and RAPD sequences were found subsequently to cross-react with *F. tricinctum* (Amoah *et al.,* 1996). Molecular analyses are also revealing, sometimes, unsuspected distinctions within species. For example, sub-groups / lineages have been recognized within *F. graminearum* (Doohan *et al.,* 1999). Sequence analysis has revealed that the product of the Fg16F/R primer pair is diagnostic of the lineage/groups tested to date and hence this primer pair may be used to detect *F. graminearum* and simultaneously determine lineage/group (O'Donnell *et al.,* 2000). Further, this assay could be used to aid the detection of migrants and monitor fungus movement. Although a number of PCR assays have been developed against mycotoxigenic species there are relatively a few reports of their use to detect mycotoxigenic species in plant material (Niller, 2000). Thus species-specific PCR-based assays to mycotoxigenic species are potentially both specific and highly sensitive.

### 4.5. Single-Strand Conformation Polymorphism (SSCP)

SSCP is another promising technique that allows efficient detection of nucleotide substitutions in short fragments (<500 bp) of DNA. SSCP analysis typically involves the amplification by PCR of a unique segment of genomic DNA, melting the PCR products, and running the single strands on a non-denaturing polyacrylamide gel (Hauser *et al*., 1998). The detection system can be accomplished by either radioactive labeling of DNA during the PCR amplification step or by silver staining of DNA after gel electrophoresis. Polymorphic differences in strand mobility result from the effects of primary sequence changes on the folded structure of a single DNA strand. The primary sequence differences alter the intra-molecular interactions that generate a three-dimensional folded structure. The molecules may thus move at different rates through a non-denaturing polyacrylamide gel. Because these conformational variations are subtle, the success of any particular SSCP experiment depends heavily on two factors: (i) the particular DNA fragments being investigated and the size of the DNA fragments and (ii) the optimization of experimental conditions to maximize differential migration among fragments. Variety of methods to improve the resolving power of SSCP including adding glycerol to polyacrylamide gels, reducing temperatures, and increasing the length of the gels or the duration of gel electrophoresis have been tried. Nonetheless, differentiation among polymorphic molecules on a polyacrylamide matrix is not entirely predictable and the method can result in false negatives, ambiguous results and experimental artifacts.

### 4.6. Random Amplified Polymorphic DNA (RAPD)

In RAPD analysis, genomic or template DNA is primed at a low annealing temperature (30–38°C) with a single short oligonucleotide (ca. 10 bases) in the PCR. Multiple PCR products of different electrophoretic mobility are typically generated (Williams *et al*., 1990). RAPD analysis detects two types of genetic variations: (i) in the length of DNA between the two primers binding sites and (ii) sequence variation at the priming regions. Nucleotide substitutions in the region of PCR primer binding, particularly at the 3′ ends, can prevent binding of the primer to the DNA template as a result this band will be missing in a PCR reaction. Similarities in banding profiles among strains (i.e., the number and mobility, but not the density of the bands) can be calculated and used to infer strain relationships. When multiple primers are screened, RAPD analysis can be very sensitive to detect variation among isolates that cannot be observed using other methods. Although technically fast and simple, there are some disadvantages of RAPD: 1. The lack of reproducibility. Consequently, RAPD analysis can detect minute variation among strains, even a single nucleotide mismatch in the priming region may prevent annealing and the absence of a characteristic band on gels. 2. In RAPDs the bands with the same electrophoretic mobility may not share the same sequence. 3. The problem of dominant and null alleles. In haploid organisms, both the dominant (presence) and recessive/null (absence) alleles can be scored.

### 4.7. Amplified Fragment Length Polymorphism (AFLP)

AFLP is a powerful method for finger printing strains and for generating a large number of dominant markers for the analysis of genetic crosses (Vos *et al*., 1995). Genomic DNA samples are first digested with two endonucleases (usually a frequent cutter and a rare cutter). Double stranded DNA adapters are then ligated to the ends of the DNA fragments to create template DNA for PCR. The adapters consist of a core sequence and an enzyme-specific sequence that allow the ligations to occur. The ligated products are then amplified with AFLP amplification primers. Amplification primers consist of the core sequence (same as the adaptor core sequence), the enzyme-specific sequence and a selective extension of one to several nucleotides, depending on the complexity of the study genome. These selective bases will allow amplification of a subset of the restriction fragments. AFLP usually involves two PCR steps. The first step is the pre-amplification step that uses unlabelled primers with a single selective nucleotide in the primer. After the first step, the reaction mixtures are diluted for second PCR amplification. In the second amplification, additional selective nucleotides are often added to enhance specificity. The selective second step often uses fluorescent or radioactively-labeled primers. AFLP has several powerful advantages over the other methods. Many more fragments can be generated and analyzed in a simple reaction. It can detect restriction site variations as well as insertions and deletions within a genomic region. Different enzymes and selective extension nucleotides can be used to create new sets of markers. Therefore, AFLP can provide an almost limitless set of genetic markers and fragments are stable and highly reproducible since they are amplified with two specific primers under stringent conditions.

## 4.8. Molecular detection of different mycotoxins in foods and feeds

**i) Aflatoxins:** The major aflatoxins consist of aflatoxins B1, B2, G1 and G2 produced by *Aspergillus flavus* or *A. parasiticus*. The aflatoxin biosynthetic pathway has been studied for over 40 years and is one of the best understood fungal secondary metabolic pathways (Fig. 4.1). Atleast 34 genes have been identified as members of the aflatoxin pathway gene cluster (Yu *et al.* 2004) Most of their functions in the pathway have been elucidated. The aflatoxin regulatory gene (*afl R*) encodes a positive regulator (AFLR), which activates aflatoxin pathway gene transcription (Chang *et al.* 1995), whereas the *afl S* (old name *afl J*) gene encodes a protein factor (AFLS), which is found to be involved in the regulation of transcription (Chang, 2003).

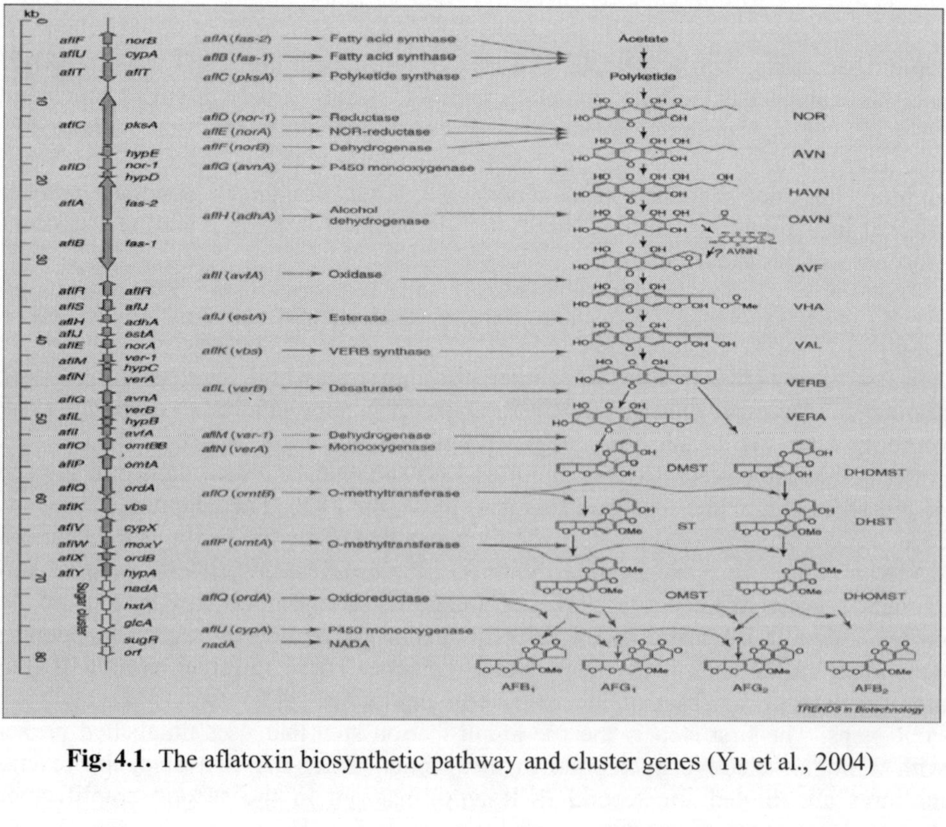


**Fig. 4.1.** The aflatoxin biosynthetic pathway and cluster genes (Yu et al., 2004)

### ii) Ochratoxin A

#### a) Identification of Ochratoxin A Gene Cluster

The PCR (Saiki *et al.*, 1988) has replaced laborious and time consuming microbiological analysis by amplification of specific genomic markers rather than growing the living organism. PCR systems have been developed for the detection and differentiation of major groups of mycotoxigenic fungi. One of the most important factors is that such methods are the reliability of the primer set designed and the targeted DNA sequence of interest organism (Saiki *et al.*, 1988). Different pairs of PCR primers were developed to set up novel diagnostic tool for identification of OTA producers among species of *Aspergillus* and *Penicillium*.

#### b) Detection of OTA producing fungi

Having established that OTA production in permissive and restrictive medium appears to be dependent on *pks* gene transcript levels expression, O'Callaghan *et al.* (2006) identified other putative OTA biosynthetic genes in *A. ochraceus* by comparing their expression profiles with that of the OTA pks gene. They also identified two putative cytochrome P450 monooxygenase genes which displayed a high degree of similarity to other monooxygenase involved in the biosynthesis of polyketide secondary metabolites and the p450-H11 gene is very similar to the averantin oxidoreductase gene involved in aflatoxin biosynthesis (Yu *et al.,* 1997). While the p450-B03 gene was not only similar to the trichodiene oxygenase gene of trichothecene biosynthetic pathway in *Gibberella zeae* (Ward *et al.,* 2002) but also displayed significant similarity to the *cyp*A gene in the aflatoxin biosynthetic pathway in *A. flavus* (Ehrlich *et al.,* 2004). Further they could establish a correlation between the amount of *pks* and cytochrome P450 transcripts and the amount of OTA production suggesting their role in OTA biosynthesis. O'Callaghan *et al*. (2006) have also cloned the full length pks gene and on the cloning part of a novel non ribosomal peptide synthetase (*nrps*) gene 900 bp upstream of the pks gene and genes transcribed in the same direction. However, they were not in the same reading frame. From this it is clear that they encoded two separate proteins rather than a single hybrid type PKS-NRPS. This finding revealed *A. ochraceus* and *A. westerdijkiae* contain similar genes. The *pks* gene also showed 60% similarity to a hypothetical PKS gene corresponding to the OTA cluster in *A. niger* (Pel *et al*., 2007) and only 38% to a *awks1* gene responsible for the biosynthesis OTA in *A. westerdijkiae* (Bacha *et al.,* 2009). Based on the similarities between the *awlc35-12* and *pks* genes, Bacha *et al.* (2009) concluded that *awlc35-12* is involved in OTA biosynthesis in *A. westerdijkiae.* They concluded that two *awks1* is different from *pks* involved in the biosynthesis of OTA in *A. westerdijkiae*. Similar PKS have been reported to be involved in the biosynthesis of zearaleone in *G. zea* (Kim *et al.,* 2005) and T2 toxin in *C. heterosphorus* (Baker *et al*., 2006).

#### c) Detection of mycotoxigenic species of *Penicillium*

Most of the gene cluster responsible for the biosynthesis of OTA in *Penicillium* has been characterized (Karolewiez and Geisen, 2005). A 10 kb genomic DNA fragment of *P. nordicum* which carried three long open reading frames has been cloned (Geisen *et al.*, 2004, Karolewiez and Geisen, 2005, Farber and Geisen, 2004). One open reading frame (*otapksPN*) is a polyketide synthase, which is different from PKS genes found in species of *Aspergillus* the second open reading frame (*npsPN)*, GeneBank accession number AY534879) has homology to non-ribosomal peptide synthetases and the third open reading frame (*aspPN*) has homology to fungal alkaline serine proteinases. Interestingly the *otapksPN* gene is present only in *P. nordicum* but not in *P. verrucosum*, indicating genetic differences also between two species of *Penicillium*

**d) Detection of ochratoxigenic fungi with ochratoxin A biosynthetic genes**

Many mycotoxin biosynthetic genes present within gene clusters of these appear to have undergone horizontal transfer from one species to another species (Sidhu, 2002, Niessen 2007). Regions of homology within mycotoxin biosynthetic gene from different species can be used to develop primers to detect the presence of relevant mycotoxigenic species. This strategy was successfully applied for aflatoxin producers (Shapira *et al.*, 1996), trichothecene-producing fungi (Niessen *et al.*, 1997, Paterson, 2006, Bluhm *et al.*, 2002), fumonisin-producing *Fusarium* species (Paterson, 2006, Bluhm *et al.*, 2002, Gonzalez-Jaen *et al.*, 2004) and also for producers of patulin (Paterson, 2006). Dao *et al.* (2005) designed two sets of specific primers, AoLC35-12L/AoLC35-12R and AoOTAL/AoOTAR, from AoLC35-12 which helped to detect *A. ochraceus* and *A. westerdijkiae* by PCR method and to detect other OTA producing *A. carbonarius*, *A. melleus*, *A. ochraceus*, *A. sulfurous* and *P. verrucosum* and citrinin producing *P. citrinum* and *Monascus ruber*.

In *A. carbonarius*, five different KS domain sequences (AcKS9, AcKS10, Ac12RL3, AcLC35-4 and AcLC35-6) were identified by Atoui *et al.* (2006). Two different PCR species specific detection systems were developed based on primers designed from AcKS10 (Selma *et al.*, 2008) and Ac12RL3 (Atoui *et al.*, 2007). PCR method also helped to differentiate to detect two ochratoxigenic *P. nordicum* and *P. verrucosum*. *P. verrucosum* gave positive reaction only with the primers for the otanpsPN gene, while *P. nordicum* is positive for both genes (otanpsPN and otanpsPN).

Advances in DNA based techniques such as real-time PCR (RT-PCR) which are new tools for detection and quantification of fungi by detecting and quantifying their DNA. RT-PCR can be performed using SYBR® Green I dye (204,205) and TaqMan® (Livak *et al.*, 1995). Both systems proved to be useful in monitoring and quantification of OTA producing fungi in many foods (Geisen *et al.*, 2004, Farber and Geisen, 2004 Schmidt *et al.*, 2004, Mulè *et al.*, 2006, Atoui *et al.*, 2007 Selma *et al.*, 2008). Assays based on OTA biosynthetic genes proved to be better than those systems based on genes unrelated to their biosynthesis. Positive correlation between OTA content and DNA quantity could be observed for *P. nordicum, A. ochraceus* (Hendrickson *et al.*, 1999 Farber and Geisen, 2004 and *A. carbonarius* (Mulè *et al.*, 2006, Selma *et al.*, 2008, Atoui *et al.*, 2007. RT- PCR quantification of *A. carbonarius* in grapes proved to be the best alternative to conventional methods (O'Callaghan *et al.*, 2003). Based on these Atoui *et al.* (2007) concluded that *A.*

*carbonarius* DNA content has to be lower than 10 ng DNA $g^{-1}$ grape berry to fulfill the maximum OTA permitted levels in the European Union.

The biosynthesis of most mycotoxins is highly complex process which is influenced by environmental factors. It needs to be understood at the transcriptional level. Further, correlation between gene expression and concentration of compounds has to be established. Farber and Geisen, (2004) could monitor growth and OTA production by *P. nordicum* in wheat by A RT-PCR system based on the *otapksPN* sequence. Strong correlation between the copy numbers of the *otapksPN* gene and the colony forming units (cfu) was observed (Karolewiez and Geisen, 2005). In addition, there was a strong congruence between *otapksPN* gene expression and OTA production. Hence, it can be used for the rapid quality assessment of food products and for HACCP. AFLP and RAPD could be the best choice to establish (1) the polymorphism of some isolated strains belonging to the same species or (2) to distinguish between relevant OTA producer species. On the other hand, detection of OTA producing fungus, the best way is to conduct conventional PCR using primers designed from housekeeping genes or mycotoxin biosynthetic genes. However, application of AFLP and RAPD for analysis of food has certain limitation such as requirement of high quality DNA. Therefore, Real time PCR technology provides better insight over AFLP and RAPD to determine mycotoxigenic status of food sample

**e) Current Status on the Biosynthesis of Ochratoxin A**

Although much information exists concerning the various toxigenic properties of OTA, unlike other important mycotoxins very information is available about OTA biosynthetic pathway in fungi. It is widely believed that the isocoumarin group is a pentaketide formed from acetate and malonate *via* a polyketide synthesis pathway (Niessen *et al.*, 2005, Edwards *et al.*, 2002, Moss, 1998) involving, a polyketide synthase (PKS) in a similar way of other polyketide mycotoxins such as fumonisins and aflatoxins. Huff and Hamilton, (1979) proposed a biosynthetic pathway based on a mechanistical model according to the structure of OTA. Three distinct steps occur in OTA biosynthesis (Fig. 4.2). The first part is polyketide synthesis of ochratoxin α *via* mellein involving a polyketide synthase. The second step includes acyl activation: mellein is methylated and oxidized to 7-Carboxy-Mellein (=OTβ). Chlorination by a chloroperoxidase leads to OTα. This component is then transformed to a mixed anhydride, an activation reaction using ATP. The second precursor phenylalanine is synthesized *via* the shikimic acid pathway followed by ethyl ester activation so that it can participate in the subsequent acyl displacement reaction. In the third step, linkage of those activated precursors *via* a synthetase takes place, generating OTC, an ethyl ester of OTA: de-esterification by an esterase or transesterification is the last step in this postulated biosynthetic pathway. This schematic pathway has been dissented by Harris and Mantle (2001) using labeled precursors to the growing cultures of OTA producing fungi *A. ochraceus*.

The heterocyclic portion of OTA is structurally similar to mellein, which produced by many OTA producing species. Mellein is also produced by non ochratoxigenic species such as *Pezicula* spp, *Botryosphaeria obtusa*, *Septoria nodorum*, *Phoma tracheiphila*, *Apiospora camptospora*, *Cercospora taiwanensis*, *C.*

*scirpicola*, *Fusarium larvarum*, *Gyrostroma missouriense*, *Pezicula livida*, *Cryptosporiopsis malicorticis*, *Cryptosporiopsis* sp. and *Plectophomella* sp. (Turner, 1983).

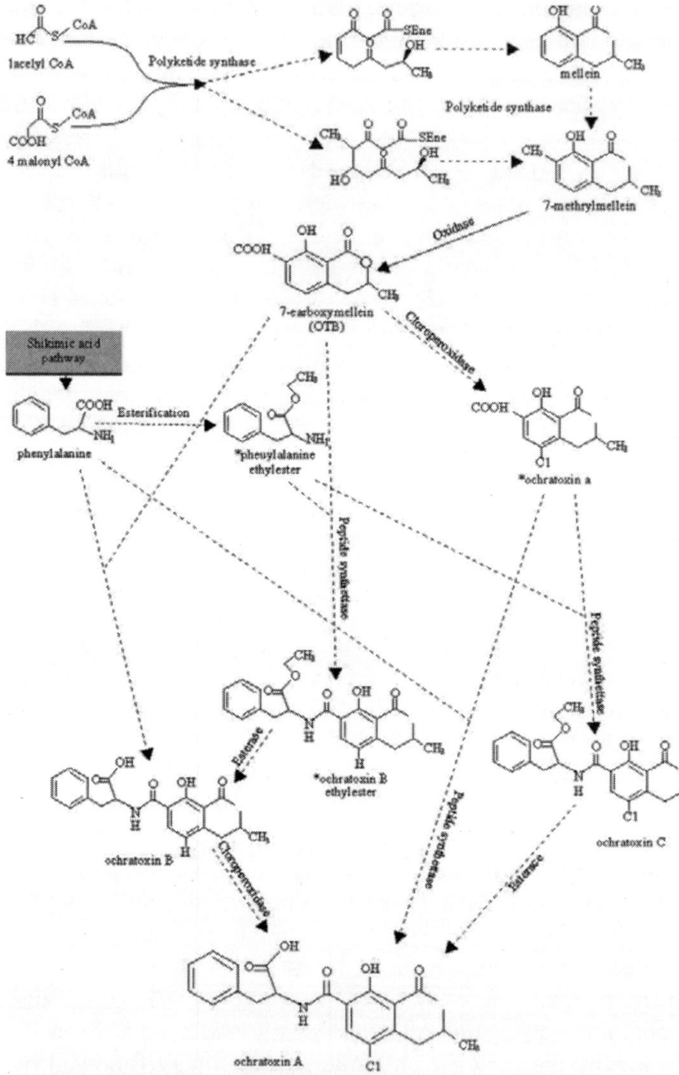


**Fig. 4.2.** Different hypotheses of biosynthesis of OTA (Harris and Mantle, 2001)

Relative incorporation of labeled putative intermediates ochtatoxin α and β and mellein indicated a strong preferential role of α, moderate role of β, and no role of mellein in OTA synthesis. The genetic nature of polyketide has been studied using molecular techniques such as Genomic DNA bank and cDNA bank construction; Polymerase Chain Reaction (PCR), subtractive PCR, gene inactivation, differential display reverse transcriptase-PCR (DDRT-PCR), microarrays *etc*. PCR and identified PKS genes responsible for the biosynthesis of various polyketides (Bingle *et al*., 1999, Liou and Khosla, 2003, Nicholson *et al*., 2001, O'Callaghan *et al*., 2003). Pairs of degenerated primers targeting KS domain; which is the most conserved domain among different PKSs, have been previously designed to amplify KS domain fragment from different types of PKS genes (Bingle *et al*., 1999, Liou and Khosla, 2003, Nicholson *et al*., 2001). These degenerated primers were successfully used in OTA producing fungi. Five different PKS genes have been identified in *A. ochraceus* by Varga *et al*. (1965) and Edwards *et al*. (2002). In addition nine different KS domains in *A. westerdijkiae* NRRL 3174 (*A. ochraceus*) as well as five KS domain sequences in *A. carbonarius* (2Mu134 = CBS 120167) (Atoui *et al*., 2006) have been isolated.

In an attempt to elucidate the molecular biosynthetic pathway of OTA, O'Callaghan *et al.* (2003) cloned part of the polyketide synthase (*pks*) gene (AY272043) required for OTA biosynthesis in *A. ochraceus*, but no information was obtained concerning the presence or absence of some metabolite like the mellein. Later on, a polyketide synthase gene *otapks*PN (AY196315) from *P. nordicum* that is essential for OTA biosynthesis was reported by Karolewiez and Geisen, (2005). Recently, Bacha *et al.* (2009) characterized a PKS gene, *awks1,* which is required for the OTA biosynthesis in *A. westerdijkiae* has no effect on mellein biosynthesis which supports Harris and Mantle (2001) hypothesis. The pathways of biosynthesis of OTA proposed by different workers are illustrated in Fig. 4.2. Recently Gallo *et al*. (2012) feel that non-ribosomal peptide synthesis (NRPS) gene play a critical role in OTA synthesis pathways proposed as illustrated in Fig.4.3.

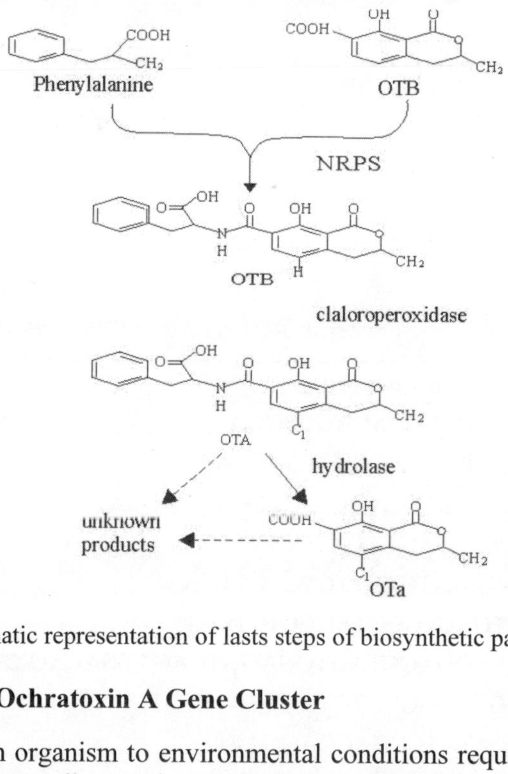


**Fig. 4.3.** Schematic representation of lasts steps of biosynthetic pathway of OTA

## f) Identification of Ochratoxin A Gene Cluster

The adaptation of an organism to environmental conditions requires the cooperation of several genes that contribute to its survival. In a metabolic chain, the product of an enzymatic reaction must be taken very quickly by the following enzyme to ensure a correct speed of formation of the final product and avoid being destroyed by the possible reactions. So, an enzymatic assembly in the form of complex or enzymatic cluster (Sidhu, 2002, Yu *et al.*, 2005) is, therefore, a necessity. In fungi, several clusters coding for polyketides for the biosynthesis of different mycotoxins (aflatoxins, fumonisins, ergot alkaloid, paxillin), antibiotics (cephal-osporin and penicillin), melanins and some pharmaceutical products (lovastatin and compactin) have been studied. In the following paragraphs we will discuss the achievement made for the identification of OTA gene cluster in *Aspergillus* and *Penicillium* species (Table 4.1).

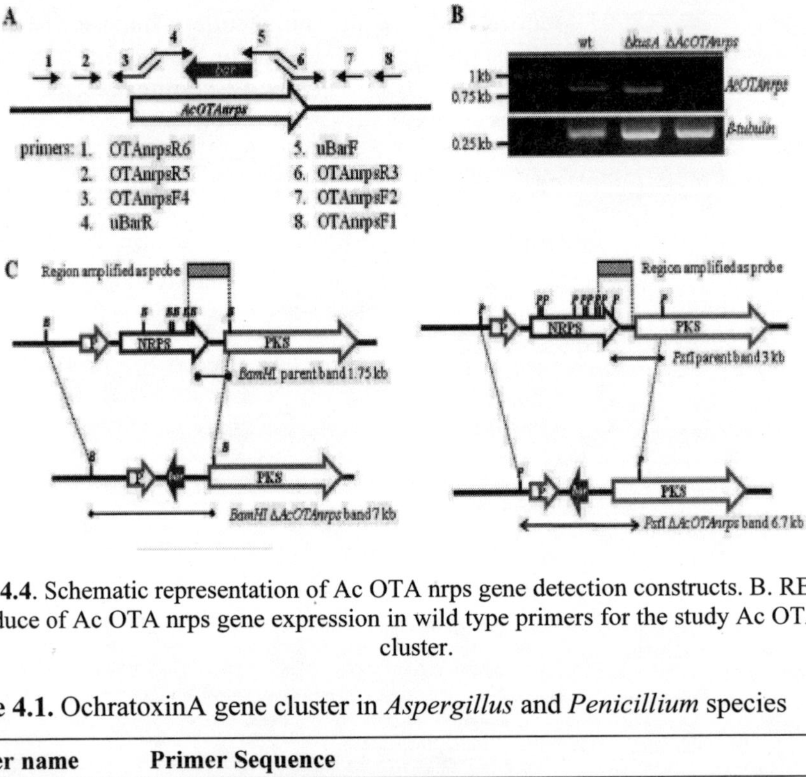


**Fig. 4.4.** Schematic representation of Ac OTA nrps gene detection constructs. B. RET-PCR produce of Ac OTA nrps gene expression in wild type primers for the study Ac OTA gene cluster.

**Table 4.1.** OchratoxinA gene cluster in *Aspergillus* and *Penicillium* species

| Primer name | Primer Sequence |
|---|---|
| OTAnrpsF1 | AATCTCCGACAGCTGGAAAA |
| OTAnrpsF2 | CTTTGGGCAGAGGTTTGAAG |
| OTAnrpsR3 | TGACCTCCACTAGCTCCAGCGCAATCCGGCTTTACTTTCA |
| OTAnrpsF4 | AATAGAGTAGATGCCGACCGGTGGAAAGTGGGAAATGGTG |
| OTAnrps5R | GGGAATGTCTTCTGCCTTGA |
| OTAnrps6R | CTGACTGCGTAGCCTCTCCT |
| uBarF | GCTGGAGCTAGTGGAGGTCAGTCGACAGAAGATGATATTG |
| uBarR | CGGTCGGCATCTACTCTATTGTCGACCTAAATCTCGGTGA |
| AcNRPS-1for | TCATCTCCGACGAGGAAC |
| AcNRPS-2rev | CAAAGGAATCCTCGTCACT |

### iii) Trichothecenes

#### a) Detection of trichothecene producing species of *Fusarium*

The trichothecene biosynthetic pathway has been well characterized in a number of species, most notably *F. graminearum* and *F. Sporotrichioides* (Brown *et al*. 2001, Kimura *et al*. 2003). Sequence from the trichothecene gene cluster, specifically the *tri4, tri5* and *tri6* genes, have been exploited in the development of generic PCR-based assays for the detection of trichothecene producers, as these genes are present in all trichothecene producing *Fusarium* species. The *tri6* gene is a transcriptional activator that can regulate expression of both *tri4* and *tri5*. Primers specific to *tri6* gene have been designed and can be used in conjunction with assays for either *tri5* or *tri4* for detection of trichothecene producers. The *tri5* gene encodes trichodiene synthase, which catalyses the first step in the trichothecene biosynthetic pathway (**Fig. 4.5**). This gene has also been targeted for the production of primers specific to trichothecene producers. The trichothecene biosynthetic pathway begins with the cyclization of the isoprenoid biosynthetic intermediate, farnesyl pyrophosphate, to trichodiene by the enzyme *trichodiene synthase*. The pathway involves a number of oxygenations, isomerizations, cyclizations and esterifications leading from trichodiene to diacetoxyscirpenol, T-2 toxin and 3-acetyldeoxyniv-alenol. The isolation of other biosynthetic enzymes proved to be more difficult, although cytochrome P- 450 monooxygenases are known to be involved (Gledhill *et al*. 1991).

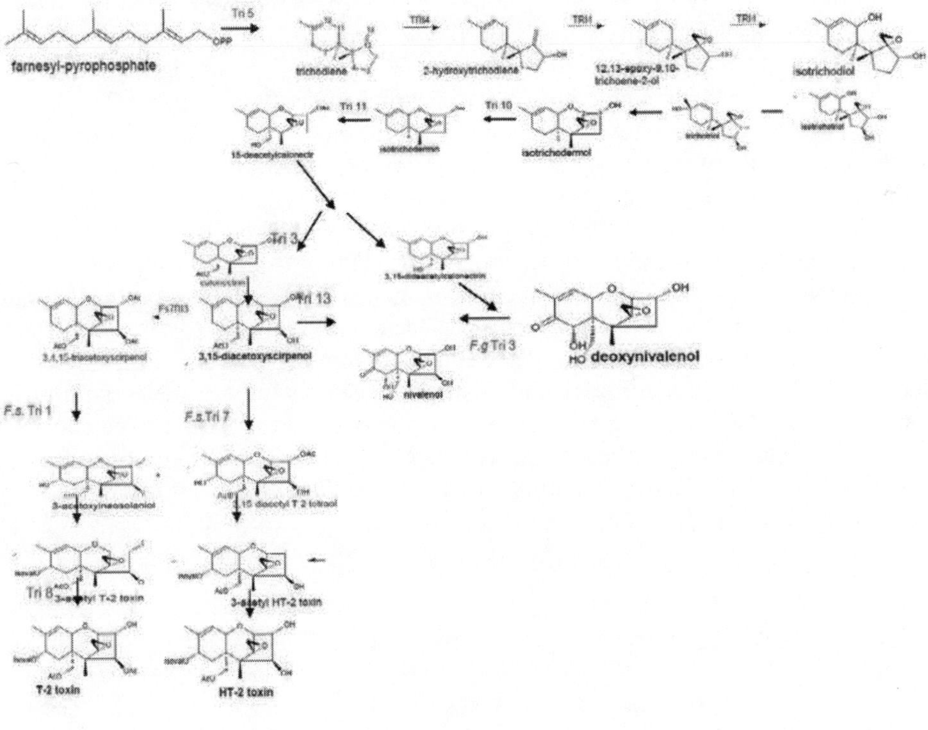


**Fig. 4.5.** Biosynthetic pathway of trichothecenes-Tri-gene products are catalyzed in the step of the pathway

### iv) Fumonisins

Out of four main groups of fumonisins, fumonisin B series occur most widely (FB1, FB2, FB3 and FB4), with the most common being FB1 and FB2 (Gelderblom *et al.*, 1988, CAST 2003, NTP, 1999). These mycotoxins proved to be strain specific (Munkvold *et al.,* 1997). Out of 23 species of *Fusarium* only *F. verticillioides, F. proliferatum* and *F. nygamai* are fumonisins producers (Nelson *et al.,* 1993, Thiel *et al.,* 1991, Rheeder *et al.,* 2002). Unfortunately, infections by fumoninsin producing *Fusarium* species are symptomless in kernels and supportive plant tissues that are unknowingly provided to animals and humans as feed and food, respectively (Munkvold *et al.,* 1997). Symptomless infections of kernels by *F. verticillioides* are generally much greater than visible infections responsible for observable tissue necrosis or damage (Headrick and Pataky, 1991).

The ribosomal DNA (rDNA) found in the nuclear genome of fungi consists of three repeating RNA genes transcribed as a single united code for the 18S, 5.8S and 28S RNA genes that are highly conserved and primers to these regions have been developed (White *et al.,* 1990). Between these coding regions, the ITS1 and ITS2 separate the three RNA genes. O'Donnell and Cigelnik (1997) and O'Donnell *et al.* (1998) recommended the use of ITS regions for species-level identification of *Fusarium* due to non-orthologous copies of the ITS2 within the *Gibberella* clade. This limitation would prevent the development of primers for identification of fumonisin producing fungi. Another approach for PCR, Random Amplified Polymorphic DNA (RAPD), or arbitrary sequences, have been shown to be an excellent methods for identifying polymorphic differences within and between fungal species (Table 4.2). This method does not require previous knowledge of the genome and uses short oligodeoxynucleotides as arbitrary primers. Bayman and Cotty (1993) used RAPD to correlate taxa relatedness based on differences in morphology, physiology, and toxigenic potential. In addition, RAPDs were employed to distinguish fungal species and pathogenicity variability within species (Zhang *et al.,* 1997, Jones and Dunkle, 1993). Preliminary studies showed that RAPD was not sensitive enough to distinguish fumonisin producing fungi. Therefore, specific primers developed from known sequences of genes for fumonisin producing fungi should be thoroughly investigated (González-Jaén *et al.* 2004, Waalwijk *et al.,* 2004). González-Jaén *et al.* (2004) compared development of primers DNA-based for identification of *F. Verticillioides* fumonisin producers and non-producers fumonisin strains. Multiplex PCR for fumonisin assay of fungi using the FUM1 gene consisting of 7.8-kb coding region or the intergenic spacer region (IGS), located between 16S-23S rRNA genes, to distinguish strains of *F. verticillioides* of fumonisin producing and non-producing strains (Patiño *et al.*, 2004).

**Table 4.2.** Primers used in detection of mycotoxin producing fungi

| Toxin | Primer | Sequence 5'------3' | References |
|---|---|---|---|
| Trichothecenes | *Tri5* | GAGAACTTTCCCACCGAATAT | Venkataramana *et al.* (2011) |
| | | GATAAGGTTCAATGAGCAGAG | |
| | *Tri5* | GCTGCT CAT CAC TTT GCT CAG | Niessen and vogel (1998) |
| | | CTG ATC TGG TCA CGC TCA TC | |

| Toxin | Primer | Sequence 5'------3' | References |
|---|---|---|---|
| | *Tri6* | GATCTAAACGACTATGAATCACC | Venkataramana *et al.* (2011) |
| | | GCCTATAGTGATCTCGCATGT | |
| | *TRI6* | CTC TTT GAT CGT GTT GCG TC | Bluhm *et al.* (2002) |
| | | CTT GTG TAT CCG CCT ATA GTG ATC | |
| | Tri13 | CATCATGAGACTTGTKCRGTTTGGG | Haratian *et al.* (2008) |
| | | TTGAAAGCTCCAATGTCGTG | |
| | TM Tri | CAGCAGMTRCTCAAGGTAGACCC | Halstensen *et al.* (2006) |
| | | AACTGTAYACRACCATGCCAAC | |
| | Tr5 | AGCGACTACAGGCTTCCCTC | Doohan *et al.* (1999) |
| | | AAACCATCCAGTTCTCCATCTG | |
| DON | *Tri13* | GGCTTACGACTCCTCAACAATGG | |
| | | AGAGCCCTGCGAAAG(C/T)ACTGGTGC | |
| | *Tri 13* | GTTGCAGTTCGCTTGATTTCG | Venkataramana *et al.* (2011) |
| | | CTGTCTATATCTTGTTTCACAG | |
| NIV | *Tri7* | GGC TTT ACG ACT CCT CAACAA TGG | Lee *et al.* (2001) |
| | | AGAGCCCTGCGAAAG(C/T)ACTGGT GC | |
| | *Tri 7* | ATAGGTACCGGATCGCAGG | Venkataramana *et al.* (2011) |
| | | CCGAAAGCCTCTAATAGTGT | |
| Fumonisins | *Fum5* | GTC GAG TTG TTGACC ACT GCG | Bluhm *et al.* ( 2002) |
| | | CGT ATC GTC AGC ATGATGTAG C | |
| | *Fum1* | ATTATGGGCATCTTACCTGGAT | Venkataramana *et al.* (2011) |
| | | ACGCAAGCTCCTGTGACAGA | |
| | *Fum13* | AGTCGGGGTCAAGAGCTTGT | Venkataramana *et al.* (2011) |
| | | TGCTGAGCCGACATCATAATC | |
| | *VERTF* | GCGGGAATTCAAAAGTGGCC | Patino *et al.* (2004) |
| | | GAGGGCGCGAAACGGATCGG | |
| | *PQF19* | ATCAGCATCGGTAACGCTTATGA | Lopez-Errasquin *et al.* (2007) |
| | | CATGTAAGTTGAGGAAGCCCTTGT | |
| | FGtub | GGTCTCGACAGCAATGGTGTT | Reischer *et al.* (2004)\ |
| | | GCTTGTGTTTTTCGTGGCAGT | |
| Aflatoxins | *Afl r1* | CGCGCTCCCAGTCCCCTTCATT | Schnerr *et al.*(2002) |
| | | CTTGTTCCCCGAGATGACCA | |
| | *Ver1* | GCCGCAGGCCGCGGAGAAAGTGGT | Geisen (1996) |
| | | GGGGATATACTCCCGCGACACAGCC | |
| | *Nor1* | ACCGCTACGCCGGCACTCTCGGCAC | Geisen (1996) |
| | | GTTGGCCGCCAGCTTCGACACTCCG | |
| | *Omt1* | GTG GAC GAA CCT AGT CCG ACA TCAC | Geisen, (1996) |
| | | GTC GGC GCC ACG CAC TGG GTT GGGG | |
| Patulin | *Idh* | CAATGTGTCGTACTGTGCCC | Paterson, (2004) |
| | | ACC TTC AGT CGC TGT TCC TC | |

| Toxin | Primer | Sequence 5'------3' | References |
|---|---|---|---|
| | *Nor1* | GTCGCTGAAGGCAGTCGGAT | Venkataramana *et al*. (2013) |
| | | TCTCAAAGTGGAACTTGC | |
| PKS | Aolc | GCCAGACCATCGACACTGCATGCTC | Dao *et al*. (2005) |
| | | CGACTGGCGTTCCAGTACCATGAGC | |
| *A. ochraceus* | AFLP | ATACCACCGGGTCTAATGCA | Schmidt *et al*. (2003) |
| | | TGCCGACAGACCGAGTGGATT | |
| | rRNA | CTTCCTTAGGGGTGGCACAGC | Patino *et al*. (2005) |
| | | GTTGCTTTTCAGCGTCGGCC | |
| | pks | CATCCTGCCGCAACGCTCTATCTTTC | Dao *et al*. (2005) |
| | | CAATCACCCGAGGTCCAAGAGCCTCG | |
| *A. westerdijkiae* | b-tubul | TGATACCTTGGCGCTTGTGACG | Morello *et al*. (2007) |
| | | CGGAAGCCTAAAAAATGAAGAG | |
| *A. niger* | RAPD | CAGTCGTCCAGTACCCTAAC | Sartori *et al*. (2006) |
| | | GAGCGAGGCTGATCTAAGTG | |
| *A. carbonarius* | rRN | GCATCTCTGCCCCTCGG | Patin˜o *et al*. (2005) |
| | | GGTTGGAGTTGTCGGCAG | |
| *A. carbonarius* | pks gene | TGGGTATGCGCGGGGTGAGGGTAT | Dobson and O'Callaghan (2004) |
| | | CCGTAGGCTTCGAAAAACTGACAC | |
| *A. carbonarius* | cmdA | CCGATG GAGGTCATGACATGA | Mule` *et al*. (2006) |
| | | AATGCGAACCGGATATTAACTTCTG | |
| *A. carbonarius* | AFLP | GAATTCACCACACATCATAGC | Schmidt *et al*. (2004a, b) |
| | | TTAACTAGGATTTGGCATTGAAC | |
| *A. carbonarius* | pks | AATATATCGACTATCTGGACGAGCG | Atoui *et al*. (2007) |
| | | CCCTCTAGCGTCTCCCGAAG | |
| | RAPD | AGGCTAATGTTGATAACGGATGAT | Fungaro *et al*. (2004a) |
| | | GCTGTCAGTATTGGACCTTAGAG | |
| | pks | CCCTGATCCTCGTATGATAGCG | Selma *et al*. (2008) |
| | | CCGGCCTTAGATTTCTCTCACC | |
| | cmdA | AAGCGAATCGATAGTCCACAAGAATAC | Perrone *et al*. (2004) |
| | | TCTGGCAGAAGTTAATATCCGGTT | |
| *P. verrucosum* | pks | TGCACGACCGGGACAACATCA | Dobson and O'Callaghan (2004) |
| | | CCGTAGGCCTCCACAAAATCTG | |
| *P. nordicum* | nrps | AGTCTTCGCTGGGTGCTTCC | Bogs *et al*. (2006) |
| | | CAGCACTTTTCCCTCCATCTATCC | |
| | pks | TACGGCCATCTTGAGCAACGGCACTGCC | Geisen *et al*. (2004) |
| | | ATGCCTTTCTGGGTCCGATA | |
| *A. flavus* | A1 | CGACGTCTACAAGCCTTCTGGAAA | Logothetiet *al*. (2009) |
| | | CAGCAGACCGTCATTGTTCTTGTC | |
| *A. niger* | A n | CAGTCGTCCAGTACCCTAAC | Logothetiet *al*. (2009) |

| Toxin | Primer | Sequence 5'------3' | References |
|---|---|---|---|
| | | GAGCGAGGCTGATCTAAGTG | |
| *A. ochraceus* | Ao | ATACCACCGGGTCTAATGCA | Schmidt *et al.* (2003) |
| | | TGCCGACAGACCGAGTGGATT | |
| *A. fumigatus* | A2 | TATGTCTTCCCCTGCTCC | Logothetiet *al.* (2009) |
| | | CTATGCCTGAGGGGCGAA | |
| *P.chrysogenum* | Pc | CTGAAGCTTACCATGTACAAC GAG | Park *et al.* (2000) |
| | | GTTCTCGAGTTTGTACTCGAAGTTCTG | |
| *F. graminearum* | Fg | CTCCGGATATGTTGCGTCAA | Mule *et al.* (2004) |
| | | GGTAGGTATCCGACATGGCAA | |
| *F. culmorum* | Fc | ATGGTGAACTCGTCGTGGC | Mule *et al.* (2004). |
| | | CCCTTCTTACGCCAATCTCG | |
| *F. verticillioides* | Fv | CTTCCTGCGATGTTTCTCC | Nicholson *et al.* (1998) |
| | | AATTGGCCATTGGTATTATATAC | |
| *F. proliferatum* | Fp | CTTTCCGCCAAGTTTCTTC | Nicholson *et al.* (1998) |
| | | TGTCAGTAACTCGACGTTG | |
| *F.sporotric-hioides* | Fs | AAAAGCCCAAATTGCTGATG | Demeke *et al.* (2005) |
| | | TGGCATGTTCATTGTCACCT | |

### a) Regulation of fumonisin biosynthesis in *Fusarium verticillioides*-maize system

Fumonisins biosynthesis are regulated by the FUM gene cluster which is approximately 42-kb in length, is known to harbor 22 Open Reading Frames (ORFs) 15 of these are known to be co-regulated (Fig. 4.6) (Proctor *et al.* 2003). Silico analysis reveals that eleven of the 15 co-regulated genes encode mono-oxygenases, dehydrogenases and fatty acyl-CoA synthetases also contain genes encoding putative transporters and proteins involved in self-protection (Proctor *et al.*, 2003). Molecular characterization of $FUM_1$, a polyketide synthase, $FUM_6$, a cytochrome $P_{450}$ monoxygenase, and $FUM_8$, an aminotransferase, revealed their critical role in $FB_1$ biosynthesis (Fig. 4.7). Disruption of $FUM_1$, $FUM_6$, and $FUM_8$ resulted in significant reduction in $FB_1$ production (Seo *et al.,* 2001). Butchko *et al.* (2003) provided the biochemical evidence linked directly to a FUM gene ($FUM_{13}$) specific reaction for fumonisin biosynthesis. The $Fum_{13}$ protein catalyzes the reduction of the C-3 carbonyl of the fumonisin backbone to a hydroxyl group. Similarly, $FUM_3$ encodes a 2-ketoglutarate dependent dioxygenase and catalyzes the hydroxylation of carbon-5 of the fumonisin backbone (Ding *et al*., 2004). Bojja *et al.* (2004) elucidated two of the early steps in fumonisin biosynthesis. Utilizing disruption mutants, Bojja *et al.* (2004) demonstrated that the enzyme encoded by $FUM_8$ catalyzes the condensation of alanine followed by the oxidation of the backbone at carbon 14 and 15 by the $P_{450}$ monooxygenase encoded by $FUM_6$. Though fumonisin biosynthesis gene cluster is nearly complete, still a critical gap exists in our understanding of the regulatory mechanism involved in fumonisin biosynthesis. Environmental as well as genetical factor play a critical role in the complex biological process. Mechanism of fumonisin biosynthesis in *F. verticillioides* is quite different from other filamentous fungi (Yu and Keller, 2005).

The FUM cluster does not contain a pathway-specific regulatory gene (Proctor *et al*., 2003) but is regulated by several genes not linked to the FUM cluster. $FCC_1$, $FCK_1$, $PAC_1$, $ZFR_1$, and $GBP_1$ are reported to regulate $FB_1$ biosynthesi in *F. verticillioides.* Detailed information on these genes is presented in Table1.1. $FCC_1$ and $FCK_1$ operate independently. Still other regulatory genes with fumonisin biosynthesis in *F. verticillioides* are yet to be identified

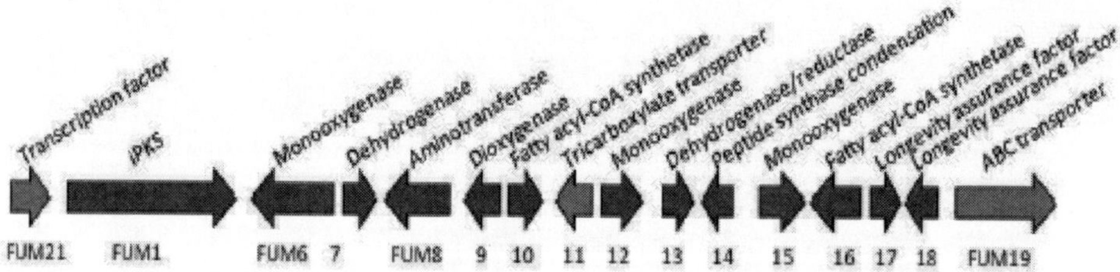


**Fig. 4.6.**The FUM gene cluster in *Fusarium verticillioides*

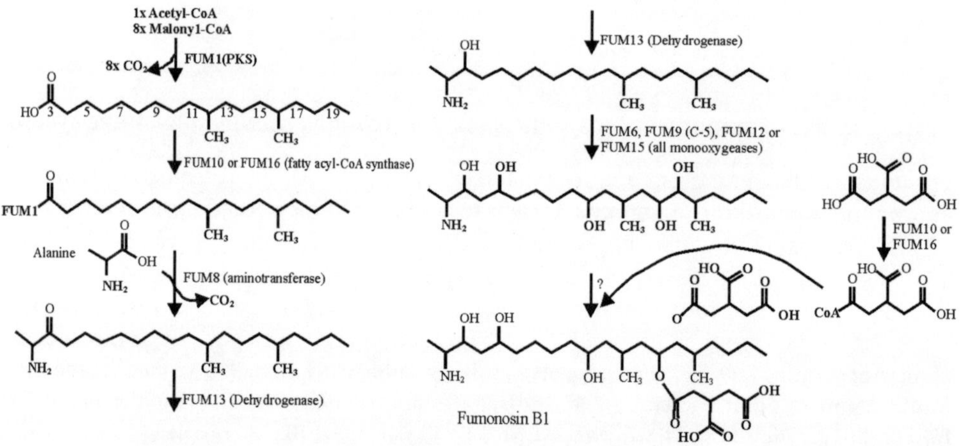


**Fig. 4.7.** Fumonisin biosynthetic pathway in *Fusarium verticillioides*

Fumonisins are believed to be synthesized through the condensation of the amino acid alanine to an acetate–derived precursor (Branham and Plattner 1993). Branched chain methyl groups are added aC-12 and C-16 by an S-adenosyl methionine transferase. The orde3r and number of the subsequent steps in the biosynthetic pathway involved in the oxygenation and subsequent esterification of the acetate derived back bone are as yet unknown. However, if fumonisin biosynthesis is similar to the synthesis of aflatoxin and trichothecenes, less oxygenated homologues, such as FB2, FB3 and FB4 are likely precursors of the most highly oxygenated homologues to FB1 (Desjardins *et al*., 1996). It is still not clear whether oxygenation and methylation occur before or after condensation with alanine. Identification of the *fum* gene cluster allowed to the development of biosynthetic model (Fig. 4.8) which also offered an explanation for the synthesis of different types of fumonisin ($B_1$, $B_2$, $B_3$ and $B_4$). At present only fum1 has been linked to the formation of the polypeptide back bone and fum9 to hydroxylation of the C5 position (Butchko *et al.,* 2003). Details of FUM Genes are described below

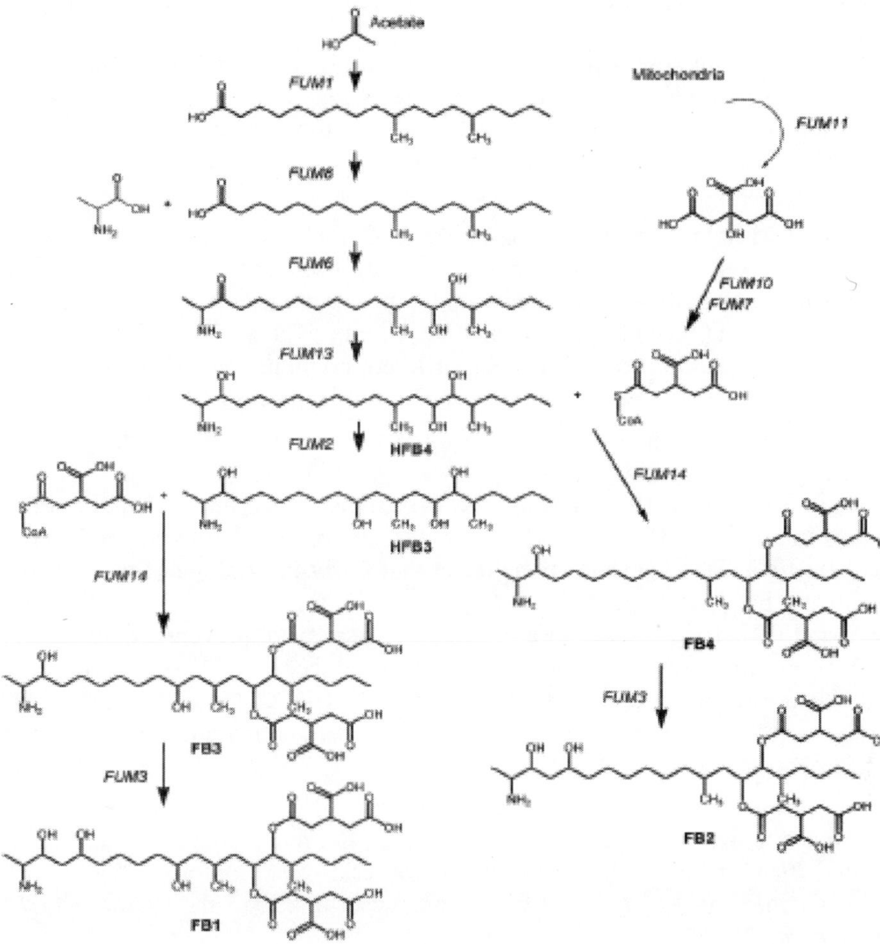


**Fig. 4.8**. Fumonisin biosynthetic pathway of FB1 in *Fusarium verticillioides*

**i) FUM1:** FUM1gene encodes the protein polyketide synthase (Fum1p), which contains the domains KS, AT, DH, MT, ER, KR and ACP (Proctor *et al*., 1999). Fum1 produces an 18 carbon backbone (C-3 to C-18) of fumonisin with a carboxylic acid at C-3 and two methyl groups at C-12 and C-16. The first carbon in the PKS (Polyketide synthase) product is designated C-3 due to the fact that the finished fumonisin backbone is a 20 carbon chain. The earlier fumonisin PKS is known as *FUM5* (Proctor *et al*. 2003), but changed to *FUM1* as this locus has initially been connected with the fumonisin PKS (Xu and Leslie 1996). Many PKSs have a Thioesterase (TE) or a TE/Cyclisation domain to stop chain elongation and facilitates the release of the polyketide chain from the PKS (Proctor *et al*., 2003). The polyketide backbone is released from the PKS $Fum_1p$ by the action of $Fum_8p$ (Gerber et al., 2009).

**ii) FUM8:** Fum8p showed similarity to a family of aminotransferases that catalyze the condensation of acyl-CoA and amino acids. Disruption of *FUM8* removed fumonisin production with no accumulation of intermediates (Seo *et al.*, 2001). Further, studies with *FUM1* and *FUM6* revealed an accumulation of compounds of high similarity to the backbone of fumonisin, suggesting that Fum8p acted before Fum6p in the synthetic pathway. It is most likely that the combination of the PKS ($Fum_1p$) product and alanine (Bojja *et al.*, 2004) was responsible for its synthesis. This was confirmed by Gerber *et al.* (2009).

**iii) FUM6:** Fum6p shares homology to Cytochrome P450 oxygenase and is fatty acid hydroxylase P450foxy from *F. oxysporum* (Seo *et al.,* 2001). Fum6p hydroxylates C-14 and C-15, whose hydroxyl groups are used for the esterification of tri carboxylic acid (TCA) to the fumonisin backbone. The action of Fum6p is similar to Fum1p and Fum8p based on growing different combinations of co-cultures of the three disrupted mutants ($FUM_1$, $FUM_6$ and $FUM_8$) (Bojja *et al.*, 2004).

**iv) FUM13:** Fum13p has been shown to be a short chain dehydrogenase/ reductase, specifically a 3-ketoreducatalase that catalyzes the reduction of the C-3 ketone at an early stage of biosynthesis. Mutants with disrupted *FUM13* produced 90% less fumonisin B than its wild counterpart and produces fumonisins 3-keto-FB3 and 3-keto-FB4. This led to the conclusion that Fum14p was one of the last steps in the synthesis but prior to the C-10 hydroxylation by $Fum_2p$ which is the branch point for $FB_3$ and $FB_4$. The presence of a ketone at C-3 seems to prevent the hydroxylation of C-5 by $Fum_3p$. However, since $FB_1$-4 was produced in 10% of WT levels there has to be a secondary and less effective 3-ketoreductase present, most likely a homologue of 3-ketosphinganine reductase (Butchko, Plattner and Proctor, 2003a).

**v) FUM2:** Fum2p is a homologue of Cytochrome P450 oxygenase and deleting this gene looses the ability to produce FB1 and FB3 (Butchko *et al.*, 2006). By comparing two naturally occurring strains which did not produce $FB_1$ and $FB_3$, was identified $FUM_2$ responsible for hydroxylation of C-10 (Proctor *et al.*, 2006). This gene was named as $FUM_{12}$ until its function was linked to the known loci $FUM_2$ (Proctor *et al.*, 2003).

**vi) FUM3:** The phenotype of Fum3 has been characterized as the FB1/FB2 producing locus *Fum3* and as the putative gene *FUM9*. Disruption of *FUM9* resulted in the loss of ability for the mutants to produce FB1/FB2, and the complementation of a *Fum3*-defective with the wild-type *FUM9* gene revealed that the *FUM9* gene and *Fum3* locus was the same (Butchko *et al.,* 2003b). The function of Fum3p was confirmed to be the hydroxylation of FB3/FB4 at C-5 to yield FB1/FB2 (Ding *et al.*, 2004). This gene is known as *FUM9* until its function was linked to the loci *FUM3* (Proctor *et al.*, 2003). All observations indicate that Fum3p is the last step in FB1 and FB2 synthesis which is supported by the observation that disruption of TCA addition yields only hydrolysed FB3/FB4 and no FB1/FB2, suggesting that Fum3p modifications take place after those of Fum14p (Zaleta-Rivera *et al.*, 2006).

**b) Tricarboxylic acid transporter**

FUM11 is not essential for the biosynthesis of fumonisins but disruption of this gene results in lower concentration of fumonisins. Since Fum11p shares homology with membrane bound tricarboxylate transporters in the mitochondria, it is likely that *FUM11* provides precursors for tricarboxylic acid formation (Proctor *et al.,* 2003).

**c) Tricarboxylic acid (TCA) formation and esterification**

*FUM7*, *FUM10* and *FUM14* are necessary for the correct assembly and attachment of the TCA side chains. The exact sequence of Fum7p and Fum10p is not clear. However, most studies show that Fum14p is responsible for the esterication of TCA and the fumonisin backbone (HFB3/ HFB4). It is possible that *FUM7*, *FUM10,* and *FUM14* together form a Non-Ribosomal Petide Synthetases (NRPS) complex (Zaleta-Rivera *et al.* 2006).

**FUM7:** Fum7p shows similarity to the dehydrogenase, which usually reduces hydroxyl groups to double bonds (Proctor *et al.* 2003). However, disruption of *FUM7* results in formation of fumonisin analogues where the TCA group had a double bond at the center carbon instead of a hydroxyl group (Butchko *et al.,* 2006). It is therefore, believed that Fum7p removes double bond thereby yielding tricarboxylic acid (Zaleta-Rivera *et al.* 2006).

**FUM10:** Disruption of *FUM10* results in accumulation of hydrolysed fumonisin, but no fumonisin B suggesting that this step is essential for attaching the TCA side group. Since Fum10p shows homology to an acyl-CoA synthase it is believed that Fum10p mediates TCA-CoA formation.

**FUM14:** Fum14p performs the esterification between TCA and hydroxyl groups at C-14 and C-15. Disruption of *FUM14* results in the accumulation of hydrolysed fumonisins, Fum14p's homology with non-ribosomal petide synthethases (NRPS) and catalyses ester formation (C-O), instead of the amide formation (C-N) which is normally expected of NRPSs (Zaleta-Rivera *et al.* 2006). These are the genes and steps directly involved in fumonisin biosynthesis in *F. verticillioides.* Based on the knowledge from *F. verticillioides* fumonisins synthesis in *A. niger* takes place as follows.

**d) Genes associated with fumonisin biosynthesis regulation**

Unlike other fungal secondary metabolite gene clusters, the *FUM* cluster does not contain a pathway-specific regulatory gene (Proctor *et al*., 2003). Fumonisin biosynthesis is regulated by several genes not linked to the *FUM* cluster. The genes that regulate $FB_1$ biosynthesis are *FCC1*, *FCK1*, *PAC1*, *ZFR1*, and *GBP1.* Mutational analysis of these genes reveals that a functional copy of these genes is necessary for proper production of $FB_1$ in *F. verticillioides*. Detailed information on these genes is presented in Table 4.3. Other than *FCC1* and *FCK1*, no definitive epistatic relationship between these regulatory genes has been established. It is likely that these genes operate independently. However, it is clear that additional regulatory genes, yet to be identified, are associated with fumonisin biosynthesis in *F. verticillioides* Table 4.4.

**Table 4.3.** Genes involved in regulation of fumonisins biosynthesis in *F. verticillioides*

| Gene Name | Gene Family | FB1 Production in mutants | Other Mutant Phenotype | Regularoty mode | Reference |
|---|---|---|---|---|---|
| FCC1 | C-Type cyclin | No FB1 When grown on maize kernel. leakly FB1 production in acidic defined media | Severe reduction in conidiation | Positive | Shim and Woloshuk, 2001 |
| FCK1 | Cyclin dependent kinase | Significant reduction in FB1 Production | Severe reduction in growth and conidiation | positive | Bluhm and Woloshnuk, 2006 |
| PAC1 | PACC Group of pH regulatory genes | Higher FB1 When grown on acidic conditions and on maize kernels | Severely impaired growth at alkaline pH | negative | Flaherty *et al.* 2003 |
| ZFR1 | Zn( )2cys6 binuclear cluster | Severe reduction in FB1 Production | No observable effect on growth and development | positive | Flaherty and Woloshuk, 2004 |
| AMY1 | Amylase | Unable to digest starch impaired in FB1 | No observable effect on growth and development | positive | Bluhm and Woloshuk, 2005 |

**Table 4.4.** Primer sequences and melting temperatures for genes involved in fumonisin biosynthesis

| Primer Name | Primer sequence | Tm °C |
|---|---|---|
| Tub2F | ACT CCG ATC TCC GAA AGC | 58 |
| Tub2R | TCT GTT GGG TCA ACT CAG G | 61 |
| Fcc1F | AGA GCA CTT CAG GAG CAG | 54 |
| Fcc1R | CTG TCA ACC AGC CAT TTC G | 60 |
| Zfr1F | TAT CAA CGA CCT CTA CAA CAG C | 57 |
| Zfr1R | TCT CAA GAA ATT CGG CAG GT | 60 |
| Fum1F | ACC ATC CCT TTC TTC TCC AG | 58 |
| Fum1R | AAA CAC CTG CTT TGA ACC AG | 58 |
| Fum6F | ACA GAA TCC CGT CGA CAG | 57 |
| Fum6R | AGA AGC AGA ATG ACA GGA GAC | 56 |
| Fum10F | TAC TTG TGC TTC CAG AGG C | 57 |
| Fum10R | CCA ACC GAG ATT CCG AGA G | 61 |
| Fum14F | CAG AAT CCT GAG CGT CCT | 56 |
| Fum14R | AGT TCA CCA CCT ACC ACA G | 60 |
| Fum15F | TGC GAA TAA CTT CAA TCC CGA | 63 |
| Fum15R | ATA GTT GCT CCG TGC CTC | 56 |
| Fum16F | GTG AAG CTC GTC TCT ATT CCC | 58 |
| Fum16R | CAT CAA ACT CGC CAA TGT CG | 63 |
| Fum19F | CAG TCA CGT CAA AGA AAC TCC | 57 |
| Fum19R | TCT GTC ATA CAC TAC TCC AAG C | 55 |

### v) Zearalenone Cluster

Zearalenone (ZEA or ZON) is a non-steroidal estrogenic mycotoxin produced by several *Fusarium* species (Caldwell *et al*., 1970). It can be found primarily on maize but can occur at lower concentration in barley, wheat, and other cereals (Caldwell *et al.,* 1970, Kuiper-Goodman *et al.,* 1987, Stob *et al.,* 1962). ZON can cause serious contamination of grains and feeds with synergistic effects in combination with other mycotoxins (Wolf and Mirocha, 1997, Placinta *et al*., 1999). ZON affects sexual reproduction in fungi. It enhances as well as inhibits sexual reproduction depending on concentration and time dependent (Wolf and Mirocha, 1973). Most ZON-producing *Fusarium* species also produce trichothecenes. It has genotoxic and carcinogenic effects in mice (El-Nezami *et al*., 2002, Pfohl-Leszkowicz *et al*., 1995).

Gene disruption and Real-Time Quantitative Expression studies revealed that four tightly linked genes are responsible for zearalenone biosynthesis (Table 4.5) (Kim, 2005, Lysøe *et al*., 2009).

**Table 4.5.** Primer sequences for genes involved in zearalenone biosynthesis

| Primer name | Sequence Forward (5'–3') | Reverse (5'–3') |
|---|---|---|
| Gene disruption | | |
| ZEA1 | GCAATGCGTCCAGCTCCAAAAG | TCGGTTCACCTCGGTCAAATCCAG |
| ZEA2 | GGAGGATATGGGCGTGGTGAAGGA | GGCTGGCTGATCTCGGGCAAAGT |
| Sequencing intercontig region | | |
| 118A | GTGGCTGATCCCGCTCTTTGACG | TACAAGCATGGCGCAACAATAGAC |
| 118B | GACGGCCTCAACCTTTTCAGTG | TACAAGCATGGCGCAACAATAGAC |
| Expression analysis | | |
| ZEA1 | GAAGAGGCCCCGGTAGCGATAAC | TGAAGCCACTCCAGCAGCAGAATT |
| ZEA2 | GTCTCACTGACTTTGTTCGCAT | TCAAAGGATGTTCCTGGTTGCT |
| AUR1 | GGATTCACCGTGCCCGACAT | TTCAACGCAGAGCTCCATTACGC |
| Cluster analysis | | |
| Fg02393.1 | ACCATCCAATCTCGGGCTTT | TAACTTCAGTCGCCTCGCAAA |
| Fg02394.1 | TCGGGTCGAGGACTTCTGTG | GGTATGCTCCTGGACAG |
| Fgd117-550 | CCAACGATTACACCAAGTCTCACTAT | TTGCGCCGTCTGAGTACCC CGT |
| Fg12015 | TGCTGGATAAGATTGCTCGTCTC | GTTGTCTGGACGAGACAGGGTAC |
| Fg12056 | AAATAATTTACCCGTTCTTCTGGGAACT | TCTCCAGTATCTGCTCCGGTCT |
| Fg02398.1 | GCTGTGGCCCAAATGCA | CTGAAACGGAGGTGTTGAGG |
| β-Tubulin | TCGACTGGATCCCCAACAAC | GCTATGATCGCCGATAGAAAGG |
| GAPDH | CTACATGCTCAAGTACGACTCTTCCC | GAACTGGTCGTCGACACGCT |
| ZEA1 | GGCACTTTGACAACCGCTTC | GCCGGTCTCGGACCACTTG |
| ZEA2 | ACATGCGTGTCGTTGTTGTAAG | TGCACCGTGAGAAGCCGT |

Two polyketide synthases genes (PKS4 and PKS13) are essential for the production of ZON. Two other genes encode a transcriptional regulator and an alcohol oxidase which controls oxidation of zearalenol to zearalenone. The analysis of the functional gene organization revealed 8 genes as zearalenone mycotoxin genes. Three additional genes are located close to the upstream region of the previously reported ZON cluster; a non-ribosomal peptide synthetase gene (NPS15, FGSG02394), a transporter gene (FGSG12124) and a putative K+ channel beta subunit (FGSG12125). It is not clear yet whether these genes are part of the same cluster. Only one gene encoding the putative K+ channel beta subunit (FGSG12125) is known to be expressed in a manner similar to that of two PKS genes in the ZON cluster (Gaffoor and Trail, 2006) and also reported to be clearly up-regulated in wheat (Lysøe *et al.*, 2009), suggesting that it may have some role in ZON production as well as in infection process. Regulatory motifs were examined based on two different sets of genes, the ZON cluster newly identified by Gaffoor and Trail, 2006 (**Fig. 4.9**). Regulatory motifs predicted from the two different sets of genes are different but both motif sets include the short promoter motif 5'CCG3' which needs to be further investigated.

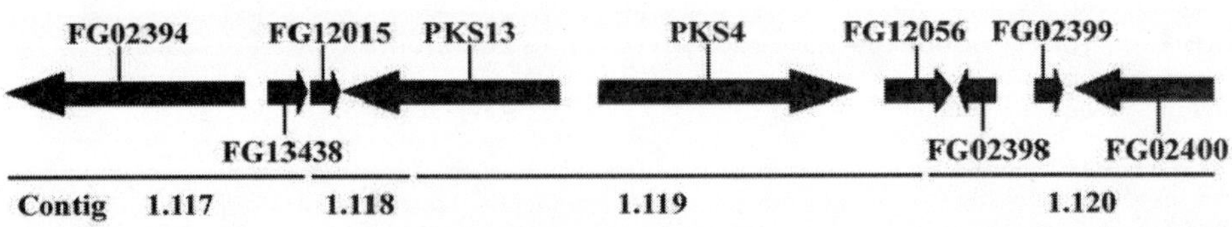


| Locus | Gene product | Contig-(coordinates) |
|---|---|---|
| FG02394 | Non-ribosomal peptide synthetase | 1.117-(125981-119004) |
| FG13438 | Monocarboxylate transporter | 1.117-(127684-128199) |
| FG12015 | $K^+$ channel β subunit | 1.118-(89-859) |
| **FG02395** | **PKS13** | **1.119-(4295-51)** |
| **FG12055** | **PKS4** | **1.119-(5336-12703)** |
| **FG12056** | **Isoamyl alcohol oxidase** | **1.120-(63-957)** |
| **FG02398** | **Transcription factor** | **1.120-(1161-2208)** |
| FG02399 | Protein kinase | 1.120-(3977-5066) |
| FG02400 | $Ca^{2+}$-transporting ATPase | 1.120-(9620-5606) |

**Fig. 4.9.** Genetic organization of the zearalenone (ZON) biosynthetic gene cluster with putative protein function and gene coordinates.

### vi) Patulin

Patulin is elaborated by several species of *Penicillium*, *Aspergillus*, *Paecilomyces* and *Byssochlamys* out of these only three species of clavate group of *Aspergillus* (*A. clavatus*, *A. giganteus* and *A. longivesica*) (Varga *et al.*, 2007) and thirteen species of *Penicillium* (*P. carneum*, *P. clavigerum*, *P. concentricum*, *P. coprobium*, *P. dipodomyicola*, *P. expansum*, *P. glandicola*, *P. gladioli*, *P. griseofulvum*, *P. marinum*, *P. paneum*, *P. sclerotigenum*, *P. vulpinum*) are real producers of patulin (Frisvad *et al.*, 2004)

Though gene clusters related to synthesis of different mycotoxins have been identified, only some genes in a gene cluster in patulin biosynthesis have been identified. The first gene, 6-methylsalicylic acid synthase (6MSAS) was isolated by Wang *et al.* (1991). The *idh* gene encoding the seventh enzyme, isoepoxydon dehydrogenase (IDH) involved in patulin synthesis was isolated by Fedeshko (1992) from *P. griseofulvum* and later from several other *Penicillium* species (Dombrink-Kurtzman, 2007, White *et al*., 2006) and *B. nivea* (Dombrink-Kurtzman and Engberg, 2006).

The biosynthetic pathway of patulin consists of about 10 steps as suggested by several biochemical studies and by the identification of several mutants that are blocked at various steps in the patulin biosynthetic pathway (Artigot *et al.,* 2009).

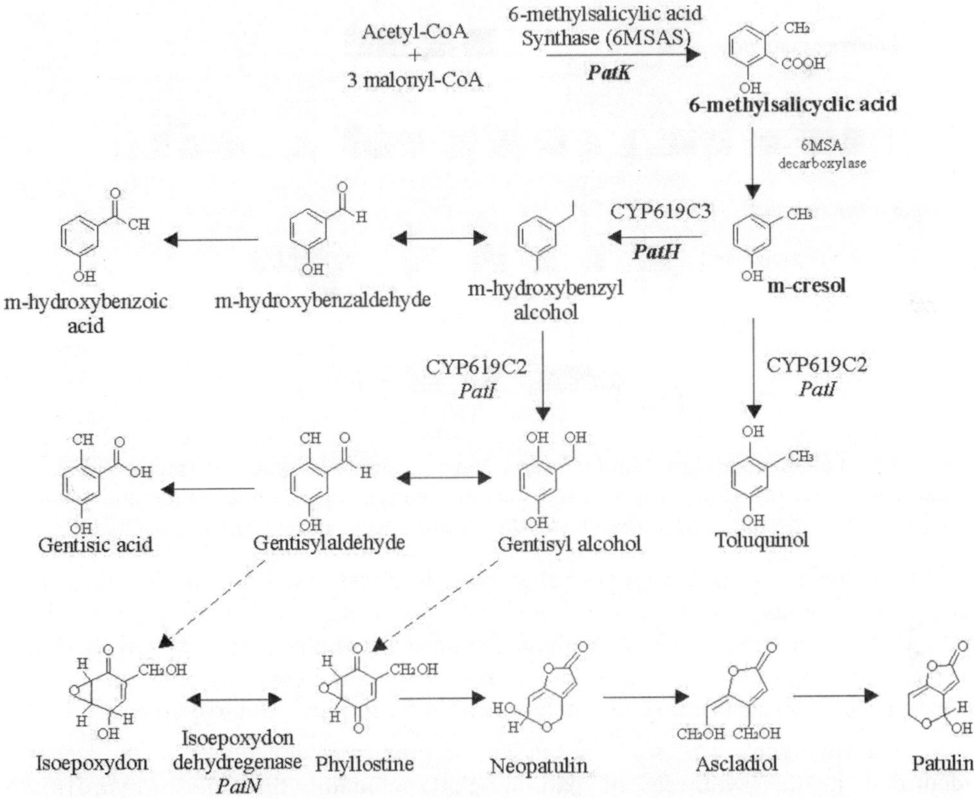


**Fig. 4.10.** Scheme of patulin biosynthetic pathways (Artigot *et al*., 2009)

A cluster of 15 genes located in 40 kb are region involved in patulin biosynthesis in *A. clavatus* was identified by Artigot *et al.* (2009). The genes encoding enzymes are necessary for the patulin biosynthesis specific regulatory factor and transporters. This cluster contains one ABC (ATP binding cassette) transporter, one MFS (Major Facilitator Superfamily) transporter and one acetate transporter genes. The cluster also contains genes responsible for putative carboxyl esterase (*PatB*), one putative Zn-dependent alcohol dehydrogenase (*PatD*), one GMC (Glucose-Methanol-Choline) oxidoreductase (*PatE*). Gene that is probably encodes a putative decarboxylase displaying an amido hydroxylase conserved domain (*PatG*), two genes (*PatH* and *PatI*) encoding cytochromes P450 (responsible for the hydroxylation of *m*-cresol to *m*-hydroxybenzyl alcohol and of *m*-hydroxybenzyl alcohol to gentisyl alcohol, respectively) and a gene encoding a putative dioxygenase

(*PatJ*) are also included. The cluster also contains the *6msas* gene (*PatK*), one putative transcription factor gene (*PatL*) and the isoepoxydon dehydrogenase gene (*PatN*) previously isolated as *idh* gene, a putative isoamyl alcohol oxidase from *P. griseofulvum* (*PatO*), and a gene of PatF which does not have any obvious function. Related clusters are also present in the genome of some species such *P. chrysogenum*, *Talaromyces stipitatus* and *Aspergillus terreus* that do not produce patulin as due to lack of some key genes such as the *6msas* gene (Fig. 4.11).

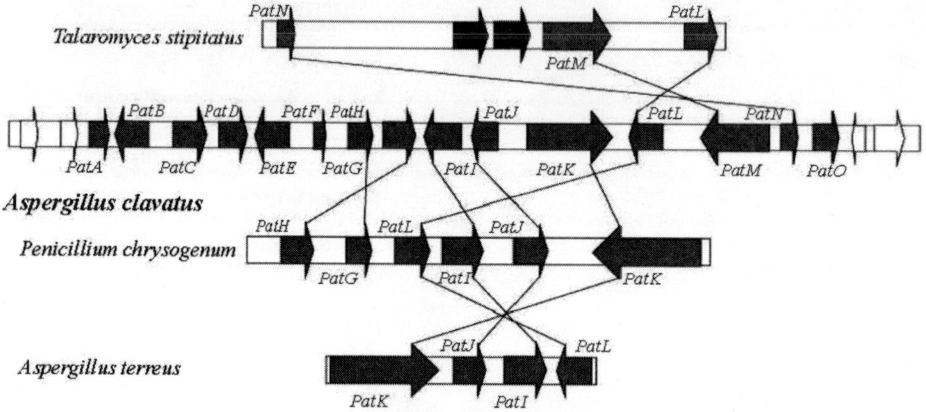


**Fig. 4.11**. Patulin gene cluster in *A. clavatus* and comparison of secondary metabolite gene clusters in three other fungal species that contain some genes related to patulin production. The black arrows in *T. stipitatus* cluster do not form gene cluster (Artigot *et al.* (2009)).

The gene encoding a functional enzyme was responsible for 6MSA formation after heterologous expression in *A. oryzae* (Fujii *et al.*, 1996). In *A. terreus* which produces terreic acid, needs 6-methylsalicylic acid synthase activity (Read *et al.*, 1969). Since there is only one in the *A. terreus* genome contains only *6msas* gene which is likely to encode an enzyme responsible for terreic acid biosynthesis.

The involvement of some genes can be predicted with regard to the steps of identified in the synthesis of patulin. PatG contains the amido hydroxylase superfamily signature sequence motifs shared by resorcylate decarboxylase and 5-carboxyvanillic acid decarboxylase (Liu and Zhang, 2006). Pat G is most likely involved in the decarboxylation of 6-methylsalicylic acid to *m*-cresol. On the basis of the high homology with VBS (59%), which catalyses the transformation of versiconal into versicolorin B in the aflatoxin pathway, and the similarity between their chemical structures, PatE could catalyze the last step of the patulin synthesis from ascladiol. The vesicles play a key role in aflatoxin biosynthesis and the hypothesis of the role of sub cellular compartmentalization could explain the presence of three transporter genes in the cluster (Chanda *et al.* 2009).

*Chapter 5*

# Pleomorphism in Mycotoxigenic Fungi

Many of the mycotoxigenic fungi exists atleast in two phases one represented by asexual reproductive phase called anamorphic phase and the other sexual phase called teleomorphic phase. Still many of the fungi exists only in anamorphic phase because either they lost sexual phase or occur rarely in their life cycle. These fungi are called differentially as fungi imperfecti, deuteromycetes, mitosporic fungi and anamorphic fungi etc. They are usually better known as fungi imperfecti. This grouping may or may not represent phylogeny and usually in reality represent ascomycota or basidiomycota. Members of this group often have two and sometimes more acceptable scientific names. The reason for existence in asexual phase may be either due to not getting a favorable conditions for production of sexual spores as a result they continue to produce asexual spores. The other possibility is that their genetic makeup and physiological conditions. In an organism the recognition is for sexual cycle as it is important for understanding relationships. Most medically important fungi are heterothallic and haploid self sterile (non-sexual) strains and identify and treat the fungus causing disease without knowing about sexual stage or how the fungus is classified. In the absence of this phase in a fungus created a problem which is solved by creating a unique phylum of fungi in which fungi are named with little regard for relationships. Identification based on morphological features without regard to classification. These fungi imperfect are named according to rules of Botanical nomenclature when perfect (Sexual) states become known. They are classified and renamed according to the rules of Botanical Nomenclature for Sexual Fungi. Thus, the concept of anamorphy (asexual), teleomorphy (sexual) and holomorphy (Both asexual and sexual) is developed to depict the correct picture of the fungus. The fungi in which connection between anamorph and teleomorph is established are precised in Table 5.1.

However, still some fungi exist in more than two phases such fungi are called synanomorphy or pleomorphic fungi. In these fungi anamorphy is represented by more than one phase. However, these fungi may become absolute because of molecular biology.

**Table 5.1.** List of anamorphic hypomycetes with teleomorphic connections in ascomycetes and basidiomycetes

| Group | Anamorph | Teleomorph | Taxonomic group |
|---|---|---|---|
| II | *Hansfordiella meliolae* (Hansf) Hughes | *Trichothyrium hansfordii* Hughes | Hemisph-aeriales |
| | *Haffansfordiella* spp. | *Trichothyrium asterolibertiae* Deighton | |
| III | *Trichothecium* spp. | *Hypomyces trichothecoides* | Hypocreales |
| IV | *Calcarisporium* spp. | *Ceratogstis floccosa* (Mathiesen) Hunt | Dothidiales |
| | *Dematophora necatrix* Hartig | *Rosellinia necatrix* Prill | Hypocreales |
| | *Diplorhinotrichum* | *Acrospermum compressum* Tode ex Fries | |
| | *Helicoma muelleri* Corda | *Lasiosphaeria pezizula* (Berk. & Curt.) Sacc. | |
| | *Helicosporium vegetum* Nees ex Fries | *Ophionectria cerea* (Berk. & Curt.) | |
| | *Verticicladium* spp. | *Ceratocystis minor* (Hedgec.) Hunt, *Desmazierella acicola* Lib. | |
| | *Virgariella spp* | *Acrospermum graminum* Lib | |
| | *Virgariella spp* | *Daldinia angolensis* (Welw. & Curr.) Sacc. | |
| V | *Cercospora* spp | Many species have perfect states referable to *Mycosphaerella* | Dothidiales |
| | *Cercosporella* spp | Perfect states are preferable to *Mycosphaerella* | Endomy-cetales |
| | *Fusicladium cerasi* (Rabenh.) Sacc | *Venturia cerasi* Aderh | Sphaeriales |
| | *Fusicladium pyrorurn* (Lib.) Fuckel | *Venturia pirina* Aderh. | |
| | *Fusicladium saliciperdum* (Ailesch. & Tub.) Lind | *Venturia chlorospora* (Ces.) Karst | |
| | *Fusicladium* spp. | *Venturia crataegia* Aderh | |
| | *Hadrotrihum phragmites* Fuckel | *Scirrhia rimosa* (Alb.& Schw) Fuckel | |
| | *Pleurophragrmium spp.* | *Melanamma subdispersum* (Karst.) Berl. & Vogl. | |
| | *Polythrincium trifolii* Kunze & Schrn. CY. Fries) | *Cymadothea trifolii* Wolf *Ascoidea rubescens* Bref. | |
| | *Polythrinium* spp | | |
| VI | *Nigrospora oryzae* (Berk.& Br.) Petch | *Khuskia oryzae* Hudson | Sphaeriales |
| VII | *Cladosporium echinulatum* (Berk.) de Vries | *Didymellina dianthi* Burt | Eurotiales |
| | *herbarum* Link ex Fries: | *Mycosphaerella tassiana* (de Not.) Johnson | Helotiales |
| | *ladinurn* Mueller: | *Leptosphaeria ladina* Mueller | |
| | *Cladosporium omithogali* (Klotzsch) de Vrics: | *Didymellina omithogali* Jacques | Sphaeriales |
| | *Cladosporium* | *Ascocybe grovesii* Wells (*Cephaloascus fragrans* Hanawa) | Endomy-cetales |
| | *Cladosorium spp.* | *Ceratocystis piceae* (Munch) | Ustilaginales |

| Group | Anamorph | Teleomorph | Taxonomic group |
|---|---|---|---|
| | *Cladosporiurn spp.* | *Ceratocystis pilifera* (Fries) | |
| | *Cladosporium spp.* | *Ceratocystis pluriannulata* (Hedgec.) Moreau | |
| | *Cladosporium* | *Graphiola cylindrica* Kobayasi | |
| | *Cladosporiurn* (*Hormodendrum pedrosoi* Brumpt) | *Microascus pedrosoi* Fuentes & Wolf | Eurotiales |
| | *Heterosporium cladosporioides* McW. | *Didymellina poecilospora* McW. | |
| | *Heterosporium gracile* Sacc. | *Didymellina macrospora* Klebh. | |
| | *Monilia cinerea* Bon. | *Sclerotinia laxa* Aderh. & Ruhl. | Helotiales |
| | *Monilia fructigena* Pers. | *Sclerotinia fructigena* Aderh. & Ruhl | |
| | *Monilia sitophila* (Mont.) Sacc. | *Neurospor sitophila* Shear & Dodge | |
| | *Monilia spp.* | *Ascobolus furfurascs ers.* (Thbaki) | |
| | *Monilia spp.* | *Ascobolus stercorarius* Rehm ex Bull | |
| | *Septoglindrium gossypii* (Speg.) Subram. | *Mycosphaerella areola* Ehrlich & Wolf | |
| | *urticae* (Ces.) Subram. | *Mycosphaerella superflua* (Auersw.) Fetrak. | |
| | *Mycosphaerella* spp. Many species of *Ramularia auct* in which conidial chains develop are better placed in *Septocylindrium*. | Several species of *Ramularia auct*. have perfect states in *Mycosphaerella.* | |
| | *Septonema secedens* Corda | *Mytilidon karstenii* Sacc. | |
| IX | *Cephalotrichum stemonites* (Pres) Link Fries | *Melanospora stysanophora matt.* | Eurotiales Hypoereales |
| | *Cephalotrichum* | *Microascus albo-nigrescens* (Sopp) Curzi | |
| | *Scopulariopsis albo-flavescens* Zach. | *Micrococcus manginii* (Loub.) Curzi | |
| | *Scopulariopsis sp.* | *Microascus cinereus* (Emile-Weil & Gaudin) Curzi | |
| | *Scopulariopsis sp.* | *Microascus cirrosus* Curzi | |
| | *Scopulariopsis sp.* | *Microascus desmosporus* (Lech.) Curzi | |
| | *Scopulariopsis sp.* | *Microascus doguetii* Moreau | |
| | *Scopulariopsis sp.* | *Microascus longirostris* Zukal | |
| | *Scopulariopsis sp.* | *Microascus pyramidus* Barron &Gilman | |
| | *Scopulariopsis sp.* | *Microascus styanophorus* (Matt.) Barron | |
| | *Scopulariopsis sp.* | *Microascus trigonosporous* Emmons &Dodge | |
| X | *Arthrinium curvatum* Kunze ex Fries | *Pseudoguignardia scirpi* Gutner | Sphaeriales |
| | *Papularia arudinis* (Corda) Fries | *Apiospora montagnei* Sacc | |
| | *Papularia vinosa* (Berk. St Curt.) Mason | *Apiospora camptospora* Penz. Sacc | |

| Group | Anamorph | Teleomorph | Taxonomic group |
|---|---|---|---|
| XI | *Septoidium sp.* | *Parodiopsis* spp | Erysiphale |
| | *Spilocae pomi* Fries ex Fries | *Venturia inaequalis (cooke)* wint | Sphaeriales |
| XIII | *Acladium capitatum* Link | *Botryobasidium candicans* Eriks | Eurotiales |
| | *Acladium sp* | *Plicaria fulva* Schneider | Helotiales |
| | *Acladium sp* | *Lamprospora trachycarpa* (Curr.) Seaver | Hypoereales |
| | *Acrostaphylus sp* | *Daldinia concentrica* (Fries) Ces | Pezizales |
| | *Acrostaphylus sp* | *Rosellinia aquila* (Fries) de Not, *Rosellinia buxi* Fabre and *Rosellinia thelena* (Fries) Rabenh | Sphaeriales |
| | *Basidiobotrys sp.* | *Hypoxylon punctulatum* (Berk. & Ray.) Cooke | |
| | *Botrytis cinerea* Pers. ex Fries | *Sclerotinia fuckeliana* (de Bary) Fuckel | |
| | *Botrytis convolute* Whetzel | *Sclerotinia convolute* Drayton | |
| | *Botrytis globose* Raabe | *Sclerotinia globosa* (Buchwald) Webster | |
| | *Botrytis narcissicola* Klebh | *Sclerotinia narcissicola* Gregory | |
| | *Bobytis polyblastis* Dowson | *Sclerotinia polyblastis* Gregory | |
| | *Botrytis sphaerosperma* Buchwald | *Sclerotinia sphaerosperma* Gregory | |
| | *Botrytis squamosa* Walker | *Sclerotinia squamosa* (Viennot-Bourgin) Dennis | |
| | *Botrytis sp.* | *Sclerotinia draytoni* Buddin & Wakcf and *Sclerotinia spermophila* Noble | |
| | *Haplographium delicatarn* Berk. & Br. | *Hyaloscypha dernatiicola* (Berk. & Br.) Nannf. | |
| | *Oedocephalum lineatum* Bakshi | *Fomes annosus* (Fries) Cooke | |
| | *Oedocephalum sp* | *Ascophanus testaceus* (Moug. ex Fries) Phill | |
| | *Oedocephalum sp* | *Corticium effuscatum* Cooke & Ellis | |
| | *Oedocephalum sp* | *Peniophora allescheri* Bres, *Peniophora heterocystidia* Burt and *Peniophora mutate* Peck | |
| | *Oedocephalum sp* | *Peziza pustulata* Pers. | |
| | *Sporotrichum sp* | *Cordyceps hokkidoensis* Y. Kob. | |
| | *Sporotrichum sp* | *Pseudoeurotium ovalis* Stolk | |
| XIV | *Isthmospora glabra* Stevens | *Trichothyrium reptans* (Berk. & Curt.) Hughes | Hemispha- |
| | *Isthmospora trichophila* (Atk.)-Damon | *Trichothyrium asterophorum* (Berk. & Br.) Hohnel | eriales |
| XVI | *Tetracriurn aurantii* P. Henn. | *Podo aurantii* (P. Henn.) Petch | Sphaeriales |
| | *Tetracrium coccicola* (Ellis & Evers.) | *Podonectria coccicola* (Ellis & Everh.) Petch | Hypocreales |
| | *Tetracrium echinatum* Petch | *Podonectria echinata* Petch | |
| | *Xenosporella thaxteri* Linder | *Acanthostigmella thaxteri* Linder | |
| XVII | *Aegerita candida* Pers. ex Fries | *Peniophora candida* Lyman | Eurotiales |
| | *Bototrichum piluliferum* Sacc. & March | *Chaetomium piluliferum* Daniels | Hypocreales<br>Sphaeriales |
| | *Mycogone* spp. | *Hypomyces* spp | Pezizales |

| Group | Anamorph | Teleomorph | Taxonomic group |
|---|---|---|---|
| | *Sepedonium chrysospermum* (Bull) Fries. | *Hypomyces chrysospermus* Tul. | |
| | *Sepedonium* sp. | *Thielavia sepedonium* Emmons | |
| XVIII | *Catenularia heimii* Mangenot | *Chaetosphaeria myriocarpa* (Fries) Booth | Sphaeriales |
| | *Catenularia* sp. | *Clzaetosphaeria innumera* Tul. | |
| | *Chalara quercina* Henry | *Ceratocystis fagacearum* (Bretz) Hunt | |
| | *Thielaviopsis paradoxa* (de Seynes) Hohnel | *Ceratocystis paradoxa* (Dade) C. Moreau | |
| | *Thielaviopsis* sp. | *Ceratocystis adiposa* (Butl.) Moreau, *Ceratocystis fimbriata* Ellis & Halst, *Ceratocystis major* (v. Beyma) C. Moreau, *Ceratocystis moniliformis* (Hedgec.) C. Moreau, *Ceratocystis radicicola* (Bliss) C. Moreau and *Ceratocystis variospora* (Davids.) C. Moreau | |
| XIX | *Aspergillus amstelodami* (Mang.) Thom & Church | *Eurotium amstelodami* Mangin | Sphaeriales Eurotiales |
| | *Aspergillus brunneus* Delacr. | *Eurotium echinulatum* Delacr. | Hypocreales |
| | *Aspergillus carnoyi* Thorn & Raper | *carnoyi* (Thom & Raper) Benjamin | |
| | *Aspergillus chevalieri* (Mang.) Thom & Church | *Eurotium chevalieri* Mang. | |
| | *Aspergillus fischeri* Wehmer | *Sartova fumigata* Vuill. | |
| | *Aspergillus medium* Meiss. | *Eurotium medius* Meiss | |
| | *Aspergillus nidulans* (Eidam) Wint. | *Emericella nidulans* (Eidam) Vuill. | |
| | *Aspergillus quadrilineatus* Thom & Raper | *Emericella quadrilineata* (Thom & Raper) Benjamin | |
| | *Aspergillus repens* (Corda) de Bary | *Eurotium repens* de Bary | |
| | *Aspergillus ruber* (Bremer) Thom & Raper | *Eurotium rubrum* Bremer | |
| | *Aspergillus rugulosus* Thom & Raper | *Emericella rugulosa* (Thom & Raper) Benjamin | |
| | *Aspergillus stellatus* Curzi | *Emericella variecolor* Berk. & Br. | |
| | *Cephalosporium salmosynnematum* Rob. | *Emericellopsis salmosynnemata* Grosklags & Swift | |
| | *Cephalosporium* sp. | *Emericellopsis minima* Stolk | |
| | *Cephalosporium* sp. | *Ceratocystis brunneo-ciliata* (Mathies.) Hunt, *Ceratocystis cana* (Munch) C. Moreau, *Ceratocystis galeiformis* Bakshi, *Ceratorystis peniciiiuta* (Grosm.) C. Moreau, *Ceratocystis perparvispora* Hunt, *Ceratocystis piceae* (Munch) Bakshi, *Ceratocystis tetropii* (Mathies.) Hunt and *Coprinus lagopus* Fries | |
| | *Cephalosporium* sp. | *Levispora terricola* Routien | |
| | *Cephalosporium sp.* | *Nectria ellisii* Booth | |

| Group | Anamorph | Teleomorph | Taxonomic group |
|---|---|---|---|
| | *Cephalosporium* sp. | *Neocosmosara vasinfecta* Smith | |
| | *Cephalosporium* sp. | *Saturnomyces humicola* Cain | |
| | *Cephalosporium* sp. | *Trichocoma paradoxa* Jungh. | |
| | *Cylindocarpon cylindrois* Wollen W. | *Nectria cucurbitula* (Tode) Fries v. mizcrospora Wollenw. | |
| | *Cylindocarpon mali* (Ailesch.) Woilenw. | *Nectria galiigena* Bres. (*Dialonectria galligena* (Bres.) Petch.) | |
| | *Cylindocarpon sp.* | *Nectria coccinea* (Pers. ex Fries), *Nectria hederae* Booth, *Nectria mammoidea* Phil- & Plowr, *Nectria mammoidea* Phil- & Plowr. v. rubi (Osterw) Weese and *Nectria veuiliotiana* Roum. & Sacc | |
| | *Cylindrocladium ilicicola* (Hawley) Boedijn & Reistma | *Calonectria ilicicola* Boedijn & Reistma | |
| | *Cylindrocladium macrosporurn* v. *hederae* Arnaud | *Calonectria hederae* Arnaud | |
| | *Dendrodochium epistroma* (Syd.) Hohn. | *Nectria magnusiana* Rehm | |
| | *Flagellospora penicillioides* Ingold. | Nectria *penicillioides* Ranzoni. | |
| | *Fusarium autrantiacum* Corda ex Fries | *Gibberella intricans* Wollen w. | |
| | *Fusarium coccidicola* P. Henn. | *Calonectria diploa* (Berk. & Curt.) Wollenw. | |
| | *Fusarium coccophilum*(Desm.) Wollenw. & Rg. | *Nectria coccophila* Wollenw. | |
| | *Fusarium coerulentum* Lib. | *Hypomysis esclepiadis* Zerova | |
| | *Fusarium cuneiforme* Sherb. | *Nectriopsis solani* (Reink. & Berth.) Booth | |
| | *Fusarium decemcellulare* Brick | *Calonectria rigidiuscula* (Berk. & Br.) Sacc | |
| | *Fusarium hippocastani* (Corda) Sacc. | *Gibberella acuminata* Wollenw. | |
| | *Fusarium kurdicum* Petrak | *Calonectria kurdica* Petrak | |
| | *Fusariurn lateritium* Nees ex Fries | *Gibberella baccata* (Wallr.) Sacc. | |
| | *Fusariurn maydis* Kalchbr. | *Gibberella pulicaris* (Fries) Sacc. | |
| | *Fusarium moniliforme* Sheld. | *Gibberella fujikuroi* (Saw.) Wollenw. | |
| | Fusarium *nivale* (Fries) Ces. | *Calonectria nivalis* Schaffn. | |
| | *Fusarium* sp. | *Ascopolyporus* | |
| | *Fusarium* sp. | *Corallomyces jatrophae* Moller | |
| | *Fusarium* sp. | *Hypocrella cavernosa* Moller | |
| | *Fusarium* sp. | *Mycocitrus aurantium* Moller | |
| | *Fusarium* sp. | *Nectria brassicae* Ellis & Sacc, *Nectria desmazierii* Beccari & de Not, *Nectria episphaeria* (Tode ex Fries) Fries, *Nectria flatio-virldis* (Fuck.) Wollenw, *Nectria haematococca* Berk. & Br, *Nectria hakonensis* | |

| Group | Anamorph | Teleomorph | Taxonomic group |
|---|---|---|---|
| | | Kob, *Nectria leptosphaeriae* Niessl, *Nectria pinea* Dingley and *Nectria purtonii* (Grey.) Berk | |
| | *Fusarium* sp. | *Peleronectria vinosa* Moller | |
| | *Gibellula leiopus* (Vuill.) Mains | *Torrubiella arachnophila* (Johnst.) Mains | |
| | *Gibellula pulchra* (Sacc.) Cav. | *Torrubiella arachnophila* (Johnst.) Mains v. pulchra Mains | |
| | *Gliocladium roseum* Banner | *Nectria gliocladioides* Smalley & Hansen | |
| | *Gliocladium* sp. | *Littiputia rufula* (Berk. & Br.) Hughes, *Nectria keithii* Berk. & Br, *Nectria pallidula* Cooke and *Nectria subquaternata* Berk & Br. | |
| | *Gonytrichum caesium* C. G. & T.F.L. Nees | *Melanopsamella inaequalis* (Grove) Hohnel | |
| | *Graphium ulmi* Schwartz | *Ceratocystis ulmi* (Buism.) C. Moreau | |
| | *Graphium* sp. | *Ceratocystis brunneo-ciliata* (Mathies.) Hunt, *Ceratocystis cana* (Munch) C. Moreau, *Ceratocystis clavata* (Mathies.) Hunt, *Ceratocystis froccosa* (Mathies.) Hunt, *Ceratocystis galeiformis* Bakshi, *Ceratocystis ips* (Rumb.) C. Moreau, *Ceratocystis obscura* (Davids.) Hunt, *Ceratocystis olivacea* (Mathies,) Hunt, *Ceratocystis piceae* (Munch) Bakshi, *Petriella guttulata* Barron & Cam, *Petriella lindforsii* Curzi and *Petriella sordida* (Zukal) Barron & Gilman | |
| | *Graphium* sp. | *Petriella boulangeri* Curzi, | |
| | *Heliscus lugdanensis* Sacc. & Therry | *Nectria lugdunensis* Webster | |
| | *Menispora caesia* Preuss | *Cluzetosphaeria pulviscula (*Curr.) Booth | |
| | *Myrothecium sp.* | *Nectria ralfsii* Berk. & Br. | |
| | *Paecilomyces varioti* Bainier | *Byssochlamys fulva* Olliver & Smith | |
| | *Paescilomyces* sp. | *Byssochlamys nivea* Westl. | |
| | *Paescilomyces* sp. | *Melanospora mangini* Vincens | |
| | *Paescilomyces* sp. The conidial state is sometimes treated as a *Scopulariopsis* | *Microascus manginii* (Loub.) Curzi | |
| | *Penicillum asperum* (Shear) Raper & Thom | *Carpenteles asperum* Shear | |
| | *Penicillium brefeldianum* Dodge | *Carpenteles brefeldianum* (Dodge) Shear | |
| | *Penicillium javanicum* v. Beyma | *Carpenteles javanicum* (v. Beyma) Benjamin n | |
| | *Penicillizan levitum* Raper & Fennell | *Carpenteles levitum* (Raper & Fennell) Benjamin | |
| | *Penicillium avellaneum* Thom & Turess. | *Taloromyces avellaneous* (Thom & Turess.) Benjamin | |

| Group | Anamorph | Teleomorph | Taxonomic group |
|---|---|---|---|
| | *Penicillium bacillosporum* Swift | *Talaromyces bacillosporus* (Swift) Benjamin | |
| | *Penicilliurn helicum* Raper & Fennell | *Talaromyces hell*icus (Raper & Fennell) Benjamin | |
| | *Penicillium luteum* Zukal | *Talaromyces luteus* (Zukal) Benjamin | |
| | *Penicilliurn rotundum* Raper & Fennell | *Talaromyces rotundus* (Raper & Fennell) Benjamin | |
| | *Penicillium spiculisporum* Lehman | *Talaromyces spiculisporus* (Lehman) Benjamin | |
| | *Penicillium stipitatum* Thom | *Talaromyces stipitatus* (Thom) Benjamin | |
| | *Penicillium striatum* Raper & Fennell | *Talaromyces striatus* (Raper & Fennell) Benjamin | |
| | *Penicilliwn vermiculatum* Dang | *Talaromyces vermiculatus* (Dang.) Benjamin | |
| | *Penicillium wortmanni* Kloecker | *Tataromyces wortmanni* (Klocker) Benjamin | |
| | *Penicillium* sp. | *Penicilliopsis calvariaeformis* Solms-Laub. | |
| | *Penicillium* sp. | *Trichocoma paradox*a Jungh | |
| | *Phialophora* sp. | *Bombardia lunata* Zickler | |
| | *Phialophora* sp. | *Lasiosphaeria hirsuta* (Fries) Ces. & de Not. | |
| | *Phialophora* sp. | *Microascus pedrosoi* Fuentes & Wolf | |
| | *Sphacelia segetum* Lev. | *Claviceps purpurea* (Fries) Tulasne | |
| | *Stachybotrys* sp. | *Melanospsamma pomiformis* (Pers.) Sacc. | |
| | *Stilbum aurantiacum* Babington | *Nectria aurantiaca* (Tul.) Jacz. | |
| | *Stilbum* sp. | *Spheaerostilbie gracilipes* Tul. | |
| | *Stibum* sp. | *Thytionectria* spp. | |
| | *Trichoderma viride* Pers. ex Fries | *Hypocrea rufa* (Pers. ex Fries) Fries | |
| | *Tubercularia vulgaris* Tode ex Fries | *Nectria cinnabarina* (Tode ex Fries) Fries | |
| | *Verticillium agaricinunz* Gorda | *Hypomyces ochraceus* (Pers.) Tul. | |
| | *Verticillium luteo-album* (Link ex Fries) Subram | *Nectria inventa Pethybr.* | |
| | *Phialospore* sp. | *Melanospora globosa* Berlese | |
| | *Phialospore* sp. | *Sphaeroderma bulbilliferurn* Berlese | |
| XXI | *Alternaria* sp. | *Clathrospora diplospora* (Ellis & Everh.) Wehm. | Sphaeriales |
| | *Alternaria* sp. | *Leptos*phaeria *heterospora* (de Not.) Sacc. | |
| | *Alternaria* sp. | *Pleospora* infectoria Fuckel | |
| | *Bipolaris homomorphus* (Luttre1 & Rogers.) Subram. c.v. | *Cochliobolus homomorphus* Luttrell & Rogers. | |
| | *Bipolaris kusanoi* (Nisik.) Shoemaker | *Cochliobolus kusanoi* (Nisik.) Drechsl. ex Dastur | |

| Group | Anamorph | Teleomorph | Taxonomic group |
|---|---|---|---|
| | *Bipolaris maydis* (Nisik.) Shoemaker | *Cochliobolus heterostrophus* Drechsl. | |
| | *Bipolaris nodulosa* (Berk. & Curt. ex Sacc.) Shoemaker | *Cochliobolus nodulosus* Luttrell | |
| | *Bipolaris oryzae* (Breda de Haan) Shoemaker | *Cochliobolus miyabeanus* (Ito & Kurib.) Dickson | |
| | *Bipolaris setariae* (Saw.) Shoemaker | *Cochliobolus setariae* (Ito & Kurib.) Drechsl. ex Dastur | |
| | *Bipolaris sorokiniana* Sacc. *in* Sorok.) Shoemaker | *Cochliobolus sativus* (Ito & Ktirib.) Drechsl. ex Dastur | |
| | *Bipolaris stenospila* (Drechsl.) Shoemaker | *Cochliobolus stenospilus* (Carp.) Matsum. & Yamam. | |
| | *Bipolaris turcicum* (Pass.)ˉ Shoemaker | *Trichometasphaeria turcica* Luttrell | |
| | *Bipolaris reicola* (Stout) Shoemaker | *Cochliobolus carbonum* Nelson | |
| | *Bipolaris* sp. | *Cochliobolus sporoboli* E. Cast. | |
| | *Dendryphiopsis atra* (Corda) Hughes. | *Amphisphaeria incrustans* Ellis & Everh | |
| | *Drechslera avenacea* (Curt. ex Cooke) Shoemaker: | *Pyrenophora avenae* Ito & Upaper | |
| | *Drechslera bromi* (Died.) Shoemaker | *Pyrenophora bromi* (Died.) Drechsl | |
| | Drechslera *graminea* (Rab. ex Schlecht.) Shoemaker | *Pyrenophora graminea* (Rah. ex Schlecht.) Ito & Kurib. | |
| | *Drechslera japonica* (Ito & Kurib.) Shoemaker | *Pyrenophora japonica* Ito & Kurib | |
| | *Drechslera siccans* (Drechsl.) Shoemaker | *Pyrenophora lolli* Dovaston | |
| | *Drechslera teres* (Sacc.) Shoemaker | *Pyrenophora teres* Drechsler | |
| | *Drechslera tritici-repentis* (Died.) Shoemaker | *Pyrenophora tritici-repentis* (Died.) Drechsler | |
| | *Drechslera tritici-vulgaris* (Nisik.) Ito | *Pyrenophora tritici-vulgaris* Dickson | |
| | *Drechslera tuberosa* (Atk.) Shoemaker | *Pyrenophora secalis* Whiteh. & Dickson | |
| | *Drechslera* sp. | *Pyrenophora phaeocomes* (Fries) Rebent. | |
| XXII | *Acrosporium* sp. | *Erysiphae, Microsphaeria, uncinula & uncinulopsin* | Erysiphales |
| | *Coniosporium* | *Hysterium isidens* Schw. | Hysteriales |
| | *Oidiopsin sicula* Scalia | *Leveillula taurica* (Ler.) Salmon | |
| XXIII | *Coreniella* | *Onygena corvine* Alb. & Schw. | Eurotiales |
| XXIV | *Monilia sp.* | *Sclerotium cydoniae* Schellenb.<br>*S. mespili Scellenb.* | Heliotiales |

| Group | Anamorph | Teleomorph | Taxonomic group |
|---|---|---|---|
| | | *S. Johansonii* (Ellis. & Everh.) Rehm, *Nectar detritus* Massee, *S. baccarum* (Schroet.) Rehm *Corticium salmonicolor* Berk. & Br. | |
| XXV | *Hirsutella confragosa* Mians | *Torrubiella confragosa* | Hypocreales Herrisphaeri |

Members of Group I, VIII, XII and XV have not been reported to produce teleomorphic states.
Group-I: Conidia blast pores usually subtended by separating cells
Group-II: Conidia blastospores, not subtended by separating cells, conidiophores with characteristic cupulate scars
Group-III: Conidiophore with multicupulate cells, sympodial growth, conidia are produced in basipetal chains
Conidiophore without multicupulate cells, sympodial growth, conidia produced singly or in acropetal chains.
Group-IV: Conidia apiculate or pointed at the base conidiophore appearing denticulate
Group-V: Conidiophores sympodial with basal scars conidia solitory or acropetal chains with flattened basal scars or flattened at the base
Group-VI: Conidiophores not showing sympodial growth, conidia produced on vesicles or sporogenous cells
Group-VII: Conidiophores not showing sympodial growth conidia are formed in acropetal chains and maturing from base to apex
Group VIII: Conidia formed in acropetal chain, maturing apex backwards
Group VIII: Conidiophores not showing sympodial growth conidia are formed in acropetal chains and maturing from apex backwards
Group-IX: Conidiophores not showing sympodial growth, conidia are formed in basipetal chains
Group-X: Conidiophores not showing sympodial growth, conidia produced singly and conidiophores elongating by basal growth
Group- XI: Conidia produced singly and conidiophore grow by apical growth and with annellations
Group XII: Conidiophores elongating from the scar of fallen conidia (proliferation). Conidia single no annulations conidia are produced repeatedly through scars in successive sometimes on sporogenous cells
Group XIII: Conidiophores elongating by apical growth
Group XIV: Conidia gangliopspores, isthamospores
Group XV: Gangliospores and subtended by separating cells
Group XVI: Gangliospores are cut off by septum early and get detached early
Group XVII: Gangliospores not cut off early and not detached early
Group XVIII: Conidia phialospores and endogenous
Group XIX: Conidia phialospores not truely endogenous expect the first formed
Group XX: Conidia porospores, acropleurogenous
Group XXI: Conidia porospores and acrogenous
Group XXII: Conidia arthorospores and formed by basipetal succession
Group XXIII: Arthrospores formed by septation and formed by fragmentations and produced endogenously
Group XXIV: Arthrospores formed either by septulation or fragmentation and not endogenous
Group XXV: Conidia are borne on spicules (Spiculospores)

Considering the limited information on teleomorph connection within their groups and enormous morphological variation within and between perfect states (teleomorphs) and imperfect states (anamorphs), we are still forced to maintain two different systems for their classification one for the perfect states (ascomycetes and basidiomycetes) arranged according to the development and the morphology of their sexual fructification and a second for the imperfect fungi including the anamorphs of ascomycetes and basidiomycetes based on their asexual fructification. The nomenclatural difficulties and consequences arising from this situation have been discussed by Hennebert and Weresub (1977). In most cases the proof that teleomorph and conidial anamorph are phases of the same life cycle can only be obtained by culture work. In some of the fungi, anamorphs have been positively connected with relatively few of known species of ascomycetes or basidiomycetes such as erysiphales (Spencer, 1978), trichocomaceae (Malloch and Cian, 1972), nectrioid fungi (Samuel and Roseman, 1979) and uredinales. Tubaki (1958) and Muller (1971) have tried to link the system for teleomorphs with that of anamorph could enable them to make general rules. De Hoog (1977) tried to approach the problem with the aid of cluster analysis. Kendrick *et al.* (1979) have tried to correlate these fungi at the genus level. They discussed the kinds and the variation of both sexual and asexual forms but also the presence and absence of anamorphs within the individual life cycles. Fungi reported to have connection between anamorph and teleomorph are listed in Table 5.2.

**Table 5.2.** Seasonal succession of states of some parasitic Ascomycetes

| | Anamorph | | | Teleomorph | | | |
|---|---|---|---|---|---|---|---|
| **Example** | **Host** | **Generic name** | **Season** | **Spermatogonia** | **Ascomata initials** | **Ascospores** | **References** |
| *Venturia inaequalis* | *Pirus or Sorbus* | *Spilocaea* | Late spring | | Autumn | Spring | Aderhold (1987) |
| *Mycosphaerella berberidis* | *Berberis* | *Septoria* | Summer | Asteromella autumn | Autumn | Spring | Von Arx (1949) |
| *Polystigma rubra* | *Prunus* | *Polystigmina* | Summer | | Late summer | Autumn, spring | Moreau (1930) |
| *Claviceps purpurea* | Gramineae | *Sphacelia* | Summer | | Spring | Early summer | Tulasne (1953) |
| *Monilinia fructigena* | Rosaceae | *Monilia* | Early summer, Autumn | Autumn, spring | Spring | Spring | Aderhold and Ruhland (1905) |
| *Diplocarpon maculatum* | Rosaceae | *Entomosporium* | Summer | Autumn | Autumn, winter | Spring | Klebahn (1918) |
| *Rhystisma acorina* | *Acer* | | | Melasmia, autumn | Autumn | Spring | Webster (1970) |

Further, it is to be noted that ascomycetous anamorphs always occur in the haploid phase, while basidiomycetous anamorphs are either in the haploid or the dikaryon mycelium and produce conidia. It has been found that the same fungus can produce conidia in both phases. In some groups conidium like spores may function as male gametes (spermatia). In the heterobasidiomycetes the formation of conidia directly from basidiospora is common to all members of the group. In rust fungi closer connection between anamorph and teleomorph is observed for *Gymnosporangium gaeumanni albertensis.* Most species of *Gymnosporangium* occurring on Rosaceae are in the haploid and cuppressoideae in the dikaryophase, which do not form the uredinial anamorph. *G. gaeumanni,* however, is an eu form. Unicellular uredinospores develop within the same sori as the bicellular Teliospores as in many rust fungi but *G. gaeumanni* also forms chimeric bicellular spores with one cell anamorphic (germinating by a germ tube) and the other cell teleomorphic (forming a basidium).

The fungi are unique in having almost uniform assimilative stages (such as unicellular or hyphae) but with diverse propagative stages. The yeast cells form which in both assimilative and propagative is better adapted to spread through a liquid medium. The filamentous form is better adapted for rapid colonization of solid substrates. Fungi which can change from mycelial to yeast form are often called dimorphic fungi.

The propagative stages of higher fungi are extremely diverse both in kinds of spores and fruiting structures produced by different fungi and in the variety of different spores produced during the life cycle of a single species. Many conidial fungi are asexual propagative stages (anamorph) in the life cycle of higher fungi. For other species no sexual stage (teleomorph) is known and it is presumed that at least some of the species have evolved a completely asexual or parasexual cycle.

In some species the propagative stage may be represented by a single type of reproductive structure and spores. For example, basidiospores (*Agaricus brunnecens*), ascospores (*Gelasinospora tetrasperma*) and only phialoconidia (*Aspergillus fumigatus*). Such fungi could be called as monomorphic in propagation but the term is rarely used. On the other hand, fungi which produce more than one type of spores are called Pleomorphic. For example the rusts (Uredinales) may produce two kinds of conidia (aeciospores and uredinospores) and three kinds of spores connected with sexual reproduction (spermatia, teliospores and basidiospores). The cup fungus, *Botryotinia fuckeliana,* produces one kind of conidia plus spermatia and ascospores. *Cephalotrichum stemonitis* produces two or three kinds of conidia but no teleomorph has yet been found. The pleomorphism in fungi can be with separate anamorphism or pleomorphism only in anamorphic phase Carmichael *et al.* (1980) discussed in detail about types of pleomorphism in anamorphic fungi. The anamorphic pleomorphism can be divided into three kinds:

1. Pleomorphism with conidia of the same ontogenic type
2. Pleomorphism with conidia of different types
3. Pleomorphism of conidiomata

### 5.1. Pleomorphism with conidia of the same ontogenic type

*Ajellomyces capsulatus* produces small, smooth aleuro-conidia and large tuberculate aleuro-conidia both of which are roughly globose in shape. Many species of *Fusarium* also have micro- and macroconidia but here the macroconidia are slimy phialoconidia. The microconidia of some species are also phialoconidia but in other species they seem to be solitary. Most conidial fungi produce spores in either powdery or slimy masses. A few phialidic species like *Acremonium alternatum, Stachybotrys echinata* and *Gliocladium roseum* are pleomorphic. In *Scopulariopsis brevicaulis*, the conidia are usually formed in long, dry chains from the annellate conidiogenous cells but under moist conditions they sometimes collapse into slimy balls.

### 5.2. Pleomorphism with conidia of different types

One of the most common kinds of anamorphic pleomorphism is the production of small, single celled phialidic conidia plus larger, thick walled, solitary conidia which may be septate and pigmented. For example, *Sepedonium* species with large, globose, echinulate or tuberculate aleuroconidia plus small slimy, hyaline phialoconidia borne on *Acremonium* or *Verticillium* like conidiophore. *Mycogone* species may have similar tuberculate microconidia and slimy phialoconidia. *Sepedonium ampullasporum* has unusual solitary bottle shaped microconidia borne terminally on penicellate conidiophores. A few species of *Chalara* type phialides produce dark, thick walled macroconidia. In conidial *Ceratocystis paradoxia,* the dark conidia are unicellular and catenate.

Hyaline phialoconidia of *Verticillium* or *Paecilomyces* type are associated with larger, thick-walled, hyaline dictyoconidia in *Diheteospora.* In some species of *Cylindrocarpon* also dictyoconidia or phragmaconidia are associated with phialidic phragmoconidia. Some species of *Phoma* also produce chains of dark dictyoconidia (*Peyronellia*) in addition to the aleuro conidia formed in pycnidial conidiomata.

### 5.3. Pleomorphism of conidiomata

Some fungi can produce their conidia diffusely on the hyphae. Even though conidium morphology and ontogeny are not altered, the general appearance is different. Fungi that are synnematous on their natural substrate often fail to produce synnemata in culture. Species of *Penicillium* (dry, blastic - phialidic conidia), *Scopulariopsis putredinis* (dry, blastic – annellidic conidia), *Beauveria densa* (dry, blastic-sympodial conidia) are examples of pleomorphism in different conidial ontogeny. Anamorphs of *Fusarium*, *Pestalotiopsis* and *Epicoccum* produce conidia in sporodochia under natural conditions but under cultural conidition the conidia are produced from scarcely differentiated mats of hyphae. In some species of *Volutella*, *Verticillium* or *Acremonium* like diffuse conidial state may be present in addition to the sporodochia. *Phoma* species generally produce conidia in a typical pycnidium but a few species do not produce pycnidium under cultural conditions but produce mass of slimy pink conidia directly from the hyphae.

Anamorphs and teleomorphs undergo completely different development and there is no reason to expect that these two states should look the same. This is true for majority of ascomycetes and basidiomycetes. However, we know many cases in which at least some characters of the teleomorph and corresponding anamorph are alike. Similarities may be found in the form of fructification and in the spore morphology. Similarities of ascomata or basidionata to conidiomata may be expected if they begin to develop before formation of the gametangia. If fertilization is suppressed for any reason, conidia may develop in the previously formed fruit bodies. For instance *Leptosphaeria doliolum* whose ascomata and pycnidia (*Phoma* sp) are identical in form, size, ostiolum and wall texture. The behavior of other *Leptosphaeria* species eg. *L. macrospora* and its *Rhabdospora* anamorph is similar to that of *L. doliolum*. Apothecia and apothecioid conidiomata of *Heterosphaeria* species and their *Heteropatela* conidial anamorphs are often macroscopically indistinguishable. We may even observe first development of conidia and later the replacement of conidial layer by a hymenium with asci and paraphyses.

Ascospores of teleomorph and conidia of anamorph may be similar. For example teleomorph *Discostroma* and *Broomella* (Amphisphaeriaceae), *Melanochaeta* (Sphaeriaceae) *Durandiella* (Leotiaceae) and *Phaeosphaeria* and *Pleospora* (Pleosporaceae). The anamorphs of *Discostroma* belong to the form genus *Sematosporium* (Conidia with hyaline, filiform appendages) or *Sporocadus* (conidia without appendages) have conidia which may be distinguished from corresponding ascospores only by the thick walled septa or by the less obvious septal pores (eg. *D. massarina, D. saccardiana*). In *Melanochaeta hemipsila* with the anamorph of *Sporoschisma saccarda* and in *Chaetosphaerlla phaeostroma*, ascospores and conidia (*Oedemium*) are transversely septate with brown central cells and paler end cells. Ascospores in both species, however, tend to be more slender or fusoid than the conidia which are cylindrical or ellipsoidal. A strict similarity in form and septation is also found in several species of *Durandiella* whose ascospores and conidia are both long fusoid or filiform and transversely pluriseptate.

A most striking likeness occurs on morphology of ascospores of *Melanochaeta* or *Chaetosphaerella* and conidia of *Pestalotia* are transversely septate and the central cells are darker than the end cells. In addition in both morphs the spores are provided with simple or branched, filiform, colorless appendages. The ascospores of *Phaeosphaeria herpotricha* are long, filiform, transversely pleuriseptate has anamorph *Rhabdospora* conidia with long slightly clavate, almost filiform transversely sepatate. *Pleospora herbarum* characterized by brown dictyospores, ascospores also produce the muriform, brown somewhat smaller conidia of *Stemphyllium* anamorph.

Sometimes a similarity in pigmentation of conidia and ascospores are also found such as *Stigmateaceae* members produce olive-green, olive-grey or olive-brown bicellular ascospores have anamorphic genera like *Fusicladium, Karakulima* (a segregate of *Cladosporium*) *Pollaccia* and *Epilocaea* whose conidia are with same pigmentation.

## 5.4. Sequence of states

Anamorphs and teleomorphs which commonly develop simultaneously are comparatively rare. In most cases a definite succession may be observed with the teleomorph following the anamorph

The sequence is logical because anamorphs are simpler and thus develop more rapidly. This is useful for propagation of its population, while teleomorph develop before the end of its life cycle and meant for rejuvenation. However, in certain bitunicate ascomycetes pycnidia of the anamorph and ascomata of the teleomorph being morphologically similar. The fruit bodies begin their development with pseudoparen-chymatous primordium and fertilization processes take place only later with in this structure. If fertilization is suppressed the multicellular structures become conidiomata, where as in *Pyrenophora* species which produce *Drechslera* anamorphs, with the failure of fertilization produces sclerotia. In many cases the occurrence of the states is governed by seasonal considerations most strikingly demonstrated by plant parasites. In these cases, the parasites often overwinter as slowly developing teleomorphs which mature in early spring so that meiospores will be available at the beginning of the growing season. The developing thallus within or on the infected host then forms the anamorphs, its conidia being immediately capable of spread of the host individuals and ensuring season (Table 5.2) pathogen during the growing.

### 5.5. Pleomorphism and Classification

Pleomorphism of higher fungi has created several taxonomists. They are mainly two kinds: 1. How to identify an anamorph state as belonging to a particular teleomorphic species; and 2. Where to classify the many species that have no known teleomorphic state. This was solved by taking anamorph as independent organism and placed them a separate Deuteromycetes or fungi imperfecti. But this violates one of the cardinal rules of biological classification that each taxon can be belong to only one taxon of the next higher rank. To avoid this anomaly the fungi imperfecti is called as a form class (or anamorph class) and form species (or anamorph species) and form genus (anamorph genera) are convenient pigeonholes for morphologically similar anamorphs. They do not necessarily correspond to the genera in the ascomycota and basidiomycota.

Two more problems are faced with anamorphic genera which include whether to recognize a double anamorph genus that is anamorph genenus based on the co-occurance of two different anamorph genera; the second problem is whether to retain anamorph genera based on bulbils, sclerotia, rhizomorphs and other non-conidial structures.

### 5.6. Pleomorphism and Nomenclature

With the acceptance of class fungi imperfecti, a single fungus can have one species name in the genus of ascomycota or basidiomycota and one or more additional species in anamorph genera of the fungi imperfecti. These different names for the same species are not taxonomic synonyms, however, since the names based on anamorphs apply only to the anamorphic state denoted by the anamorph generic name. The ICBN provides that epithets proposed for anamorphs are not transferable to genera based on teleomorphic names. Mycologists all over the globe increasingly came to favor the use of cross-reference names to refer to the anamorphs of species with known teleomorphs. For example, an anamorphic name *Monilia sitophila* is now usually replaces by the cross reference name *Monilia* anamorph of *Neurospora sitophila* or *Neurospora sitophila* anamorph.

*Chapter 6*

# Ecology of Mycotoxigenic Fungi

Variety of environmental conditions influences the growth of mycotoxigenic moulds both in the pre -and post- harvest stage. Factors affecting mold invasion of the developing corn kernel include the amount of spore inocula stress factors on the developing corn kernel on the growing plant, insect and mite populations, damage from other fungi, varietal resistance or susceptability, mechanical damage during farming, storm damage, bird damage, mineral nutrition of plant and temperature. During harvest the grain is exposed to mechanical injury and extensive spore inoculation. After harvest mold growth and mycotoxin production can be affected by the moisture level, temperature, aeration, microbiological ecosystem, storage insects, blending of corn, amount of chaff and dist, chemical treatment, amount of damage, chemical composition of the substrate and heating. The behaviour of mycotoxigenic fungi is also variable factor. Some fungi elaborate mycotoxins while they are on the plant, while others on the harvested plant material and still others in the storage (Table 6.1). Various physical, chemical and biological factors influence the mould growth and mycotoxin elaboration (Table 6.2).

**Table 6.1.** Location of mycotoxin producing fungi

| Name of the fungus | Living plant | Harvested plant material | Stored plant material |
|---|---|---|---|
| *Claviceps purpurea* | + | - | - |
| *Sclerotinia sclerotim* | + | - | - |
| *Fusarium graminarum* | + | - | + |
| *F. sporotrichoides* | - | + | - |
| *Aspergillus flavus* | + | - | + |
| *A. parasiticus* | - | - | + |
| *A. Versicolor* | - | - | + |
| *A. ochraceous* | - | - | + |
| *A. clavatus* | - | - | + |
| *A. fumigatus* | - | - | + |
| *A. rubrum* | - | - | + |
| *A. chevaleri* | + | - | + |

| Name of the fungus | Living plant | Harvested plant material | Stored plant material |
|---|---|---|---|
| *Rhizoctonia leguminicola* | + | - | + |
| *Helminthosporium biseptatum* | - | - | - |
| *Pithomyces charatum* | - | + | - |
| *Stachybotrys atra* | - | + | - |
| *Periconia minutissima* | - | + | - |
| *Cladosporium cladosporoides* | - | + | - |
| *Alternaria longipes* | - | + | - |
| *Chaetomium globosum* | - | + | - |
| *Dendrodochium toxicum* | - | + | - |
| *Myrithecium verucaria* | - | + | - |
| *Trichothecium roseum* | - | + | - |
| *Trichoderma viridi* | - | + | - |
| *Pencillium* sp | - | - | + |
| *P. islandicum* | - | - | + |
| *P. citrinum* | - | - | + |
| *P. rubrum* | - | - | + |
| *P. cyclopium* | - | - | + |
| *P. viridicatum* | - | - | + |
| *P. urtica* | - | - | + |
| *P. verrucosum* | - | - | + |
| *P. puberulum* | - | - | + |
| *P. expansum* | - | - | + |
| *P. rugulosum* | - | - | + |
| *P. palitans* | - | - | + |
| *P. roqueforti* | -- | -- | + |
| *Fusarium tricintum* | - | - | + |
| *F.nivale (Gerlachia nivale)* | - | - | + |
| *F. verticilloides* | - | - | + |

**Table 6.2.** Factors affecting mycotoxin production by moulds

| Factors | In field | At harvest | In storage |
|---|---|---|---|
| **I Physical** | | | |
| Moisture | + | + | + |
| Rapid of drying | - | + | + |
| Rewetting | - | + | + |
| Relative humidity | + | + | + |

| Factors | In field | At harvest | In storage |
|---|---|---|---|
| Temperature | + | + | + |
| Mechanical damage | + | + | + |
| Blending of grain | - | + | + |
| Hot spots | - | + | + |
| Time | + | + | + |
| **II Chemical** | | | |
| $CO_2$ | - | - | + |
| $O_2$ | - | - | + |
| Nature of substrate | + | - | + |
| Mineral nutrition | + | - | + |
| Chemical treatment | - | - | + |
| **III Biological** | | | |
| Plant stress | + | - | + |
| Invertebrate vectors | + | - | + |
| Fungus infection | + | - | + |
| Plant varietal differences | + | - | + |
| Fungal strain differences | + | - | + |
| Spore load | + | + | + |
| Microbial ecosystem | + | - | + |

### 6.1. Effect of Physical factors

**i. Moisture:** If other factors are equal, the saprophytic fungi that grow in the field require higher moisture level than those of storage fungi. Jarvis (1971) recorded the growth of storage fungi when substrate moisture is 13-18%, while saprophytic fungi grow only when the moisture level is 22-25%. Moisture level for spore germination and growth of these moulds for instance spores of *Fusarium* and *Myrothecium* will not germinate even if moisture level is optimum as they require exogenous energy source for their germination. On the other hand, spores of *Helminthosporium* will readily germinate in distilled water (Table 6.3). Moisture level required for the growth of mycotoxigenic fungi varied with the incubation temperature. Strains are reported to vary in their moisture requirement. When growth was most rapid, spore germination was also rapid. Similarly range of minimal moisture level required by different fungi varies greatly. The moisture required for growth and aflatoxin production by *A. flavus* in peanuts is 20-25%, while *A. glaucus* can grow well at 15-20% moisture level. The time required to dry a commodity is a critical aspect in preventing mycotoxin formation. For instance aflatoxin production by *A. flavus* is not there until at least 48 hrs after spore germination i.e. moisture for optimal growth of *A. flavus* will not be dangerous even when they are heavily infested, if the corn can be dried to below 13% moisture within 48 hrs. Thus peanuts can be saved if the peanut dugged are placed on the top exposing to sun enabling them to be dried quickly when dried and stored products are remoistered by leaks in the bins, floods or condensation.

**ii. Temperature:** Temperature exerts significant influence on growth and mycotoxin production by fungi. It also affects on the type of mycotoxin production. For instance *A. flavus* produces aflatoxin $B_1$ was comparatively more than aflatoxin $G_1$ at an incubation temperature between 11°C and 36°C. In nature daily temperature and moisture levels vary consideribly which is likely to influence the fungal growth but it will not kill the fungus. Infact the infected corn can be dried and held for several months without any reduction in the viability of typical storage moulds. Once this corn is remoistered, growth resumes rapidly. Similarly *Gibberella zeae* does not produce zearalenone on the ears of corn while they are in the field but it produces the mycotoxin during spring. But subsequently when the temperature is above freezing and moisture increases, the toxin is produced in significant amount on corn within the storage bins or in the field.

**iii. Mechanical injury:** Injury to plant parts promoted the growth of mycotoxigenic fungi. This was observed in peanut pods injured were colonized redily and contaminated with aflatoxin more than uninsured. Similarly, the behaviour of 4 different varieties of maize were infested more than healthy maize by *A. flavus* from two or more sources to improve condition.

Blending of grain is a common practice. Corn with high moisture is mixed with dry grain to make its moisture level too low for storage mold to develop. Sometimes lower quality grain is blended with good grade (an older) so that final dry product keep well and have good quality. However, if improperly blended, the good corn will also deteriorate if the moisture level is above that required for mold growth. Storability of grains depends not only on the average moisture level in a bin but also on the moisture concentration in individual bins. Comingling (blending) of grain to reach an average safe moisture level result in a heterogeneous blend of a seed in which some kernels contain sufficient moisture for mold growth. Also high moisture grain such as corn infected with fungi when blended may continue to develop during moisture equilibrium or drying and therefore become a source of inoculation for an entire bin. It is appears that distinctly lower levels of moisture will support *A.flavus* growth atleast when high moisture corn and dry kernels are blended.

**iv. Hot spots:** Hot spots are those areas in which a small area of only a few kernels with high moisture encourages mold growth consequently a large becomes molded. Moisture is generated from dry substrate due to mold activity and not from external source. This was repeatedly proved that hot spots developed in different bins spontaneously and aflatoxin was present in all the cases. In contrast infestation of *Fusarium* and zearalenone contamination was spread uniformly in such areas and not localized like *A. flavus.*

Storage time becomes important when we consider the fungus in relation to the substratum. For instance, 1. grain can be safely stored for a short time at high moisture levels 2. Mold grows slowly at temperatures near minimum, 3. The age of fungal spores is important because older spores require longer period to germinate and consequently to develop mycelium and to form toxin, 4. Over long periods of grain storage under conditions that inhibit mold growth, viable microbial inoculum will gradually decreases because of the death of spores, 5. Even under optimum conditions, some time is required for germination, growth of mycelial penetration of kernel will take longer time. 6. Competition among spermosphere fungi also delays the infestation and mycotoxin production and if the substrate is not ideal also delays the fungal invasion and toxin formation. Thus several physical factors influence the activity of mycotoxigenic fungi and mycotoxin elaboration.

### 6.2. Effect Chemical factors

Besides physical environmental conditions, the chemical factors such as $CO_2$, $O_2$, chemical nature of substratum, mineral nutrients available will influence the activity of mycotoxigenic fungi (Koteswara Rao *et al.*, 2013). *A. flavus* infested peanuts were exposed to different levels of $CO_2$, aflatoxin production was reduced with increased concentration of $CO_2$. Sanders *et al.* (1968) recorded decreased aflatoxins production at 20-40% $CO_2$ concentration at low temperature (17°C) or reduced humidity. Indian way of storage of food grains below ground is scientifically sound way of protecting food grains from molds and insects. Carbon dioxide at high concentration reduces mold growth even at high RH. $CO_2$ concentration of 20-40% prevented mold growth at reduced temperature (17°C) or low RH or both.

The inhibitory effect of $CO_2$ on molds is used in farm storage in India where cereal grains are stored below ground in a gowdown which is then sealed with a mixture of mud and cow dung which results in the accumulation of $CO_2$ and prevents moulds and insects. In USA storage of grains will be done in large concrete trenches.

**i. Oxygen:** Since moulds are strongly aerobic, they fail to grow and sporulate at low $O_2$ levels. Making lack of $O_2$ is not sufficient, but raising $CO_2$ concentration will make grain free from mould growth. As all mycotoxigenic moulds are aerobic reduced availability of $O_2$ will delay spore germination, mycelial growth and mycotoxin production. The nature of substratum is likely to influence mould growth and mycotoxin contamination. Rice is an ideal substratum for aflatoxin production in comparison to soybean which they attributed to chemical composition of soybean. Similarly mycotoxin formation is likely to be influenced by mineral composition of substrate food grains. The absence of Zinc, Copper and Iron resulted in the failure of *A. ochraceous* in production of Ochratoxin A. Similarly treatment of food grains with certain chemicals such as insecticides is likely to adversely affect the mycotoxin production either by direct acting on mycotoxin producing fungus or control insect population indirectly affecting mycotoxin elaboration. On the other hand, treatment of food grains with carbontetrachloride or carbondisulfide resulted in the increase of Ochratoxin A production by *P. viridicatum* and aflatoxin by *A. flavus*. However, *A. parasiticus* was inhibited for aflatoxin production with the treatment of phosphine.

### 6.3. Influence of Biological factors

Plant growth conditions are likely to influence the growth and mycotoxin production by infesting moulds. The date of sowing, density of plants, water stress conditions and humidity are likely influence the growth and mycotoxin formation in different crop plants. Presence of variety of insects on the crop is likely influence the mycotoxin contamination. Some of these insects act as vectors, provide moisture, create portal for growth of the fungus. Infection by one fungus is likely predisposes to other.

**i. Time:** Time becomes a factor when we consider the fungus in relation to substratum. Grain can be safely stored for a short time at high moisture levels. Christensen and Kaufmann (1972) demonstrated that grain can be stored safely for a few weeks at higher moisture and temperature levels than grains stored for months or years at lower levels. Molds grow slowly at temperature near their minimum and takes additional time for toxin to form. Similarly old spores of the fungus takes more time to germinate and consequently to develop mycelium and to form toxin. Under the conditions that inhibit mold growth over long periods of grain storage, viable spores of the mold decrease because of the death of spores. Even in optimum conditions some time is required for germination, growth of mycelium, penetration of the substratum and toxin production. Similarly penetration of kernel will take longer time if it is intact rather than cracked or broken. Under highly competitive conditions also mold growth and invasion of the substratum is reduced. Similarly if the substratum is not ideal, the quantity of mycelium is reduced although the nature of the substratum will not necessarily reduce the speed of the fungal invasion.

**ii. Plant relation:** Exploitation of generically variety of crop plants in the management of mycotoxin problem may be helpful. It has been successful in the control of plant pathogen from causing disease. Out of 1406 varieties of peanut varieties showed resistance against *A. flavus* infestation and aflatoxin elaboration. The hilum and seed coat are reported to play significant role in peanut infestation and aflatoxin production. Similarly fungal strains are reported to differ in their mycotoxigenicity. Therefore, an effort should be made to find out nonmycotoxigenic strains so that they can be exploited in the control of mycotoxins have successfully controlled. The aflatoxin contamination of peanut could be controlled by employing a toxigenic strains *A. flavus*.

Fungal spore's load of mycotoxigenic fungi is likely to influence their success in mycotoxin elaboration. If many infective units of inoculum exist, extensive infection occurs in the host or stored grain because of the additional growth sites. Longevity of the inoculum and environment prevailing are likely to influence mould infestation and mycotoxin contamination. Sources of mould inoculum is important in preventing mycotoxigenic fungi and mycotoxin elaboration. For instance, *A. flavus* spores carried by invertebrate vectors to variety of substrates far from the site of original infestation. Similarly, *Fusarium* species causing ear-rot of corn and spread by air and insects are attracted by the honeydew produced by the ergot fungus. An inoculum of storage moulds may develop from molds growing in pockets in conveyor pipes and elevators and may infect other grain. Similarly, some infection results from viable spores that remain in bins after removal of grain.

**iii. Interaction of microorganisms:** It is generally observed that the production of mycotoxins in controlled conditions is more than what is observed in nature. This is mainly attributed to the competition from natural spermosphere mycoflora either for space or nutrients. It may also be due to mycotoxic substances released by co-existing fungus. This was demonstrated by Boller and Schroeder (1974) in aflatoxin production by *A. parasiticus.* Aflatoxin production was considerably reduced either in the presence of *Aspergillus chevalleri* which might have metabolized aflatoxin thus reducing its level.

Thus, it is clear that mycotoxigenic fungi on the seed surface does not grow in isolation and several factors like amount of inoculum, temperature, substrate moisture, physical conditions of the substratum and growth of other microorganisms will interact resulting in the change of these conditions greatly get altered. Certain factors having more influence than others, due to dynamic nature of environment. Thus, it is clear that the ways of regulating variety of factors (both abiotic and biotic) must be considered and the range of limitations must be known.

# Glossary

**Acerose:** Needle shaped

**Acervalus**: Cushion or line shaped asexual reproductive body producing conidia

**Acicular:** Needle like tapering gradually to a point

**Acicular:** Needle-like, tapering gradually to a point

**Acleistothecial:** Not producing cleistothecia

**Acroauxic**: Growing and enlarging at the apex

**Acrogenous**: Borne at the tip.

**Acropetal**: Chain of conidia in which youngest is at the tip of distalend of the chain

**Acropleurogenous:** Formed at the end and on the sides

**Acrospore**: An apical spore

**Acuminate:** Gradually narrowing into a point

**Allantoid**: Sausage shaped and slightly curved

**Amphigenous**: Making growth all around or on two sides

**Ampullae:** A vesicular swelling on conidiophores formed at the apex or at any point along its length.

**Ampulliform:** Swelling from a narrow base, then tapering to a relatively blood apex, like a Greek ampulla or a ten pin bowling pin

**Ampulliform:** Ten pin bowling pin

**Amyloid:** Stain blue black with iodine

**Anamorph**: An asexual stage in the life cycle of the fungus

**Anamorphic**: Refers to the state of a fungus which bears asexual spores

**Anastamose**: Froming net-work of hyphae by joining irrgularly

**Annellate**: With ring like markings on the wall showing where the recurrent growth has taken place repeatedly

**Apiculate**: Pointed at both ends

**Apiculate:** Pointed at both ends; lemon-shaped

**Apothecium:** An open ascocarp

**Appendiculate:** With one or more appendeges

**Archanoid:** Covered with or formed of delicate hairs or fibers

**Archuate**: Arch like

**Arthoroconidia:** Conidium produced by a thallic conidigenous cell often cyhndrical with runcate ends.

**Ascocarp:** Friut body covering asci and ascospores

**Ascogenous:** Non-spore bearing

**Aseptate:** Lacking cross walls

**Asexual:** Reproducing without involving karyogamy and meiosis

**Asperulate:** Covered with tiny projections or points

**Attenuate:** Tapering, drawnout

**Basauxic:** Growing and elongating at the base.

**Basipetal:** A chain of spores with the youngest at the base or proximal ends of the chain

**Biseriate:** In two lines or series

**Biverticellate**: Two branch points between conidia and stipe

**Biverticillate:** Refers to a penicillus with two branch points between stipe and conidium, i.e. one bearing metulae and phialides

**Blastic conidium**: A conidium arising from a narrow region of the conidiogenous cells with elongation and swelling occurring before delinitaous of the conidium

**Blastic:** A mode of conidium development in which there is marked enlargement of a recognisable conidium initial before the initial is delimited by a septum

**Botryose**: Clustered like grapes

**Caadate**: With a tail at the base

**Caelomycetes**: A fruit body producing conidia either pycnidium or acervulus

**Caespitose**: Aggregated in tufts

**Calvate: C**lub shaped narrowing in the direction of the base

**Candate**: Having a tail

**Carmiculate:** Having horn like projections

**Caronaceous:** Black and brittle

**Catenate**: Developing in chains

**Catenate:** Formed in succession and adhering in a chain

**Centrum:** A structure with an ascomata

**Cheiroid:** Rougly the shape of a hand

**Chlamidospore**: A thick walled 1-2 celled sterile structure generally functions as a resting structure

**Cicatrized** (Conidiogenous cell, conidiophores): Bearing scars

**Circinate:** Coiled inwards

**Cirrhus:** A ribbon like cylinder or columns of of opens hold together by mucus

**Citriform**: Lemon shaped

**Clansconnection:** A Bridge like hypal connection characterstcis of secondary mycelium of many basidiomycota

**Clistothecium:** A closed ascomata

**Collula** : Apex of phialides

**Collulum:** The neck of a phialide or annelide

**Conidiamata:** Asexual reproductive body producing conidia

**Conidiogenous cells**: A cell producing conidia

**Conidiogenous cells**: A cell which produces a conidium or several conidia.

**Conidiomata** - Frit body producing conidia (pycnidia, acervulus, synnemata, sporodochium)

**Conidium**: Any asexual spore which rohen mature in liberated from a conidiophores or conidiogenous cell

**Coremium**: Aggregation of fascicles

**Corniform:** Shaped like horn

**Cruciate:** A form of cross

**Crustose:** Conidial masses in the form of crust

**Cuneiform**: Wedge shaped

**Cupulate:** Cup like

**Cynibiform:** Boat shaped

**Dactyloid**: Finger like

**Deliguescing:** Dissolving

**Demataceous:** Darkly pigmented

**Denticulate** (Conidiogenous cell, conidiophores): Bearing deuticle

**Determinate** (Conidiogenous cell, conidiophores): Ceasing to grow in length on production of terminal conidium or chains of conidia.

**Dictyoseptate**: A spore with both transverse and longitudinal septa

**Discrete (conidiogenous cells):** Separate from the main axis or branches of a conidiophore and often having distinct shape.

**Doliform:** Barrel shaped

**Endogenous:** Produced entirely within a phialide

**Enteroblastic** (Conidogenous cells): Blastic with inner wall only or neither wall contributing towards the formation of conidial wall.

**Exogenous:** Formed on the surface of a protoplasmic mass protruding from the end of a phialide.

**Extrolites** : Secondary metabolites produced by fungi

**Exudates** (Extrolites): Droplets excreted by the mycelium

**Falaform:** Formed like a bill hook

**Fasicle:** Aggregation of conidiophores

**Floccose**: Loose aerial mycelium

**Funicle**: Rudimentary form of synnemata

**Funiculose**: Occurring in ropes or bundles

**Holoblastic**: Blastic with both the outer and inner walls contributing towards the formation of the conidia wall.

**Hyphopodium:** A short, lateral, sometimes lobed appressorium which adheres firmly to the surface of a host plant; a pale spot in the centre often indicates the point of origin of a narrow filament which penetrates the cuticle. Indeterminate fertile, hyphae bearing conidiophores along their entire length.

**Infundibuliform**: Funnel shaped

**Integrated**: Incorporated in the main axis or branch of a conidiophore where it is either terminal or intercolony.

**Lageneform**: Shaped like a long flask

**Lenticular:** Lens-shaped; biconvex

**Limaniform**: Lemon shaped

**Lunate:** Crescent shaped

**Macronomators** (Conidiophore): Morphologically very different from a vegetative hyphae and usually erect.

**Mesistematic** (Conidiophore) : Increasing in length with the laying down of cross walls and elongation of cells.

**Metula:** Aspecialised cell giving rise to a verticil of phialides

**Metulae** : A sporephore branch having phialides

**Micronematous** (conidiophore): Morphologically resembling a vegetative hyphae but bearing conidia

**Monoblastic** (conidiogenous cell): Holoblastic and blowing out at only one point

**Mononematous** (Conidiophore) : Composed of a single thread or filament secondary conidiophores

**Monophyladic:** Phialide with only one opening

**Monotretic** (Conidigenous cell): Tretic with a single channel or pose

**Monoverticellate:** One branch point between conidia and stipe.

**Monoverticellate:** Refers to a penicillus with one branch point between stipe and conidium, i.e. composed of phialides only

**Muriform**: Longitudinally and transversely septate

**Mycotoxin:** Toxins produced by fungi which are toxic to warm blooded animals including man.

**Navicular**: Boat shaped

**Obovoid:** Egg shaped with the broad end at the base

**Obturbinate**: The shape of a top upside down with the broad end at the base.

**Pendulus**: Hanging down

**Penicillate**: A paint brush like arrangement

**Peridium:** The wall of an ascocarp or other fruiting body

**Phialide** (Conidiogenous cells) : A discrete or integrated phialidic conidiogenous cells

**Phialides**: Conidiogenous cells

**Phialoconidium:** A conidium produced by a phialide

**Pigments**: Water soluble colours produced by fungus

**Polyblastic:** Holoblastic and blowing out at several points.

**Polyphyldic**: Phialide with more than one opening.

**Polytretic:** Tretic with several channels or pores

**Pseudoparenchymatous: R**efers to cells which, although formed from hyphae, have the appearance of parenchymatous (plant) cells

**Pulvinate:** Cushion shaped

**Punctiform:** Dot like

**Quarterverticellate**: Four branch points between conidia and stipe.

**Quaterverticillate:** Refers to a penicil us with four branch points between stipe and conidium; i.e. bearing rami, ramuli like, metulae and phialides

**Ramulus:** An element similar to a ramus, interposed between ramus and metulae in a quaterverticillate penicillus.

**Rugale** : Finely wrinkled

**Schizogenous:** Formed by cleavage or splitting

**Sclerotigenic:** bearing sclerotia

Sclerotium: A firm mass of aggregated hyphae ofen with a rind of rather small dark, thick walled cells

**Setulae**: A delicate appendage arising from the surface of a conidium

**Solitary**: A rising singly at one point.

**Sporodichium**: A pulrinate stroma with closely packed, relatively short conidiophores.

**Sporodochia:** A pulvinate stroma with closely packed relatively short conidiophores.

**Stellate**: Star shaped

**Stipe** (Conidiophore): The main, stont axis or unbranched lower part.

**Stipe:** A stalk, in the present context a specialised hypha arising from a fertile hypha and supporting a pcnicillus

**Stipe:** Fertile hyphae

**Stomatopodium:** An apresorium produced from a lateral branch above or in a stroma.

**Sympodial:** Proliferating, the main axis elongating but growth of a succession of apices each of which develops both in and to one side of presiomapex, where growth has ceased with the production of conidium.

**Synnematous** (Conidiophore): Composed of numerous threads or filaments tightly adpressed and in some cares fased along most of their length.

**Synoptic key:** This is a key which allows entry at more than one point, and *in toto* provides a synopsis of the significant characters of each organism keyed

**Teleomorphic:** Refers to the state of a fungus which bears sexual spores such as ascosporcs or basidiospores

**Terverticillate:** Refers to a penicillus with three branch points between stipe and conidium, i.e. hearing rami, metulae and phialides

**Terverticillate:** Three branch points between conidia and stipe.

**Thallic:** In which conidium development takes place without enlargement of the recomsable conidium, initial

**Thallic**: Conidium develops without enlargement of recognizable conidium initial.

**Thallocomiola:** A conidium produced by thallic conidiogenous cell

**Type specimen:** The holotype, lectotype or neotype, as the case may be, of a validly described taxon

**Unitunicate:** Refers to an ascus with a fused or single wall

**Velutinous**: Velvet like texture

**Verruculose**: Finely warted

**Verticil:** A cluster of metulae or phialides with a common origin

# List of Mycotoxigenic species

*Acremonium species*
*Alternaria alternata*
*Apiospora campactospora*
*Arachinotus aureus*
*Ascochyta*
*Aspergillus*
*A.alliaceus*
*A.amstelodami*
*A.aurantiobrunneus*
*A.awamori*
*A.brevipes*
*A.caespitosus*
*A.candidus*
*A.chevalieri*
*A.clavatus*
*A.cucumerina*
*A.dauci*
*A.echinulatus*
*A.ficuum*
*A.fischeri*
*A.flavus*
*A.fumigatus*
*A.glaucus*
*A.kikuchina*
*A.longipes*
*A.mali*
*A.melleus*
*A.nidulans*
*A.niger*
*A.ochraceus*
*A.oryzae*
*A.oryzae var.microsporus*
*A.ostianus*
*A.parasiticus*
*A.puniceus*
*A.ruber*
*A.sclerotiorum*
*A.sulphureus*
*A.tamarii*
*A.tenuis*
*A.tenussima*
*A.terreus*
*A.ustus*
*A.versicolor*
*A.viridi-nutans*
*A.wentii*
*Bipolaris sorokiniana*
*B.botrytis allii*
*Beauveria bossiania*
*Byssochlamis nivea*
*Chaetomium cerotocinigenum*
*C. cochliodes*
*C. globosum*
*C. minutum*
*Cladosporoium*
*Clavicep paspali*
*C. purpurea*
*Chalara trilaterale*

*Calonectria nivalis*
*Ceratocystis fimbriata*
*Curvularia lunata*
*Cylinbrocarpon species*
*Diplodia maydis*
*D. macrospora*
*Emericellopsis*
*Fusarium culmorum*
*F. diversisporum*
*F. equiseti*
*F. gibbosum*
*F. graminearum*
*F. lateritium*
F. moniliforme var. subglutinans
*F. moniliforme*
*F. nivale*
*F. oxysporum*
*F. poae*
*F. roseum*
*F. sambucinum*
*F. sambucinum var. coeruleum*
*F. scirpi*
*F. semitectum*
*F. solani*
*F. sporotrichioides*
*F. sulphureum*
*F. tricinctum*
*F. avenaceum*
*Gibberella fujikuroi*
*G. intricans*
*G. virens*
*G. zeae*
*Gliocladium fimbriatum*
*Helminthosporium victoriae*
*H. dematioideum*
*Hormiscium species*
*Hyalodendron diddens*
*Lasiodiplodia theobromae*
*Myrothecium roridum*
*M. verrucaria*
*Metarshzium anisopliae*
*Nigrosabulum*
*Phoma sorghina*
*P. terrestris*
*P.brefeldianum*
*P.brevicompactum*
*P.chrysogenum*
*P.citro-viridae*
*P.citrinum*
*P.claviforme*
*P.crustosum*
*P.cyclopium*
*P.expansum*
*P.griseofulvum*
*P.aurantiogriseum*
*P.islandicum*
*P.italicum*
*P.janthinellum*
*P.lanosum*
*P.lapidosum*
*P.lividum*
*P.martensii*
*P.nigricans*
*P.novae-zeelandieae*
*P.oxalicum*
*P.palitans*
*P.puberulum*
*P.purpurogenum*
*P.roqueforti*
*P.rubrum*
*P.rugulosum*
*P.simplicissimum*
*P.spinulosum*
*P.thomii*
*P.toxicarium*
*P.verrucosum*
*Penicillium verruculosum*
*P. viridicatum*

*P. wortmanni*
*Phomopsis paspali*
*Piricauda chartarum*
*Pithomyces chartarum*
*Pyricularia oryzae*
*Rhizoctonia leguminicola*
*Rosellinia necatrix*
*Scheleobrachea echinulata*
*Sepedonium ampullosporum*
Solorina crocea
*Sporidesmium bakeri*
*Sporidesmium chartarum*
*Stachybotrys alternans*
*S. atra*
*S. echinulatum*
*Trichoderma viride*
*Trichothecium roseum*
*Verticillium* species
*Verticimonosporium* species
*Zygosporium masonii*

# References


Abildgren, M.P., F. Lund, U. Thrane and S. Elmhot 1987 Czapek-Dox agar containing Iprodione and Dichloran as a selective medium for the isolation of *Fusarium* species. *Lett. Appl. Microbiol.* **5**: 83-86.

Aderhold, R. 1897 Revision der Species Venturia chlorospora, inaequalis und ditricha autorum. Hedwigia. **36**: 67–83.

Aderhold, R. and W. Ruhland 1905 Zur Kenntnis der Obstbaum-Sklerotinien. *Arbeit. Biol. Abt. Land und Forst.* **4**:427-442.

Ainsworth, G.C 1976 Introduction to the History of Mycology, Cambridge, Cambridge University Press.

Alexander, M., P.M. Daggett, R. Gherna, S.C. Jong, F. Simione and H. Hatt 1980 American type culture collection method I. Laboratory manual on preservation freezing and freeze drying. Rockville, Maryland, American type culture collection.

Alexander, N.J., R.H. Proctor, S.P. McCormick and R.D. Plattner 1997 Genetic and molecular aspects of the biosynthesis of trichothecenes by *Fusarium. Cer. Res. Comm.* **25**:315–20.

Ames, L.M., 1963 A monograph of the Chaetoiaceae. *United State Army Research and Development series.*2:1-125.

Amoah B.K, M.V. Mac Donald, H.N. Rezanoor and P. Nicholson 1996 The use of RAPD technique to identify mating groups in *Fusarium* section Liseola. *Plant pathology.* **45**: 115-125.

Andersen, B. and U. Thrane 1996 Differentiation of Alternaria infectoria and *Alternaria alternata* based on morphology, metabolite profiles, and cultural characteristics. *Can J. Microbiol.* **42**: 685–689.

Andrews, S and J.I. Pitt 1986 Selective medium for isolation of *Fusarium* species and dematiaceous Hypomycetes from cereals. *Appl. Environ. Microbiol.* **5**: 1235-1238.

Andrews, S. 1986 Optimization of conditions for the surface disinfection of sorghum and sultanas using sodium hypochlorite solutions. In: King, A.D., J.I. Pin, L.R. Beuchat and J.E.L. Corry ed., 1986 Methods for the mycological examination of food. New York, Plenum Press, pp. 28-32.

Artigot, M.P., N.Loiseau, J. Laffitte, L. Mas-Reguieg, S. Tadrist, I.P. Oswald and O. Puel 2009 Molecular cloning and functional characterization of two CYP619 cytochrome P450s involved in biosynthesis of patulin in *Aspergillus clavatus. Microbiology.* **155:** 1738–1747.

Arx JA von, J. Guarro and M.J. Figueras 1986 The ascomycete genus *Chaetomium*. *Nova Hedwigia Beihefte*. **84**: 1−162.

Atoui, A. F. Mathieu and A. Lebrihi 2007 Targeting a polyketide synthase gene for *Aspergillus carbonarius* quantification and ochratoxin A assessment in grapes using real-time PCR. *Int. J. Food Microbiol.* **115**: 313–318.

Atoui, A. P. Dao., F.Mathieu and A. Lebrihi 2006 Amplification and diversity analysis of Ketosynthase domains of putative polyketide synthase gene in *Aspergillus ochraceus* and *Aspergillus carbonarius* producers of ochratoxin A. *Mol. Nutr. Food Res*. **50:** 448–493.

Attanayake RN, D.A. Glawe, F.M. Dugan and W. Chen 2009 *Erysiphe trifolii* causing powdery mildew of lentil (*Lens culinaris*). *Plant Disease* **93:** 797-803.

Bacha N., A. Atoui, F. Mathieu, T. Liboz and A. Lebrihi 2009 *Aspergillus westerdijkiae* polyketide synthase gene "aoks1" is involved in the biosynthesis of ochratoxin A. *Fungal Genetics and Biology*. **46:** 77–84.

Baird R, H.K.Abbas, G.Windham, P. Williams, S. Baird, P. Ma, R. Kelley, L. Hawkins 2008 Identification of select fumonisin forming Fusarium species using PCR applications of the polyketide synthase gene and its relationship to fumonisin production in vitro. *Int. J. Mol. Sci*. **9:**554–570.

Bakan B, C., Giraud-Delville, L. Pinson, D. Richard-Molard, E. Fournier E and Y. Brygoo 2002 Identification by PCR of *Fusarium culmorum* strains producing large and small amounts of deoxynivalenol. *Appl. Environ. Microbiol.***68**:5472–5479.

Baker, S.E., J. Thykaer, W.S. Adney, T.S. Brettin, F.J. Brockman, P. D'Haeseleer, A.D. Martinez, R.M. Miller, D.S. Rokhsar, C.W. Schadt, T. Torok, G. Tuskan, J. Bennett, R.M. Berka, S.P. Briggs, J. Heitman, J. Taylor, B.G. Turgeon, M. Werner-Washburne and M.E. Himmel 2008 Fungal genome sequencing and bioenergy. *Fungal Biol. Rev*. **22**: 1–5.

Baker, S.E., S. Kroken, T. Asvarak, P. Inderbitzin, B.Y. Li, L. Shi, O.C. Yoder and B.G. Turgeon 2006 Two polyketide synthase-encoding genes are required for biosynthesis of the polyketide virulence factor, T-toxin, by *Cochliobolus heterostrophus*. *Mol. Plant Microbe. Interact*. **19**:139–149.

Bayman, P. and P.J. Cotty 1993 Genetic diversity in *Aspergillus flavus*: association with aflatoxin production and morphology. *Can. J. Botany*. **71**: 23-34.

Bennett JW 2010 An overview of the Genus *Aspergillus*. In: Aspergillus Molecular Biology and Genomics, 238 pp. Edited by Masayuki Machida and Katsuya Gomi. Caister Academic Press. Norfolk, UK. pp. 1-17.

Beuehat, L. R and Hocking, A.D 1990 Some considerations when analyzing foods for the presence of xerophilic fungi. *J. Food Prot*. **53**: 984-989.

Bezuidenhout C.C., M. Prinsloo and A.M. Van der Walt 2006 Multiplex PCR based detection of potential fumonisin-producing *Fusarium* in traditional African vegetables. *Inc. Environ. Toxicol*. **21**:360–366.

Bilai V.I. 1950 The Fusaria (biology and systematics). Akademia Naukwe Ukrinskii SSR, Kiev.

Bilai, V.I. 1950 The Fusaria (biology and systematics)" Akademia Naukwe Ukrinskii SSR, Kiev.319P. (Abstract in *Rev.Appl. Mycology* .**36**:731-732.1957).

Bingle, L.E. T.J. Simpson and C.M. Lazarus 1999 Ketosynthase domain probes identify two subclasses of fungal polyketide synthase genes. *Fungal Genet. Biol.* **26**: 209–223.

Bluhm, B.H. and C.P. Woloshuk 2005 Amylopectin induces fumonisin B1 production by *Fusarium verticillioides* during colonization of maize kernels. *Mol. Plant-Microbe. Interact.* **18**: 1333-1339.

Bluhm, B.H. and C.P. Woloshuk 2006 Fck1, a C-type cyclin-dependent kinase, interacts with Fcc1 to regulate development and secondary metabolism in *Fusarium verticillioides*. *Fungal. Gen. Biol.***43**: 146–54.

Bluhm, B.H., J.E. Flaherty, M.A. Cousin and C.P. Woloshuk 2002 Multiplex polymerase chain reaction assay for the differential detection of trichothecene- and fumonisin-producing species of *Fusarium* in cornmeal. *Food Prot.* **65:**1955–1961.

Bojja, R.S., R.L.Cerny, R.H. Proctor and L. Du 2004 Determining the biosynthetic sequence in the early steps of the fumonisin pathway by use of three gene-disruption mutants of *Fusarium verticillioides. J. Agri. Food Chem.* **52**: pp. 2855-2860.

Boller, R.A. and H.W. Schroeder 1974 Influence of relative humidity on production of aflatoxin in rice by *Aspergillus parasiticus*. *Phytopathol.* **64:** 17-21.

Booth, C. 1971. The genus *Fusarium*. Commonwealth Mycological Institute, Kew, Surrey, England. pp. 237.

Bottalico, A. and A. Logrieco 1998 Toxigenic *Alternaria* species of economic importance. In Mycotoxins in Agriculture and Food Safety (K.K. Sinha & D. Bhatnager, eds) Marcel Dekker, New York. 65–108.

Brayford, D 1990 Reflections on the use of physiological and molecular technique in *Fusarium* taxonomy. In: *Fusarium* Notes, an International news letter.

Brewer D, W. Duncan, C. Jerram, K. Leach, S. Safe, A. Taylor, L.C. Vining, R. Archibald, R.G. Stevenson, C.J. Mirocha and C.M. Christensen 1972 Ovine ill-thrift in Nova Scotia. 5. The production and toxicology of chetomin, a metabolite of *Chaetomium* spp. *Can. J. Microbiol.* **18**: 1129-1137.

Brewer, D and A. Taylor 1978 The production of toxic metabolites by *Chaetomium* spp. isolated from soil of permanent pasture. *Can. J Microbiol.* **24:** 1082-1086.

Brewer, D., W.A. Jerram and A. Taylor 1968 The production of cochliodinol and a related metabolite by *Chaetomium* species. *Can. J. Microbiol.***14**: 861–866.

Bridge, P. D., D. L. Hawksworth, Z. Kozakiewicz, A.H.S. Onions, R.R. M. Paterson and M.J. Sackin 1986. An integrated approach to *Penicillium* systematics. In Advances in *Penicillium* and *Aspergillus* Systematics, Edited by R.A. Samson & J.I. Pitt. New York: Plenum Press. pp. 28 1-309.

Brown D.W, R.A. Butchko, M. Busman and R.H. Proctor 2007 The *Fusarium verticillioides* FUM gene cluster encodes a Zn (II) 2 Cys6 protein that affects FUM gene expression and fumonisin production. *Eukaryot Cell.* **6**:1210–1218.

Brown D.W., S.P. McCormick, N.J.Alexander, R.H. Proctor and A.E. Desjardins 2001 A genetic and biochemical approach to study trichothecene diversity in *Fusarium sporotrichioides* and *Fusarium graminearum*. *Fun. Gen. Biol.* **32:** 121–33.

Bulock, J.D 1980 In The biosynthesis of mycotoxins. A study in secondary metabolism (ed. P.S. Steyn) Academic Press, New York.p.1-16.

Burlakoti, R.R. S. Ali, G.A. Secor, S.M. Neate, M.P. McMullen and T.B. Adhikari 2008 Genetic relationships among populations of *Gibberella zeae* from barley, wheat, potato, and sugar beet in the Upper Midwest of the United States. *Phytopathol.* **98**: 969–976.

Butchko R.A, R.D.Plattner, R.H. Proctor 2003 FUM9 is required for C-5 hydroxylation of fumonisins and complements the meitotically defined Fum3 locus in *Gibberella moniliformis*. *Appl. Environ. Microbiol.* 69:6935–37.

Butchko, R.A., R.D. Plattner and R.H. Proctor 2003b "FUM9 is required for C-5 hydroxylation of fumonisins and complements the meitotically defined Fum3 locus in *Gibberella moniliformis*". *Appl. Environ. Microbiol.* **69:** 6935-6937.

Butchko, R.A., R.D. Plattner and R.H. Proctor 2006 Deletion analysis of FUM genes involved in tricarballylic ester formation during fumonisin biosynthesis. *J. Agri. Food Chem.* **54**. 9398-9404.

Butchko, R.A., R.D.Plattner and R.H.Proctor 2003a "FUM13 encodes a short chain dehydrogenase/reductase required for C-3 carbonyl reduction during fumonisin biosynthesis in *Gibberella moniliformis*". *J. Agri. Food Chem.* **51**:3000-3006.

Caldwell, R.W., J. Tuite, M. Stob and R. Baldwin 1970 Zearalenone production by *Fusarium species*. *Appl. Microbiol.* 20: 31-34.

Carmichael, J.W., W.B. Kendrick, I.L. Conners, and L. Sigler 1980 Genera of Hyphomycetes. The University of Alberta Press, Edmonton, Alberta, Canada. pp.386.

Carter J.P, H.N. Rezanoor, D.Holden, A.E. Desjardins, R.D. Plattner and P. Nicholson 2002 Variation in pathogenicity associated with the genetic diversity of *Fusarium graminearum*. *Eur. J. Plant .Pathol.***108:**573–583.

Carter, A. 1981 Ph.D. Thesis, University of Toronto.

Chanda, A., L.V. Roze, S. Kang, K.A. Artymovich, G.R. Hicks, N.V. Raikhel, A.M. Calvo and J.E. Linz 2009 A key role for vesicles in fungal secondary metabolism. *Proc. Natl. Acad. Sci. USA.* **106**: 19533–19538.

Chang, P.K 2003 The *Aspergillus parasiticus* protein AFLJ interacts with the aflatoxin pathway-specific regulator AFLR". *Mol. Gen. Genomics.* **268**: 711-719.

Chang, P.K., K.C. Ehrlich, J. Yu, D. Bhatnagar and T.E. Cleveland 1995 Increased expression of *Aspergillus parasiticus aflR*, encoding a sequence-specific DNA-binding protein, relieves nitrate inhibition of aflatoxin biosynthesis. *Appl. Environ. Microbiol.* **61**: 2372– 2377.

Christensen C.M., G.H. Nelson and R.J.Townsend 1996 Recent advances in mycotoxicoses, *Soc. Chem. Ind. (London), Monograph.* **23**: 31-43.

Christensen, C. M. and H. H. Kaufmann 1972 In Soybeans: Chemistry and Technology, Vol. 1, Proteins, A. K. Smith and S. J. Circle, Eds., Avi Publishing Co., Westport, Conn., U.S.A.

Cole, G.T. and R.A. Samson 1979 Patterns of development in conidial fungi. Pitman Publishing Limited. pp.190.

Cole, R.J. and R.H. Cox 1981 Handbook of Toxic Fungal Metabolites, Academic Press, New York, N.Y.

Cole, R.J., J.W. Kriksey and G. Morgan-Joneswells 1974. *Can. J. Microbiol.***20**:1159-1162.

Council for Agriculture Science and Technology 2003 Mycotoxins risks in plant, animal, and human systems. CAST, Task Force Report, Ames, IA. 139.

Cruickshank, R.H. and J.I. Pitt 1987 Identification of species in *Penicillium* subgenus *Penicillium* by enzyme electrophoresis. *Mycologia*: **79**: 614-620.

Cruickshank, R.H. and J.I. Pitt 1990 Speciation and synonymys in *Penicillium* subgenus *Penicillium* towards a definitive taxonomy. In: Samson, R.A., and Pitt, J.I., ed., Modern concepts in *Penicillium* and *Aspergillus* classification. New York; *Plenum Press, pp*. 103-119.

Cruickshank, R.H., and J.I. Pitt 1990 Isoenzyme patterns in *Aspergillus flavus* and closely related taxa. In Modern Concepts in *Penicillium* and *Aspergillus* Systemics (R.A. Samson and J.I. Pitt eds.) Plenum Press New York.pp.259-264.

Dao, H.P., F.Mathieu and A. Lebrihi 2005 Two primer pairs to detect OTA producers by PCR method. *Int. J. Food Microbiol.* **104**: 61–67.

Daub, M.E 1982 Cercosporin, a photosensitizing toxin from *Cercospora* species. *Phytopathology*. **72**:370–374.

De Hoog, G. S 1977 Rhinocladiella and allied genera. *Stud. Mycol*. 15**:**1-144.

Desjardins, A. E., R.D. Plattner. and R.H. Proctor 1996 Linkage among genes responsible for fumonisin biosynthesis in *Gibberella fujikuroi* mating population A. *Appl. Environ. Microbiol*. 62: 2571–2576.

Dev Rao and P. Raghuveer Rao 1964 Some Allied Dematiaceae-Amerosporae from India. *Trans. Am. Microsc*. Soc.**83:** 399–406.

Ding, Y., R.S. Bojja, and L. Du 2004 Fum3p, a 2-ketoglutarate-dependent dioxygenase required for C-5 hydroxylation of fumonisins in *Fusarium verticillioides*. *App. Environ. Microbiol*. **70**:1931-1934.

Dombrink-Kurtzman, M.A 2006 The isoepoxydon dehydrogenase gene of the patulin metabolic pathway differs for *Penicillium griseofulvum* and *Penicillium expansum*. *Antonie Van Leeuwenhoek*. **89**: 1–8.

Doohan, F.M., G.Weston, H.N. Rezanoor, D.W. Parry, P. Nicholson 1999 Development and use of a reverse transcription-PCR Assay to study expression of *Tri5* by *Fusarium* Species in vitro and in planta. *Appl.Environ .Microbiol.* **65**:3850–3854.

Dreyfuss, M. and O. Petrini 1984 Further investigations on the occurrence and distribution of endophytic fungi in tropical plants. *Bot Helvet*. 94:33–40.

Druzhinina, I.S., A.G. Kopchinskiy, M. Komoń, J. Bissett., G. Szakacs and C.P. Kubicek 2005 An oligonucleotide barcode for species identification in Trichoderma and Hypocrea. *Fungal Genet Biol*. 42: 813–828.

Edwards, S.G., J. O'Callaghan and A.D.W. Dobson 2002 PCR-based detection and quantification of mycotoxigenic fungi. *Mycol. Res*. 106: 1005–1025.

Ehrlich, K.C., P.K. Chang, J. Yu and P.J. Cotty 2004 Aflatoxin biosynthesis cluster gene cypA is required for G aflatoxin formation. *Appl. Environ. Microbiol*. **70**: 6518-6524.

Ellis, J.J 1979 Preserving fungus strains in sterile water. *Mycologia*. **71**: 1072-1075.

Ellis, M.B. 1976 More dematiaceous hyphomycetes .Commonwealth Mycological Institute, Kew, Survey. England.

El-Nezami, H., N. Polychronaki, S. Salminen and H. Mykkänen 2002 Binding rather than metabolism may explain the interaction of two food-Grade Lactobacillus strains with zearalenone and its derivative alpha-earalenol. *Appl. Environ. Microbiol.* **68**: 3545-3549.

Farber, P. and R. Geisen 2004 Analysis of differentially expressed ochratoxin A biosynthesis genes of *Penicillium nordicum*. *Eur. J. Plant Pathol.* **110**: 661–669.

Fedeshko, R.W 1992 Polyketide enzymes and genes in *Penicillium urticae*. PhD thesis, University of Calgary, Calgary, Alberta, Canada

Flaherty, J.E. and C.P. Woloshuk 2004 Regulation of fumonisin biosynthesis in *Fusarium verticillioides* by a zinc binuclear cluster-type gene, ZFR1. *Appl. Environ. Microbiol.* **70:** 2653–59.

Flaherty, J.E., A.M. Pirttilä, B.H. Bluhm and C.P. Woloshuk 2003 PAC1, a pH–regulatory gene from *Fusarium verticillioides*. *Appl. Environ. Microbiol.* 69: 5222–27.

Freeman, G.G. 1965 Isolation of alternariol and alternariol monomethyl ether from *Alternaria dauci* (Kükn) Groves and Skolko. Phytochemistry. **5**: 719-721.

Frisvad, J.C 1983 A selective and indicative medium for groups of *Penicillium viridicatum* producing different mycotoxins in cereals. *J. Appl. Bacteriol.* **54**: 409-416.

Frisvad, J.C 1989 The connection between the Penicillia and *Aspergillus* and mycotoxins with special emphasis on misidentified isolates. *Arch. Environ. Contamin. Toxicol.* **18**: 452-467.

Frisvad, J.C. 1985 Creatine sucrose agar, a differential medium for mycotoxin producing terverticillate *Penicillium* species. *Lett. Appl. Microbiol.* **1**: 109–113.

Frisvad, J.C. 1986 In modern methods in analysis and structural Elucidation of mycotoxins, (ed.R.J.Cole), *Academic Press, Orlando, FL*.pp.415-457.

Frisvad, J.C., D.L. Hawksvorth, Z. Kozakiewicz, J.I. Pitt, R.A. Samson and A.C. Stalk 1990 Proposal to conserve important names in *Aspergillus* and *Penicillium*. In: Samson, R.A., and Pitt, J.I., ed., Modern concepts in *Penicillium* and *Aspergillus* classification, New York, *Plenum Press*, pp. 83-90

Frisvad, J.C., J. Smedsgaard, T.O. Larsen, R.A. Samson 2004. Mycotoxins, drugs and other extrolites produced by species in *Penicillium* subgenus *Penicillium*. *Stud. Mycol.* **49**: 201–242.

Frisvad, JC 1985a Creatine sucrose agar, differential medium for mycotoxin producing terverticillate *Penicillium* species. *Lett. Appl. Microbiol.* **1**: 109-113.

Fujii, I., Y. Ono, H.Tada, K. Gomi, Y. Ebizuka and U. Sankawa 1996 Cloning of the polyketide synthase gene at X from *Aspergillus terreus* and its identification as the 6-methylsalicylic acid synthase gene by heterologous expression. *Mol. Gen. Genet.* **253***:* 1–10.

Gaffoor, I. and F. Trail 2006 Characterization of two polyketide synthase genes involved in zearalenone biosynthesis in *Gibberella zeae*. *Appl. Environ. Microbiol.* **72**:1793-1799.

Gagkaeva T and T. Yli-Mattila 2004 Genetic diversity of populations of *Fusarium graminearum* in Europe and Asia. *Eur. J. Plant .Pathol* **.110**:551–562

Gale, L.R, L.F. Chen, C.A. Hernick, K. Takamura and H.C. Kistler 2002 Population analysis of *Fusarium graminearum* from wheat fields in eastern China. *Phytopathol.* **92**:1315–1322.

Gallo, A., K.S. Bruno, M. Solfrizzo, G. Perrone, G. Mulè, A. Visconti, E. Scott and Baker 2012 New Insight into the Ochratoxin A Biosynthetic Pathway through

Deletion of a Nonribosomal Peptide Synthetase Gene in *Aspergillus carbonarius*. *Appl. Eniron. Microbiol.* **23**: 8208-8218.

Gams, W, M. Christensen, A.H.S. Onions, J.I. Pitt and R.A. Samson 1985 Infrageneric taxa of *Aspergillus*. In: Advances in *Penicillium* and Jurjevic, Peterson & Horn ARTICLE 78 ima f UN G U S *Aspergillus* Systematics (RA Samson & JI Pitt, eds): New York: Plenum Press. 55–64.

Gams, W. and E. Muller 1980 Conidiogenesis of *Fusarium nivale* and *Rhynchosporium oryzae* and its taxonomic implications. *Neth. J. Plant Path*. **86**:45-53.

Geisen, R., Z. Mayer, A. Karolewiez and P. Farber 2004 Development of a real time PCR system for detection of *Penicillium nordicum* and for monitoring ochratoxin A production in foods by targeting the ochratoxin polyketide synthase gene. *Syst. Appl. Microbiol.* **27:** 501–504.

Geiser, D.M., R.A. Samson, J. Varga, A. Rokas and S.M. Witiak 2008 A review of molecular phylogenetics in *Aspergillus*, and prospects for a robust genus-wide phylogeny. In: *Aspergillus* in the Genomic Era, Varga, J., and Samson, R.A., eds. (*Netherlands: Wageningen Academic Pubs).* pp. 17–32.

Gelderblom, W.C.A., K. Jaskiewicz, W.F.O. Marasas, P. G. Thiel, M.J. Horak, R. Vleggar and N.P.K. Kriek 1988 Fumonisins novel mycotoxins with cancer promoting activity produced by *Fusarium moniliforme*. *Appl. Environ. Microbiol.* **54:** 1806-1811.

Gerber, R., L. Lou and L. Du 2009 A PLP-Dependent Polyketide Chain Releasing Mechanism in the Biosynthesis of Mycotoxin Fumonisins in *Fusarium verticillioides*. *J. Am. Chem. Soc.* **131:** 3148–3149.

Gerlach, W. 1970 Suggestions to an acceptable modern *Fusarium* system. Helsinki. *Ann. Acad. Sci. Jenn. A, IV Biologica*. **168**:37–49.

Gerlach, W. and H. Nirenberg 1982 The genus *Fusarium* a pictorial atlas, Mitt. Biol. Bundesanst. Land Forstwirtsch. *Berl. Dahlem*. **209**: 1-406.

Gledhill, L., A.R. Hesketh , B.W. Bycroft, P.M. Dewick and Gilbert 1991 Biosynthesis of trichothecenes mycotoxins: cell free epoxydation of thricode derivatives. *FEMS Lett.* 81:241-245.

Gonzalez-Jaen, M.T., S. Mirete, B. Patino, E. Lopez-Errasqum and C. Vazquez 2004 Genetic markers for the analysis of variability and for production of specific diagnostic sequences in fumonisinproducing strains of *Fusarium verticillioides*. *Eur J Plant Pathol*. **110**:525–532.

Gordon, W. L. 1960 The taxonomy and habitats of *Fusarium* species from tropical and temperate regions. *Can. J. Bot*. 38:643–658.

Harris, J.P. and P.G. Mantle 2001 Biosynthesis of diaporthin and orthosporin by *Aspergillus ochraceus.Phytochem*. **57:**165–169.

Hauser, M.T., F. Adhami, M. Dorner, E. Fuchs, J. Glössl 1998 Generation of co-dominant PCR-based markers by duplex analysis on high resolution gels. *The Plant Journal*. 16:117–125

Hawksworth, D and J.I. Pitt 1973 A new taxonomy for Monascus species based on cultural and microscopical characters. *Aust. J. Bot*. **31**: 51-61.

Hawksworth, D. L. and J.I. Pitt 1976 Typification of the generic name *Penicillium*. Taxon, in press.

Hawksworth, D.L and B.C. Sutton 1974 Article 59 and names of perfect state taxa in imperfect genera. *Taxon*. **23**: 563-568.

Hawksworth, D.L and B.E. Kirsop 1988 Living resources for biotechnology. Filamentous fungi eds. Cambridge, Cambridge University Press.

Hawksworth, D.L. and H.I. Wells 973 Ornamentation on the terminal hairs in *Chaetomium Kunze* ex Fr. and some allied genera. *Mycol Pap*. 134: 1–24

Headrick, J.M. and J.K. Pataky 1991 Maternal influence on the resistance of sweet corn lines to kernel infection by *Fusarium moniliforme*. *Phytopathol.* **81**: 268-274.

Heitz, J.R. and K.R. Downum 1995 Light-activated pest control. Ame. Chem. Soc, Washington, D.C

Hendrickson, L., C.R. Davis., C. Roach, D.K. Nguyen, T. Aldrich, P.C. Mcada, C.D. Reeves 1999 Lovastatin biosynthesis in *Aspergillus terreus*: characterization of blocked mutants, enzyme activities and a multifunctional polyketide synthase gene. *J. Chem. Biol.* **6:** 429 439.

Hennebert, G.L. and L.K. Weresub 1977 Terms for states and forms of fungi, their names and types. *Mycotaxon*. **6**: 207-211.

Henry T., P.C. Iwen and S.H. Hinrichs 2000 Identification of *Aspergillus* species using internal transcribed spacer regions 1 and 2. *J. Clin. Microb.* **38:**1510–1515

Hocking, A.D 1981 Improved media for enumeration of fungi from foods. *CSIRO Food Research Quarterly.* 44: 73-82.

Hocking, A.D 1991 Effects of fumigation and modified atmosphere storage on growth of fungi and production of mycotoxins. Champ, B.R., Highley, E., Hocking, A.D., and Pitt, 1.1., eds. Fungi and Mycotoxins in Stored Products: Proceedings of an international conference, Bangkok. Thailand. ACIAR Proceedings No. 36, 270p. Pp.145-156.

Hocking, A.D. and J.I. Pitt 1980 Dichloran-glycerol medium for enumeration of xerophilic fungi from low-moisture foods. *Appl. Environ. Microbiol*. 39:488-492.

Hoffmeister, D., and N.P. Keller 2007 Natural products of filamentous fungi: enzymes, genes, and their regulation. *Nat. Prod. Rep.* **24**: 393–416.

Hradil, C.M. Hallo I.F, J. Clardy, D.S. Kenfield and G. Strobek 1989 Phytotoxins of *Alternaria cassiae*. *Phytochem*. 28: 73-75.

Huff, W.E. and P.B. Hamilton 1979 Mycotoxins their biosynthesis in fungi: ochratoxins: metabolites of combined pathways. *J. Food. Prot.* **42**:815–820.

Hughes, S.J 1953 Conidia, conidiophores and classification. *Canadian. J. Bot.* **31**:577–659.

Ishikawa, M. T. and F. Oki Sawada 1973 *Rept. Tottori, Mycol.Inst.*10:637-645.

Jarvis, B 1971. Factors affecting the production of mycotoxins. *J. Appl. Bacteriol.* **34**: 199-213.

Joffe, A.Z., 1974 A modern system of *Fusarium* taxonomy. *Mycopathol. Mycol. Appl.* 53: 201-228.

Joffe, A.Z. 1978 In Mycotoxic Fungi, Mycotoxin, Mycotoxicoses. An Encyclopedia Handbook, Vol. 3, (eds.T.D.Wyllic and L.G. Morehouse), Marcel Dekker, New York, pp.21-86.

Joffe, A.Z. and J. Palti 1975 Taxonomic study of Fusaria of the sporotrichiella section used in recent toxicological work. *Appl. Microb*. 29:575-579.

Joffe, A.Z. and J. Palti 1977 Species of *Fusarium* found in uncultivated desert type soils in Israel. *Phytoparasitica*. **5**:119-121.

Joffe, A.Z., J. Palti and R. Arbel-Sherman 1973 *Fusarium moniliforme* Sheldon in Israel (*Gibberella fujikuroi* (Saw.) Wollenw.). *Mycopathol. Mycol. Appl*. 50:85–107.

Jones, D., B.R. Sexton and G. Rebell 1969 Mycotic keratitis in south Florida: a review of 39 cases. Trans. Opthalmol. Soc. U.K. 89:781-797.

Jones, M.J. and L.D. Dunkle 1993 Analysis of *Cochliobulus carbonum* races by PCR amplification with arbitrary and gene-specific primers. *Phytopathol.* **83***:* 366-370.

Jurado, M., C. Vazquez, S. Marin, V. Sanchis, M.T. Gonzalez-Jaen 2006 PCR-based strategy to detect contamination with mycotoxigenic *Fusarium* species in maize. *Syst. Appl. Microbiol.* **29**:681–689.

Kameda, K., H. Aoki, H.Tanaka and M. Namiki 1973 Studies on metabolites of *Alternaria kikuchiana* Tanaka, a phytopathogenic fungus of Japanese pear. *Agri. Biol. Chem*. **37**: 2137-2146.

Karolewiez, A. and R. Geisen 2005 Cloning a part of the ochratoxin A biosynthetic gene cluster of *Penicillium nordicum* and characterisation of the ochratoxin polyketide synthase gene. *Syst. Appl. Microbiol.* **28***:* 588–595.

Karugia, G.W., H. Suga, L.R. Gale, T. Nakajima, A. Ueda and M. Hyakumachi 2009 Population structure of *Fusarium asiaticum* from two Japanese regions and eastern China. *J. Gen. Plant Pathol*. **75**: 110–118.

Kellerman, T. S., W. F. 0. Marasas, J.G. Pienaar and T.W. Naude 1972 A mycotoxicosis of Equidae caused by *Fusarium moniliforme* Sheldon. A preliminary communication. *Onderstepoort J. Vet. Res*.**39**:205-208.

Kendrick, B. (Ed.) 1971 The Taxonomy of the Fungi Imperfecti. Proceedings of the International Specialists' Workshop Conference on Criteria and Terminology in the Classification of Fungi Imperfecti. University of Toronto Press, Toronto.

Kendrick, B. (ed.) 1979 The Whole Fungus: The Sexual-Asexual Synthesis. Volume 1, National Museums of Canada, Ottawa.

Kim, J.E., K.H. Han, J. Jin, H. Kim, J.C. Kim, S.H. Yun and Y.W. Lee 2005 Putative polyketide synthase and laccase genes for biosynthesis of aurofusarin in *Gibberella zeae*. *Appl. Environ. Microbiol*. **71**: 1701–1708.

Kim, Y. u. a. 2005Two different polyketide synthase genes are required for synthesis of zearalenone in *Gibberella zeae*. *Mol. Microbiol.***58:** 1102-1113

Kimura, M., T. Tokai, K. O'Donnell, T.J. Ward, M. Fujimura, H. Hamamoto, T. Shibata and I. Yamaguchi 2003 The trichothecene biosynthesis gene cluster of *Fusarium graminearum* F15 contains a limited number of essential pathway genes and expressed non-essential genes. *FEBS.* Lett. **539**:105–110.

King, A.D., A.D. Hocking and J.I. Pitt 1979 Dichloran-rose bengal medium for enumeration and isolation of molds from foods. *Appl. Environ. Microbiol*. **37:** 959-964.

Kinoshita, T., Y. Renbutsu, I. Khan, K. Kohmoto and S. Nishimura 1972 Distribution of tenuazonic acid production in the genus Alternaria and its phytopathological evaluation. *Annals. Phytopathol. Soc. Japan.* **38**: 397-404.

Kirsnop, B.E and J.J.S. Snell 1984 Maintenance of microorganisms. London, *Academic Press*.

Klebahn, H 1918 Haput-und Nebenfruchtformen von Ascomyceten" Leipzig.

Klich M.A and J.I. Pitt 1988 A Laboratory Guide to Common *Aspergillus* Species and Their Teleomorphs. North Ryde, NSW: CSIRO.

Klich, M.A. 2002 Identification of common *Aspergillus* species. First Ed., Utrecht, The Netherlands, *Centraalbureau voor Schimmelcultures* pp. 122 pp.

Klich, M.A., E.J. Mullaney, C.B. Daly J.W. Cary 2000 Molecular and physiological aspects of aflatoxin and sterigmatocystin biosynthesis by *Aspergillus tamarii* and *A. ochraceoroseus*. *Appl. Microbiol. Biotechnol.* **53**: 605-609.

Knoche, H.W. and J.P. Duvick 1987 The role of fungal toxins in plant disease In: Fungal Infection of Plants (Pegg, G.F. and 204 M.E. Daub et al. / *FEMS Microbiology Letters 252* (2005) 197–206 Ayres, P.G., Eds.), Cambridge University Press, Cambridge, pp. 158–191.

Knutsen, A.K., and A. Holst-Jensen 2004 Phylogenetic analyses of the *Fusarium poae, Fusarium sporotrichioides* and *Fusarium langsethiae* species complex passed on partial sequences of the translation elongation factor-1 alpha gene. *Int. J. Food. Prot.* **95**: 287–295.

Konstantinova, P. and T. Yli-Mattila 2004 IGS-RFLP analysis and development of molecular markers for identification of *Fusarium poae*, *Fusarium langsethiae, Fusarium sporotrichioides* and *Fusarium kyushuense*. *Int. J Food. Microbiol.* **95**: 321–331.

Koteswara Rao,V., M.V. Ramana,S. Girisham and S.M. Reddy 2013 Culture media and factors influencing ochratoxin A production by two species of *Penicillium* isolated from poultry feeds. *Natl. Acad. Sci. Lett.* 36:101–110.

Kozakiewicz, Z 1989 *Aspergillus* species on the stored products. *Mycological Papers*. **161**:1-188

Krysinska-Traczyk, E., J. Perkowski and J. Dutkiewicz 2007 Levels of fungi and mycotoxins in the samples of grain and grain dust collected from five various cereal crops in eastern Poland. *Ann. Agric. Environ. Med.* **14**:159–167.

Kuiper-Goodman, T., P.M. Scott and H. Watanabe 1987 Risk assessment of the mycotoxin zearalenone. *Regul. Toxicol. Pharmacol.* **7:** 253-306.

Kulik, T 2008 Detection of *Fusarium tricinctum* from cereal grain using PCR assay. *J. Appl. Genet.* **49**: 305–311.

Kuraishi, H., M. Itoh, N. Tsuzaki, Y. Katayama, T. Yokoyama and J. Sugiyama 1990 The ubiquinone system as a taxonomic aid in *Aspergillus* and its teleomorphs. In Modern Concepts in *Penicillium* and *Aspergillus* Classification, ed. by Samson, R. A. and Pitt, J. I., Plenum Press, New York, pp. 407–421.

Kurtzman, C.P., M.J. Smiley, C.J. Robnett and D.T. Wicklow 1986 DNA relatedness among wild and domesticated species in the *Aspergillus flavus* group. *Mycologia.* **78**: 955–959.

Laday, M., A. Juhasz, A. Moretti, A. Szecsi and A. Logrieco 2004a Mitochondrial DNA diversity and lineage determination of European isolates of *Fusarium graminearum*. *Eur J Plant Pathol.***110**: 545–550.

Laday, M.G., A. Mule, Z. Moretti Ham, A. ari Juhasz, A. Szecsi A. and Logrieco 2004b Mitochondrial DNA variability in *Fusarium proliferatum* (*Gibberella intermedia*). *Eur. J. Plnt. Pathol.***110:** 563–571.

Lee,T., D. Oh, J. Kim, T. Lee, S. Kim, Yun and Y. Lee 2001 Identification of deoxynivalenol and nivalenol-producing chemotypes of *Gibberella zeae* by using PCR. *Appl. Environ. Microbiol.* **67:** 2966–2972.

Li, S.G.L. Hartman, L.L. Domier and D. Boykin 2008 Quantification of *Fusarium solani* f. sp. glycines isolates in soybean roots by colony-forming unit assays and real-time quantitative PCR. *Theor. Appl. Genet.* **117**:343–352.

Lieckfeldt, E. and KA. Seifert 2000 An evaluation of the use of ITS sequences in the taxonomy of the 1-lypocreales. *Stud. Mycol.* **45:**35-44.

Liou, G.F. and C. Khosla 2003 Building-block selectivity of polyketide synthases. *Curr. Opin. Chem. Biol.* **7:** 1–6.

Liu, A. and H. Zhang 2006 Transition metal-catalyzed nonoxidative decarboxylation reactions. *Biochem.* **45**: 10407–10411.

Livak, K.J., S.J. Flood, J. Marmaro, W. Giusti, and K. Deetz 1995 Oligonucleotides with fluorescent dyes at opposite ends provide a quenched probe system useful for detecting PCR product and nucleic acid hybridization. *PCR Methods Applic*. 4: 357-362.

Lysøe, E., K.R. Bone and S.S Klemsdal 2009 Real-time quantitative expression studies of the zearalenone biosynthetic gene cluster in *Fusarium graminearum*. *Phytopathol.* **99:** 176-184.

Mach, R.L., CM. Kullnig-Gradinger, A.H. Farnleitner, G. Reischer, A. Adler and CP. Kubicek 2004 Specific detection of *Fusarium langsethiae* and related species by DGGE and ARMS-PCR of a b-tubulin (tub1) gene fragment. *Int. J Food. Prot.* **95**:333–339.

Malloch, D. and R.F. Cain 1972 New species and combinations in cleistothecial Ascomycetes. *Can. J. Bot.* **50**: 61–72.

Marasas, W.F.O., F.C. Wehner, S.J. van Rensburg and D.J. van Schalkwyk 1981 Mycoflora of corn produced in human esophageal cancer areas in Transkei, southern Africa. *Phytopathol.* **71**: 792796.

Marasas, W.F.O., P.E. Nelson and T.A. Tousson 1984 Toxigenic *Fusarium* Species: Identity and Mycotoxicology. University Park, Pennsylvania: Pennsylvania State University Press.

Marcia McMullen M, R. Jones and D. Gallenberg 1997 Scab of wheat and barley: a re-emerging disease of devastating impact. *Plant Dis* **81**:1340–1348.

Mason, E.W. and M.B. Ellis 1953 British species of Periconia. *Mycol. Pap.* 56. Pp. 126.

Millner, P.D. 1977 Radial growth responses to temperature by 58 *Chaetomium* species, and some taxonomic relationships. *Mycologia.* 69:492–502.

Minter, D.W., P.M. Kirk and B.C. Sutton 1982 Holoblastic phialides. Trans. *Brit. Mycol. Soc.* **79**: 75-93.

Minter, D.W., P.M. Kirk and B.C. Sutton 1983 Thallic phialides. Trans. *Brit. Mycol. Soc.***80**: 39-66.

Mirete, S., C. Vazquez, G. Moule, M. Jurado and M.T. Gonzales-Jaen 2004 Differentiation of *Fusarium verticillioides* from banana fruits by IGS and EF-1a sequence analysis. *Eur. J. Plant. Pathol.* **110**:525–523.

Mislivec and Bruce 1977 Incidence of toxic and other mold species and genera in soybeans. *J. Food. Prot.* **40**: 309–312.

Miyake, Naito, H. and H. Sumeda 1940 Japanese title. Report of the Research Institute for Rice Improvement, 1: 1.

Moreau, M. and F. Moreau 1930 Le development du perithece quelques ascomycetes. *Rev.Gen. Bot.* **42**: 1-34.

Moss, M.O. 1998 Recent studies of mycotoxins. Symp. *Ser. Soc. Appl. Microbiol.* **27**: 62S–76S.

Mule, G., A. Susca, G. Stea and A. Moretti 2004 A species specific PCR assay based on the calmodulin partial gene for identification of *Fusarium verticillioides, F. roliferatum* and *F. subglutinans*. *Eur. J. Plant. Pathol.* **110**:495–502

Mulè, G.A., A. Susca, G. Logrieco, Stea and A.Visconti 2006 Development of a quantitative real-time PCR assay for the detection of *Aspergillus* carbonarius in grapes. *Int. J. Food Microbiol.* **111**: 28–34.

Mullaney, E.J and M.A. Klich 1990 In modern Concepts in *Penicillium* and *Aspergillus* Classification. (eds. R.A. Samson and J.I. Pitt), Plenum, NewYork.pp.321-334.

Muller, E 1971 Imperfect-perfect connections in Ascomycetes. In "Taxonomy of fungi imperfecti" (W.B. Kendrick, ed.) Univ. of. Toronto Press, Toronto. pp. 184-201.

Munkvold, G. P. R. L. Hellmich and W.B. Showers 1997 Reduced *Fusarium* ear and symptomless infection in kernels of maize genetically engineered for European cornborer resistance. *Phytopathol.* **87:** 1071-1077.

National Toxicology Program 1999 Toxicology and carcinogenesis studies on fumonisin B1 in F344/N rats and B6CF1 mice (feed studies). Technical Report Series, n.496. NIH Publication No. 99-3955. U.S. Department of Health and Human Services, National Institutes of Health: Research Triangle Park, NC.

Nelson, P., E.A.E. Desjardins and R. D. Plattner 1993 *Fumonisins*, mycotoxins produced by *Fusarium* species: Biology, chemistry and significance. *Annu. Rev. Phytopathol.* **31:** 233-252.

Nelson, P.E., T.A. Toussoum and W.F. Marasas 1983 *Fusarium* species: An illustrated Manual for Identification, P. 203. The State University Press, Penn *USA*. PP.193.

Nelson, P.E., T.A. Toussoun and W.F.O. Marasas 1985 *Fusarium* species, an illustrated Manual for identification. Penn. State. Univer. Press, Univ. Pak.

Nganje, W.A., D.A. Bangsund, F.L. Leistritz, W.W. Wilson and N.M. Tiapo 2004 Regional economic impacts of *Fusarium* head blight in wheat and barley. *Rev. Agr. Econ.* **26**:332–337.

Nicholson, P., D.R. Simpson, A.H. Wilson, E. Chandler and A. Thomsett 2004 Detection and differentiation of trichothecene and enniatin-producing *Fusarium* species on small-grain cereals. *Eur. J. Plant. Pathol.* **110:** 503–514.

Nicholson, T.P. B.A.M. Rudd, M. Dawson, C.M. Lazarus, T.J. Simpson and R.J. Cox 2001 Design and utility of oligonucleotide gene probes for fungal polyketide synthases. *Chem. Biol.* **8:** 157–178.

Niessen, L 2007 Current trends in molecular diagnosis of ochratoxin A producing fungi. In Mycotechnology Present Status and Future Prospects; Rai, M.; Ed.; I.K. International Publishing House: New Delhi, India. p. 320.

Niessen, L., H. Schmidt, E. Muhlencoert, P. Farber, A, Karolewiez and P. Geisen 2005 Advances in the molecular diagnosis of ochratoxin A producing fungi. *Food .Addit. Contam.* **22**:324–334.

Niessen, M.L. and R.F. Vogel 1997 Specific identification of *Fusarium graminearum* by PCR with gaoA targeted primers. *Syst. Appl. Microbiol.* **20:** 111–123.

Niessen, M.L. and R.F. Vogel 1998 Group specific PCR-detection of potential trichothecene-producing *Fusarium*-species in pure cultures and cereal samples. *Syst. Appl. Microbiol.* **21**:618–631.

Niller, J.D 2000 Development of a PCR based assay for rapid and reliable identification of pathogenic Fusaria. *Plant Breed.* 19–36.

Nirenberg, H.I 1976 Untersuchungen über die morphologische und biologische Differenzierung in der *Fusarium*-Sektion Liseola. Mitteilungen aus der Biologischen Bundesanstalt für Land- Forstwirtschft Berlin-Dahlem 169: 1-117

Nirenberg, H.I 1990 Recent advances in the taxonomy of *Fusarium*. Studies in Mycology, Baarn. **32**: 91-101.

Njambere, E.N., W. Chen, C. Frate, B.M. Wu, S. Temple, and F Muehlbauer 2008 Stem and crown rot of chickpea in California caused by *Sclerotinia oiJliorum*. *Plant Dis.* **92**:917-922.

O'Brian, G.R., A.M. Fakhoury and G.A. Payne 2003 Identification of genes differentially expressed during aflatoxin biosynthesis in *Aspergillus flavus* and *Aspergillus parasiticus. Fun. Gen. Biol.* **39**: 118–27.

O'Callaghan, J., M.X. Caddick and A.D.W. Dobson 2003 A polyketide synthase gene required for ochratoxin A biosynthesis in *Aspergillus ochraceus*. *Microbiol.* **149**: 3485–3491.

O'Callaghan, J., P.C. Stapleton and A.D. Dobson 2006 Ochratoxin A biosynthetic genes in *Aspergillus ochraceus* are differentially regulated by pH and nutritional stimuli. *Fungal Genet Biol.* **43**:213–221.

O'Donnell, K. and E. Cigelnik 1997 Two divergent intragenomic rDNA ITS2 types within a monophyletic lineage of the fungus *Fusarium* are nonorthologous. *Mol. Phylogenet. Evol.* **7**: 103-116.

O'Donnell, K., E. Cigelnik and H. Nirenberg 1998 Molecular systematics and phylogeography of the *Gibberella fujikuroi* species complex. *Mycologia.* **90:** 465-493.

O'Donnell, K., H.I. Nirenberg, T. Aoki, E. Cigelnik 2000 A multigene phylogeny of the *Gibberella fujikuroi* species complex: detection of additional phylogenetically distinct species. *Mycosci.* **41**:61-78.

Onions, A.H.S. 1983 Preservation of fungi. In: Smith, J.E., Barry, D.R., and Kristiansen, B., ed., The filamentous fungi 4, Fungal technology, ed. Smith, J.E. Berry, D.R., and Kr Kristiansen B. London, *Edward Arnold*. 373-390.

Osbourn, A. 2010 Secondary metabolic gene clusters: evolutionary toolkits for chemical innovation. *Trends. Genet*. 26:449–457.

Parry, D.W. and P. Nicholson 1996 Development of a PCR assay to detect *Fusarium poae* in wheat. *Plant. Pathol*. **45**: 383- 391.

Paterson, R.R.M 2006 Identification and quantification of mycotoxigenic fungi by PCR. *Process Biochem.* **41**: 1467–1474.

Patiño, B., S. Mirete, M.T. González-Jaén, E.G. Mulé and C. Vázquez 2004 PCR detection assay of fumonisin-producing *Fusarium verticillioides* stains. *J. Food Protect*. **67:** 1278-1283.

Pel, H.J. et al. 2007. Genome sequencing and analysis of the versatile cell factory *Aspergillus niger* CBS 513.88. *Nature Biotech.* **25**: 221–231.

Pfohl-Leszkowicz, A., L. Chekir-Ghedira and H. Bacha 1995 Genotoxicity of zearalenone, an estrogenic mycotoxin: DNA adduct formation in female mouse tissues. *Carcinogenesis.* **16**: 2315-2320.

Pitt, J.I 1996 What are mycotoxins? Aust. *Mycotoxin Newsl.* 7:1–3.

Pitt, J.I and A.D. Hocking 1985 Fungi and food spoilage. Sydney, *Academic Press*.

Pitt, J.I. 1987. Penicillium viridicatum, Penicillium verrucosum, and Production of Ochratoxin A. Appl. Environ. Microb. 53: 266-269.

Pitt, J.I. 1979 The Genus *Penicillium* and its Teleomorphic States *Penicillium* and Talaromyces. London, Academic Press.

Pitt, J.I. 1984 The significance of potentially toxigenic fungi in foods. Food Technology in Australia. **36**: 218-219.

Pitt, J.I. 1988 A laboratory guide to common *Penicillium* species, 2nd ed. CSIRO Division of Food processing, North Ryde, N.S.W., Australia

Pitt, J.I. and A.D. Hocking 1985 "Fungi and Food Spoilage". Sydney, N.S.W.: Academic Press

Pitt, J.I. and A.D. Hocking 1985b Fungi and Food Spoilage. Sydney: Academic Press. p. 52.

Pitt, J.I., A.D. Hocking and D.R. Glenn 1983 An improved medium for the detection of *Aspergillus flavus* and *A.parasiticus*. *J. Appl. Bacteriol.* **54**:109-114.

Placinta, C.M., J.P.F. D'Mello and A.M.C. Macdonald 1999 A review of worldwide contamination of cereal grains and animal feed with *Fusarium* mycotoxins. *Anim. Feed. Sci. Techn.* **78**: 21-37.

Proctor, R.H., A.E. Desjardins, R.D. Plattner, T.M. Hohn 1999 A polyketide synthase gene required for biosynthesis of fumonisin mycotoxins in *Gibberella fujikuroi* mating population A. *Fungal Genet. Biol.* **27**:100–12.

Proctor, R.H., D.W. Brown, R.D. Plattner, A.E. Desjardins 2003 Co-expression of 15 contiguous genes delineates a fumonisin biosynthetic gene cluster in *Gibberella moniliformis*. *Fungal Genet. Biol.* **38**:237–249.

Proctor, R.H., R.D. Plattner, A.E. Desjardins, M. Busman and R.A. Butchko 2006 Fumonisin production in the maize pathogen *Fusarium verticillioides*: genetic basis of naturally occurring chemical variation. *J. Agri. Food Chem.* **54**: 2424-2430.

Raistrick, H., C.E. Stickings and R. Thomas 1953 Studies in the biochemistry of microorganisms. Alternariol and alternariol monomethyl ether, metabolic products of *Alternaria tenuis*. *Biochem. J.* **55**: 421-433.

Ramana, M.V., K. Balakrishna, H.S. Murali and H.V. Batra 2011 Multiplex PCR-based strategy to detect contamination with mycotoxigenic Fusarium species in rice and fingermillet collected from southern India. *J. Sci. Food .Agri.* **91**: 1666- 1673.

Raper, K.B 1957 Nomenclature in *Aspergillus* and *Penicillium*. *Mycologia*. **49**:644-662.

Raper, K.B and C. Thom 1949 A manual of the Penicillia. Baltimore: Williams & Wilkins Co. p.875.

Raper, K.B. and D.I. Fennell 1965 The Genus *Aspergillus*, Baltimore: Williams & Wilkins Company. pp 686.

Read, G., D.W.Westlake, L.C. Vining 1969 Quinone epoxides V The biosynthesis of terreic acid. *Can. J. Biochem.* **47:** 1071–1079.

Ren, P, T.M. Jankun and B.P. Leaderer 1999 Comparisons of seasonal fungal prevalence in indoor and outdoor air and in house dusts of dwellings in one Northeast American county. *J. Expo. Anal. Environ. Epidemiol.* **9**: 560-568.

Rheeder, J.P., W.F.O Marasas, H.F.Vismer 2002 Production of fumonisin analogs by *Fusarium* species. *Appl. Environ. Microbiol.* **68:** 2101-2105.

Saccardo, P.A 1886 Sylloge Fungorum. Vol. 4. Johnson Reprint Corporation

Saiki, R.K., D.H. Gelfand, S. Stoffel, S.J. Scharf, R. Higuchi, G.T. Horn, K.B. Mullis and H.A. Erlich 1988 Primer-directed enzymatic amplification of DNA with a thermostable DNA polymerase. *Science.* **239:** 487-491.

Samson, R. A. and W. Gams 1985 Typification of the *Aspergillus* species and associated teleomorphs. In Advances of *Penicillium* and *Aspergillus* systematics (eds. R.A.Samson & J.I. Pitt), Plenum Publ. p. 31-54.

Samson, R. A., A.C. Stolk and R. Hadlock 1976 Revision of the subsection *Fasciculata* of *Penicillium* and some allied species. *Stud. Mycol.***11**:1–47.

Samson, R.A. and J.I. Pitt 1990 Modern Concepts in Penicillium and Aspergillus Classification. NATO ASI series eds. Plenum Press, New York. pp.478.

Samson, R.A. 1994 Current Systematics of the genus *Aspergillus*. In The genus *Aspergillus*: From Taxonomy and Genetics to Industrial Application (Powell, K.A., A. Renwick Peberdy, J.F, eds.). Plenum Press, London. pp. 261-276.

Samson, R.A. 1994 Taxonomy Current concepts in *Aspergillus*. In Biotechnology Handbooks: *Aspergillus* (J. E. Smith editor). *Plenum Publishing* Co. pp. 1-22.

Samson, R.A. and J. Varga 2008 *Aspergillus* in the Genomic Era. Wageningen Academic Publishers. The Netherlands.

Samson, R.A. and J.I. Pitt. eds. 1990 Modern Concepts in *Penicillium* and *Aspergillus* Classification. Plenum Press: New York and London, pp.478.

Samson, R.A., and E.S. Van Reenen-Hoekstra 1988 Introduction to food-borne fungi. Third edition. *Centraalbureau voor Schimmelcultures.* pp. 299.

Samson, R.A., and J.I. Pitt 1990 Approaches to *Penicillium* and *Aspergillus* systematics. *Stud. Mycol., Baarn.* **32**: 77-91.

Samson, R.A., and J.I. Pitt 2000 Types for accepted species in *Penicillium, Aspergillus* and their teleomorphs In Integration of modern taxonomic methods for *Penicillium* and *Aspergillus* classification (Eds. Samson, R.A. and Pitt, J.I). Harwood Publishers, Amsterdam, pp 51-72

Samson, R.A., E.S. Hoekstra, J.C. Frisvad and O. Filtenborg 1995 Methods for the detection and isolation of food-borne fungi. In Introduction of Food borne Fungi (eds. Samson *et al.)* p. 235-242.

Samson, R.A., J.I. Pitt, and J.C. Frisvad 2000 List of accepted species and synonyms in the family Trichocomaceae. In: Samson RA, Pitt JI (Eds.). Integration of Modern taxonomic methods for *Penicillium* and *Aspergillus* classification. 510 pp. Singapore: Harwood Academic Publishers. pp. 9-49.

Samson, R.A., M.C. Rombach and K.A. Seifert 1984 *Hirsutella guiguardii* and *Stilbella kervellei*, two troglobiotic entomogenous hyphomycetes. *Persoonia.* **12**: 123-134

Samuel, G.J. and A.Y. Roseman 1979 Conidia and classification of the nectriod fungi. In 'The Whole Fungus" (W.B. Kendrick, ed.). National Museum of Canada, Ottawa. pp. 167-182.

Samuels, G.J. and I.C. Hallett 1983 *Microdochium stoveri* and *Monographella stoveri* new combinations for *Fusarium stoveri* and *Micronectriella stoveri.* Trans. Brit. Mycol. Soc. **81**:473-483.

Sanchez-Ballesteros, J., V. Gonzalez, O. Salazar, J. Acero, M.A. Portal, M. Julian, V. Rubio, GF. Bills, J.D. Polishook, G. Platas, S. Mochales, F. Pelaez 2000 Phylogenetic study of Hypoxylon and related genera based on ribosomal ITS sequences. *Mycologia.* **92**:964–977.

Sanders, T.H., N.D. Davis and V.L. Diener 1968. Effect of carbon dioxide, temperature, and relative humidity on production of aflatoxins in peanuts. *J. Ame. Oil.Chem. Soc.* 45: 683-685.

Sargeant, K. A., J. Sheridan, O'Kelly and R.B.A. Carnaghan 1961 Toxicity associated with certain samples of groundnuts. *Nature.* **192**:1096-1097.

Schmale, D.G, III., J.F. Leslie, K.A. Zeller, A.A. Saleh, E.J. Shields and G.C. Bergstrom 2006 Genetic structure of atmospheric populations of *Gibberella zeae. Phytopathol.* **96**: 1021–1026.

Schmidt, H., M. Bannier, R.F. Vogel, L. Niessen 2004 Detection and quantification of *Aspergillus ochraceus* in green coffee by PCR. *Lett. Appl. Microbiol.* **38**: 464–469.

Schmidt, H., M. Ehrmann, RF.Vogel, M.H. Taniwaki, L. Niessen 2003 Molecular Typing of *Aspergillus ochraceus* and construction of species specific SCAR-primers based on AFLP. *Syst Appl Microbiol.* **26**:138–146.

Schneider, J.H.M., O. Salazar, V. Rubio and J. Keijer 1997 Identification of *Rhizoctonia solani* isolates associated with field-grown tulips using ITS rDNA polymorphism and pectic zymograms. *Euro. J. Plant. Pathol.* **103**: 607–622.

Selma, M.V., P.V. Martínez-Culebras and R. Aznar 2008 Real-time PCR based procedures for detection and quantification of *Aspergillus carbonarius* in wine grapes. *Int. J. Food Microbiol.* **122:** 126–134.

Seo, J.A., R.H. Proctor and R.D. Plattner 2001 Characterization of four clustered and coregulated genes associated with fumonisin biosynthesis in *Fusarium verticillioides*. *Fungal. Genet. Biol.* **34**:155–165.

Seth, H. K 1972 A monograph of the genus *Chaetomium*. Beihefte zur Nova Hedwigia heft 37.

Shapira, R., N. Paster, O. Eyal, M. Menasherov, A. Mett and R. Salomon 1996 Detection of aflatoxigenic molds in grains by PCR. *Appl. Environ. Microbiol.* **62**: 3270-3273.

Shim, W. B. and C.P. Woloshuk 2001 Regulation of fumonisin B biosynthesis and conidiation in *Fusarium verticillioides* by a cyclin-like (C-type) gene, FCC1. *Appl. Environ. Microbiol.* **67**: 1607–1612.

Sidhu, G.S 2002 Mycotoxin genetics and gene clusters. *Eur. J. Plant Pathol.* **108**: 705–711.

Smalley, E. B. and F.M. Strong 1974 Toxic trichothecenes, In I.F.H. Purchase (ed.), Mycotoxins. Elsevier Scientific Publishing Co., Amsterdam. pp. 199-228.

Snyder, W. C. and H.N. Hansen 1945 The species concept in *Fusarium* with reference to Discolor and other sections. *Amer. J. Bot.* **32**:657–666.

Snyder, W. C. and T.A. Tousson 1965 Current status of taxonomy in *Fusarium* species and their perfect stages. *Phytopathol.* **55**:833–837.

Solfrizzo, M., A. D. Girolamo, C. Vitti, K. Tylkowska, J. Grabarkiewicz Szczesna, D. Szopinska and H. Dorna 2005 Toxigenic profile of *Alternaria alternata* and *Alternaria radicina* occurring on umbelliferous plants. *Food. Addit. Contam.* **22**: 302-308.

Sörgel, G. 1960 Zum Problem der Trennung von Arten bei Pilzen, dargestellt am Beispiel der Ascomycetengattung *Chaetomium*. *Arch. Mikrobiol.* **36**:51–66.

Spencer, S.M. ed 1978 The "Powdery Mildew" Academic Press., New York

Stark A.A., B.D. Kobbe, G. Matsuo, G.N. Buchi, A.L. Wogan and Demain 1978 Mollicellins: mutagenic and antibacterial mycotoxins. *Appl. Environ. Microbiol.* **36:** 412-420.

Stob, M., R.S. Baldwin, J. Tuite, F.N. Andrews and K.G. Gillette 1962 Isolation of an anabolic, uterotrophic compound from corn infected with *Gibberella zeae*. *Nature.* **196**: 1318.

Stolk, A.C. and D.B. Scott 1967 Studies on the genus *Eupenicillium* Ludwig 1. Taxonomy and nomenclature of Penicillia in relation to their Sclerotioid ascocarpic states. *Persoonia.* 4:391-405.

Subramanian, C.V. 1952 Studies on South Indian Fusaria. II. Fusaria isolated from black cotton soils. Proc. Nat. Inst. Sci. India 18:557–584.

Taylor, J.W., J.I. Pitt and A.D. Hocking 1990 Ribosomal DNA restriction studies of *Talaromyces* species with *Paecilomyces* and *Penicillum* anamorphs. In: Samson, R.A. and Pitt, J.I. ed., Modern concepts in *Penicillium* and *Aspergillus* classification, *New York Plenum Press*. pp. 357-369.

Thiel, P.G. W.F.O. Marasas, E.W. Sydenham, G.S. Shepard, W.C.A. Gelderblom and J.J. Nieuwenhuis 1991 Survey of fumonisin production by *Fusarium* species. *Appl. Environ. Microbiol.* **57**: 1089-1093.

Thom, C. and K.B. Raper 1954 A Manual of the Aspergilli (Baltimore: Williams & Wilkins)

Thom, C. and M.B. Church 1926 The Aspergilli. Baltimore, Maryland, *Williams and Wilkins*.

Torzilli, A.P. and J.D. Isbister 1994 Comparision of coal solubilizarion by bacteria and fungi. *Biodegradation*. S. 55-62.

Tubaki, K 1958 Studies on the Japanese Hypomycetes. *V.J. Hattori Bot. Lab.* **20**:142-244.

Tubaki, K. 1981 Hyphomycetes their perfect-imperfect connexions. *J. Cramer, Vaduz*.1-181.

Tulanse, L.R 1953 Memorie sur 1'ergot des Glumacees. *Ann. Sci. Nat., Bot. Biol. Veg.* 20: 5-56.

Tulloch M. 1972 The genus Myrothecium Tode ex. *Fr Mycol Pap*. 130:1–42.

Turner, W.B. and D.C. Albridge 1983 Fungal Metabolites. Academic Press Inc., London, UK.

Ueno, Y. and I. Ueno 1972 Isolation and acute toxicity of citreoviridin, a neurotoxic mycotoxin of *Penicillium citreoviride* Biourge. *Jap. J. Exp. Med.* **42**: 91-105.

Ueno, Y., N. Sato, K. Ishii, K. Sakai and M. Enomoto 1972 Toxicological approaches to the metabolites of Fusaria. V. Neosolaniol, T-2 toxin and butenolide, toxic metabolites of *Fusarium sporotrichioides* NRRL 3510 and *Fusarium poae* 3287. *Jpn. J. Exp. Med.* **42**:461-472.

Umeda M, K. Ohtsubo, M.Saito, S. Sekita, Y.Yoshira, S. Natori, S.Udagawa, H. Sakabe, Kurata 1975 Cytotoxicity of new cytochalasans from *Chaetomium globosum*. *Experientia*. **31**: 435- 438.

Van Reenen-Hoekstra, E.S., J.C. Frisvad, R.A. Samson and A.C. Stalk 1990 The *Penicillum funiculosum* complex: well defined species and problematic taxa. In: Samson, R.A. and Pitt, J.I., ed., Modern concepts in *Penicillium* and classification. New York, *plenum Press*. 173-191.

Varga, J., M. Due, J.C. Frisvad, R.A. Samson 2007 Taxonomic revision of *Aspergillus* section *Clavati* based on molecular, morphological and physiological data. *Stud. Mycol.* **59:** 89–106.

Von Arx, J.A 1949 Beitrage zur Kenntnis der Gattung *Mycosphaerella. Sydowia* **3**: 28-100.

Vos, P., R. Hogers, M. Bleeker, M. Reijans, T. Van de Lee, M. Hornes, A. Frijters, J. Pot, J. Peleman, M. Kuiper, and M. Zabeau 1995 AFLP: a new technique for DNA fingerprinting. *Nucl. Acids. Res.* **23**:4407–4414.

Voss, E. G., H. M. Burdet, W.G. Chaloner, V. Demoulin, P. Hiepko, J. McNeill, R.D. Meikle, D.H. Nicolson, R.C. Rollins, P.C. Silva and W. Greuter 1983 International Code of Botanical Nomenclature (Sydney Code). Adopted by the Thirteenth International Botanical Congress, Sydney, August 1981. Bohn, Scheltema & Holkema, Utrecht. *Regnum Veg*. **111**: 1–472.

Waalwijk, C., R. van der Heide, I. deVries, T. van der Lee, C. Schoen, G. Costrel-de Corainville, I. Häuser-Hahn, P. Kastelein, J. Köhl, P. Lonnet, T. Demarquet and G.H.J. Kema 2004 Quantitative detection of *Fusarium* species in wheat using TaqMan. *Eur. J. Plant Pathol.* **110:** 481-494.

Wang, I.K., C. Reeves and G.M. Gaucher 1991 Isolation and sequencing of a genomic DNA clone containing the 3' terminus of the 6-methylsalicylic acid polyketide synthetase gene of *Penicillium urticae. Can. J. Microbiol.* **37:** 86–95.

Ward, T.J., J.P. Bielawski, H.C. Kistler, E. Sullivan and K. O'Donnell 2002 Ancestral polymorphism and adaptive evolution in the trichothecene mycotoxin gene cluster of phytopathogenic *Fusarium*. *Proc. Natl. Acad. Sci. USA*. **99**:9278–9283.

Webster, J. 1970 Introduction to Fungi. Cambridge Univ. Press, London and New York.

Wei, D.L., and S. Jong 1986 Production of aflatoxin strains of the *Aspergillus flavus* group Maintained in ATTC. *Mycopathologia*. **93**: 19-24.

White, S., J. O'Callaghan and A.D. Dobson 2006 Cloning and molecular characterization of Penicillium expansum genes upregulated under conditions permissive for patulin biosynthesis. FEMS. *Microbiol. Lett.* 255: 17–26.

White, T.J., T. Bruns, S. Lee and J. Taylor 1990 Amplification of direct sequencing of fungal ribosomal RNA genes for phylogenetics. PCR Protocols: A guide to method and applications. Innis, D.; Gelfand, H.; Sninsky, J.J., Eds. Academic Press, Inc.: New York.

Wicklow, D.T. and R.J. Cole 1984 Citreoviridin in standing corn infested by *Eupenicillium ochrosalmoneum*. *Mycologia*. **76**: 959-961.

Williams, J.G.K., A.R. Kubelik, K.J. Livak, J.A. Rafalski and S.V. Tingey 1990 DNA polymorphisms amplified by arbitrary primers are useful as genetic markers. *Nucleic. Acids. Res.* **18**:6531–6536.

Wilson B.J, and R.R. Maronpot 1971 Causative fungal agent of leucoencephalomalacia in equine animals. *Vet. Rec.* **88**:484–485.

Wolf, J.C. and C.J. Mirocha 1973 Regulation of sexual reproduction in *Gibberella zeae* (*Fusarium roxeum "graminearum"*) by F-2 (Zearalenone). *Can. J. Microbiol.* **19:** 725-734.

Wolf, J.C. and C.J. Mirocha 1977 Control of Sexual Reproduction in *Gibberella zeae, Fusarium roseum "Graminearum"*). *Appl. Environ. Microbiol.* **33**: 546-550.

Wollenweber, H.W. and O.A. Reinking 1935 Die Fusarien, ihre Beschreibung, Schadwirkung und Bekämpfung. *Paul Parey, Berlin.* pp. 355.

Wolpert, T.J., L.D. Dunkle and L.M. Ciuffetti 2002 Host selective toxins and a virulence determinants: Whats in a name. *Annu. Rev. Phytopathol.* **40**: 251–285.

Xu, J., and T. G. Mitchell 2003 Comparative gene genealogical analyses of strains of serotype AD identify recombination in populations of serotypes A and D in the human pathogenic yeast Cryptococcus neoformans. *Microbiol.* **149**: 2147–2154.

Xu, J.R. and J.F. Leslie 1996 A genetic map of *Gibberella fujikuroi* mating population A (*Fusarium moniliforme*). *Genetics.* **143**:175-189.

Yli-Mattila, T., R.L. Mach, I.A. Alekhina, S.A. Bulat, S. Koskinen, C.M. Kullnig-Gradinger, C.P. Kubicek, S.S. Klemsdal 2004a Phylogenetic relationship of *Fusarium langsethiae* to *Fusarium poae* and *Fusarium sporotrichioides* as inferred by IGS, ITS, b-tubulin sequences and UP-PCR analysis. *Int. J. Food. Microbiol.* **95**:267–285.

Yli-Mattila,T., S.Paavanen-Huhtala, P. Konstantinova, T.Y. Gagkaeva 2004b Molecular and morphological diversity of *Fusarium* species in Finland and north-western Russia. *Eur. J. Plant. Pathol.* **110**:573–585.

Youssef, S.A., M. Maymon, A. Zveibil, D. Klein-Gueta, A. Sztejnberg, A.A. Shalaby and S. Freeman 2007 Epidemiological aspects of mango malformation disease caused by *Fusarium mangiferae* and source of infection in seedlings cultivated in orchards in Egypt. *Plant. Pathol.* **56**:257–263.

Yu, J., D. Bhatnagar and T.E. Cleveland 2004 Completed sequence of the aflatoxin pathway gene cluster in *Aspergillus parasiticus*. *FEBS Lett.* **564**: 126–130.

Yu, J., P.K. Chang, J.W. Cary, D. Bhatnagar and T.E. Cleveland 1997 avnA, a gene encoding a cytochrome P-450 mono-oxygenase, is involved in the conversion of averantin to averufin in aflatoxin biosynthesis in *Aspergillus parasiticus. Appl. Environ. Microbiol.* **63:** 1349-1356.

Yu, J., P.K. Chang, K.C. Ehrlich, J.W. Cary, D. Bhatnagar, T.E. Cleveland, G.A. Payne, J.E. Linz, C.P. Woloshuk and J.W. Bennett 2004 Clustered pathway genes in aflatoxin biosynthesis. *Appl. Environ. Microbiol.* **70:** 1253–1262.

Yu, J.H. and N. Keller 2005 Regulation of secondary metabolism in filamentous fungi. *.Annu. Rev. Phytopathol.* **43**:437–58.

Zaleta-Rivera, K., C. Xu, F. Yu, R.A. Butchko, R.H. Proctor, M.E. Hidalgo-Lara, A. Raza, P.H. Dussault, and L. Du 2006 A bidomain non-ribosomal peptide synthetase encoded by FUM14 catalyzes the formation of tricarballylic esters in the biosynthesis of fumonisins. *Biochem.* **45:** 2561-2569.

Zeller, K., B.A.Summerell, S. Bullock and J.F. Leslie 2003 *Gibberella konz* (*Fusarium konzum*) sp. nov., a new species within the *Gibberella fujikuroi* complex from native prairie grasses. *Mycologia*. **95**:943–954

Zeller, K.A., R.L. Bowden and J.F. Leslie 2004 Population differentiation and recombination in wheat scab populations of *Gibberella zeae* from the United States. *Mol. Ecol.* **13**: 563–571.

Zhang, A.W., G.L. Riccioni, L. Chen, W. D. Ma and W.L. Petersen 1997 Using PCR to distinguish *Diaporthe phaseolorum* and *Phompsis longicolla* from other soybean fungal pathogens and to detect them in soybean tissues. *Plant. Dis.* **81:** 1143-1149.

Zhang, Y.Q., H.Wilkinson, N.P. Keller and D. Tsitsigiannis 2004 Secondary metabolite gene clusters. In: An Z (ed) Handbook of Industrial Mycology. Marcel Dekker, New York, pp 355–386.